VOLUME FIVE HUNDRED AND FIFTY FOUR

METHODS IN ENZYMOLOGY

Hydrogen Sulfide in Redox Biology Part A

METHODS IN ENZYMOLOGY

Editors-in-Chief

JOHN N. ABELSON and MELVIN I. SIMON
Division of Biology
California Institute of Technology
Pasadena, California

ANNA MARIE PYLE
Departments of Molecular, Cellular and Developmental Biology and Department of Chemistry Investigator
Howard Hughes Medical Institute
Yale University

Founding Editors

SIDNEY P. COLOWICK and NATHAN O. KAPLAN

VOLUME FIVE HUNDRED AND FIFTY FOUR

METHODS IN ENZYMOLOGY

Hydrogen Sulfide in Redox Biology Part A

Edited by

ENRIQUE CADENAS

Pharmacology & Pharmaceutical Sciences,
School of Pharmacy,
University of Southern California, USA

LESTER PACKER

Department of Molecular Pharmacology and Toxicology,
School of Pharmaceutical Sciences,
University of Southern California, USA

ELSEVIER

AMSTERDAM • BOSTON • HEIDELBERG • LONDON
NEW YORK • OXFORD • PARIS • SAN DIEGO
SAN FRANCISCO • SINGAPORE • SYDNEY • TOKYO
Academic Press is an imprint of Elsevier

Academic Press is an imprint of Elsevier
225 Wyman Street, Waltham, MA 02451, USA
525 B Street, Suite 1800, San Diego, CA 92101-4495, USA
125 London Wall, London, EC2Y 5AS, UK
The Boulevard, Langford Lane, Kidlington, Oxford OX5 1GB, UK

First edition 2015

ISBN: 978-0-12-801512-4
ISSN: 0076-6879

For information on all Academic Press publications
visit our website at store.elsevier.com

CONTENTS

Section II
Hydrogen Sulfide Donors and Their Pharmacological Activity

Section III
Hydrogen Sulfide Metabolism in Mammalian Tissues

CONTRIBUTORS

Abbas Abou-Hamdan
Inserm U1016; CNRS UM8104, and Université Paris Descartes UMR-S1016, Institut Cochin, Paris, France

Mireille Andriamihaja
INRA-CRNH-IdF-AgroParisTech, UMR 914 Nutrition Physiology and Ingestive Behavior, Paris, France

Ruma Banerjee
Department of Biological Chemistry, University of Michigan Medical School, Ann Arbor, Michigan, USA

Jin-Song Bian
Department of Pharmacology, Yong Loo Lin School of Medicine, National University of Singapore, Singapore

François Blachier
INRA-CRNH-IdF-AgroParisTech, UMR 914 Nutrition Physiology and Ingestive Behavior, Paris, France

Frédéric Bouillaud
Inserm U1016; CNRS UM8104, and Université Paris Descartes UMR-S1016, Institut Cochin, Paris, France

Christopher J. Chang
Department of Chemistry; Department of Molecular and Cell Biology, University of California, Berkeley, California, and Howard Hughes Medical Institute, Chevy Chase, Maryland, USA

Taurai Chiku
Department of Biological Chemistry, University of Michigan Medical School, Ann Arbor, Michigan, USA

Brian W. Dymock
Department of Pharmacy, National University of Singapore, Singapore

Hala Guedouari-Bounihi
Inserm U1016; CNRS UM8104, and Université Paris Descartes UMR-S1016, Institut Cochin, Paris, France

Takaaki Ito
Department of Pathology and Experimental Medicine, Graduate School of Medical Science, Kumamoto University, Kumamoto, Japan

Michael R. Jackson
Department of Biochemistry and Molecular Biology, College of Medicine, Drexel University, Philadelphia, Pennsylvania, USA

Marilyn Schuman Jorns
Department of Biochemistry and Molecular Biology, College of Medicine, Drexel University, Philadelphia, Pennsylvania, USA

Omer Kabil
Department of Biological Chemistry, University of Michigan Medical School, Ann Arbor, Michigan, USA

Christopher G. Kevil
Department of Pathology, Louisiana State University Health Sciences Center–Shreveport, Shreveport, Louisiana, USA

Gopi K. Kolluru
Department of Pathology, Louisiana State University Health Sciences Center–Shreveport, Shreveport, Louisiana, USA

Véronique Lenoir
Inserm U1016; CNRS UM8104, and Université Paris Descartes UMR-S1016, Institut Cochin, Paris, France

Zhong-Guang Li
School of Life Sciences; Engineering Research Center of Sustainable Development and Utilization of Biomass Energy, Ministry of Education; Key Laboratory of Biomass Energy and Environmental Biotechnology, Yunnan Normal University, Kunming, Yunnan Province, PR China

Marouane Libiad
Department of Biological Chemistry, University of Michigan Medical School, Ann Arbor, Michigan, USA

Vivian S. Lin
Department of Chemistry, University of California, Berkeley, California, USA

Alexander R. Lippert
Department of Chemistry, and Center for Drug Discovery, Design, and Delivery (CD4), Southern Methodist University, Dallas, Texas, USA

Scott L. Melideo
Department of Biochemistry and Molecular Biology, College of Medicine, Drexel University, Philadelphia, Pennsylvania, USA

Philip K. Moore
Neurobiology Program, Life Science Institute and Department of Pharmacology, National University of Singapore, Singapore

Nicole Motl
Department of Biological Chemistry, University of Michigan Medical School, Ann Arbor, Michigan, USA

Noriyuki Nagahara
Isotope Research Center, Nippon Medical School, Tokyo, Japan

Masatoshi Nagano
Department of Pharmacology, Nippon Medical School, Tokyo, Japan

Péter Nagy
Department of Molecular Immunology and Toxicology, National Institute of Oncology, Budapest, Hungary

Chung-Min Park
Department of Chemistry, Washington State University, Pullman, Washington, USA

Bo Peng
Department of Chemistry, Washington State University, Pullman, Washington, USA

Michael D. Pluth
Department of Chemistry and Biochemistry, Institute of Molecular Biology, Materials Science Institute, University of Oregon, Eugene, Oregon, USA

Peter Rose
University of Lincoln, Lincoln, Lincolnshire, United Kingdom

Xinggui Shen
Department of Pathology, Louisiana State University Health Sciences Center–Shreveport, Shreveport, Louisiana, USA

T. Spencer Bailey
Department of Chemistry and Biochemistry, Institute of Molecular Biology, Materials Science Institute, University of Oregon, Eugene, Oregon, USA

Hidenori Suzuki
Department of Pharmacology, Nippon Medical School, Tokyo, Japan

Victor Vitvitsky
Department of Biological Chemistry, University of Michigan Medical School, Ann Arbor, Michigan, USA

Ming Xian
Department of Chemistry, Washington State University, Pullman, Washington, USA

Xue Xue
Department of Pharmacology, Yong Loo Lin School of Medicine, National University of Singapore, Singapore

Pramod K. Yadav
Department of Biological Chemistry, University of Michigan Medical School, Ann Arbor, Michigan, USA

Shuai Yuan
Department of Pathology, Louisiana State University Health Sciences Center–Shreveport, Shreveport, Louisiana, USA

PREFACE

Hydrogen sulfide is viewed as the third gasotransmmitter, gaseous signaling molecule, together with nitric oxide and carbon monoxide. The cellular sources of hydrogen sulfide involve enzymes of the trans-sulfuration pathway CBS (cystathionine β-synthase), CSE (cystathioine γ-lyase), and 3MST (3-mercaptopyruvate sulfur-transferase). Storages of hydrogen sulfide occur in mitochondria (iron–sulfur clusters of enzymes) and cytosol (bound sulfane sulfur). Of course, the release of hydrogen sulfide from these storages is tightly regulated by several pathophysiological processes.

In addition to the myriad of effects arising from hydrogen sulfide as a signaling molecule, it also protects against oxidative stress and glutamate toxicity, inhibits the release of insulin, preserves mitochondrial function, and is a modulator of inflammatory responses. These pleiotropic effects of hydrogen sulfide have been the subject of numerous investigations in the last years and are largely accounted for by as its role as a gaseous signaling molecule. Hydrogen sulfide may act alone or in conjunction with other gasotransmitters and, in doing so, it regulates a number of physiological processes and is involved in some stages of the pathogenesis of several diseases.

These volumes of *Methods in Enzymology* were designed as a compendium for hydrogen sulfide detection methods, the pharmacological activity of hydrogen sulfide donors, the redox biochemistry of hydrogen sulfide and its metabolism in mammalian tissues, the mechanisms inherent in hydrogen sulfide cell signaling and transcriptional pathways, and cell signaling in specific systems, such as cardiovascular and nervous system as well as its function in inflammatory responses. Three chapters are also devoted to hydrogen sulfide in plants and a newcomer, molecular hydrogen, its function as a novel antioxidant.

In bringing these volumes of *Methods in Enzymology* to fruition, credit must be given to the experts in various aspects of hydrogen sulfide research, whose thorough and innovative work is the basis of these *Methods in Enzymology* volumes. We hope that these volumes will be of help to both new and established investigators in the field.

Enrique Cadenas
Lester Packer

SECTION I

Hydrogen Sulfide Detection Methods

CHAPTER ONE

Mechanistic Chemical Perspective of Hydrogen Sulfide Signaling

Péter Nagy[1]
Department of Molecular Immunology and Toxicology, National Institute of Oncology, Budapest, Hungary
[1]Corresponding author: e-mail address: peter.nagy@oncol.hu

Contents

Abstract

Hydrogen sulfide is now a well-appreciated master regulator in a diverse array of physiological processes. However, as a consequence of the rapid growth of the area, sulfide biology suffers from an increasing number of controversial observations and interpretations. A better understanding of the underlying molecular pathways of sulfide's actions is key to reconcile controversial issues, which calls for rigorous chemical/biochemical investigations.

Protein sulfhydration and coordination/redox chemical interactions of sulfide with heme proteins are the two most extensively studied pathways in sulfide biochemistry. These pathways are important mediators of protein functions, generate bioactive

Methods in Enzymology, Volume 554
ISSN 0076-6879
http://dx.doi.org/10.1016/bs.mie.2014.11.036

sulfide metabolites, contribute to sulfide storage/trafficking and carry antioxidant functions. In addition, inorganic polysulfides, which are oxidative sulfide metabolites, are increasingly recognized as important players in sulfide biology.

This chapter provides an overview of our mechanistic perspective on the reactions that govern (i) sulfide's bioavailability (including the delicate enzyme machineries that orchestrate sulfide production and consumption and the roles of the large sulfide-storing pools as biological buffers), (ii) biological significance and mechanisms of persulfide formation (including the reduction of disulfides, condensation with sulfenic acids, oxidation of thiols with polysulfides and radical-mediated pathways), (iii) coordination and redox chemical interactions of sulfide with heme proteins (including cytochrome *c* oxidase, hemoglobins, myoglobins and peroxidases), and (iv) the chemistry of polysulfides.

ABBREVIATIONS

3MST 3-mercaptopyruvate sulfurtransferase
AAT aspartate/cysteine aminotransferase
CBS cystathionine-β-synthase
CcO cytochrome *C* oxidase
CSE cystathionine-γ-lyase
CySOH Cys sulfenic acids
CySSH and GSSH, respectively Cys- and GSH-persulfides
DTNB 5,5′-dithiobis-(2-nitrobenzoic acid)
ER endoplasmic reticulum
ERK1/2 extracellular signal-regulated kinases ½
Ero1 ER oxidoreductin
Erv1 essential for respiration and vegetative growth sulfhydryl oxidase
GAPDH glyceraldehyde-3-phosphate dehydrogenase
GOR glutathione oxido-reductase
Hb Hemoglobin
$HSSH^{-\bullet}$ disulfide anion radical species
Keap1 Kelch-like ECH-associated protein 1
LPO lactoperoxidase
Mb myoglobin
MEK1 mitogen-activated protein kinase kinase
MPO myeloperoxidase
NfκB nuclear factor kappa-light-chain-enhancer of activated B cells
Nrf2 nuclear factor (erythroid-derived 2)-like 2
PARP-1 poly [ADP-ribose] polymerase 1
PDI protein disulfide isomerase
PLP pyridoxal phosphate
PTEN phosphatase and tensin homolog
PTP1B protein-tyrosine phosphatase 1B
roGFP reduction–oxidation sensitive green fluorescent protein
ROS reactive oxygen species
RPS3 ribosomal protein S3
SAM *S*-adenosylmethionine

SDO sulfide dioxigenase
SO sulfite oxidase
SQR sulfide quinone reductase
SUMO small ubiquitin-like modifier
TNB 5-thio-2-nitrobenzoic acid
TPO thyroid peroxidase
Trx thioredoxin
TST thiosulfate:glutathione sulfurtransferase

1. INTRODUCTION

Sulfide biology has expanded very rapidly in the past decade with a great number of groundbreaking fundamental discoveries (see, e.g., Abe & Kimura, 1996; Blackstone, Morrison, & Roth, 2005; Coletta et al., 2012; Elrod et al., 2007; Flannigan et al., 2014; Mustafa et al., 2009; Papapetropoulos et al., 2009; Suzuki et al., 2011; Szabo et al., 2013; Yang et al., 2008). However, the field suffers from a side effect common to many extensively growing areas, that of generating a large number of controversial reports. Better understanding of the underlying molecular mechanisms of sulfide actions is a fundamental requirement for reconciling these controversial physiological observations. Therefore, this review is intended to provide our mechanistic chemical perspective on what are currently considered to be the most important reactions in sulfide biology.

2. BIOAVAILABILITY OF SULFIDE—THE SIGNAL

In considering the mechanistic details of sulfide signaling, the first obvious parameter to address is the physiological concentration of sulfide, i.e., the signal. Moreover, in dynamic, living systems, it is rather the change in sulfide concentration over time that triggers the cascade of biochemical reactions that constitute the signaling process.

Sulfide concentrations can change: (i) as a result of physiological or pathophysiological events (for example, in some cancer cells the sulfide-producing enzymes, *cystathionine-β-synthase* (CBS), and/or *cystathionine-γ-lyase* (CSE) are overexpressed and could increase sulfide levels, but in inflammation or at sites of oxidative stress, sulfide levels could be significantly diminished), (ii) due to administration of authentic sulfide or slow releasing sulfide donor molecules, or (iii) simply as a result of normal cellular functions. The latter can be demonstrated by the fact that in cellular systems metabolic pathways work in parallel to strictly govern free sulfide levels, and a small perturbation of sulfide-producing or consuming reactions can have a profound effect on steady-state concentrations (Vitvitsky, Kabil, & Banerjee, 2012).

2.1. Endogenous sulfide production

Endogenous sulfide production via cysteine metabolism is catalyzed by at least three different enzymatic systems, the main ones being the two pyridoxal phosphate (PLP) dependent CBS and CSE enzymes and the cooperative actions of *aspartate/cysteine aminotransferase* (AAT) and *3-mercaptopyruvate sulfurtransferase* (3MST) (reviewed recently in Kabil & Banerjee, 2014; Nagahara, 2013). Although these enzymes all use cysteine as their biological substrate, their individual enzyme kinetic properties are very different. Differences include the involvement of (i) a variety of co-substrates, (ii) parallel enzymatic activities, and (iii) mechanisms of inhibition/potentiation (Fig. 1).

Therefore, the dominant cysteine metabolism-mediated sulfide-producing pathway in a certain biological situation depends not only on the relative CSE: CBS:AAT/3MST concentrations on site, but also on many other components including the bioavailability of cysteine, homocysteine, α-ketoglutarate, and other enzyme activity modulating factors (Chiku et al., 2009; Yadav, Yamada, Chiku, Koutmos, & Banerjee, 2013). The primary sulfide-producing reactions catalyzed by these enzymes are also very different chemically: CSE and CBS catalyze α,β-elimination and β-replacement reactions of cysteine, respectively, while 3MST produces sulfide via reductive elimination from a preformed cysteine-persulfide intermediate species (see Fig. 1).

Two additional mechanisms have the potential to produce sulfide indirectly, from Cys- and GSH-persulfides (CySSH and GSSH, respectively). These are CSE (Yamanishi & Tuboi, 1981) and CBS (Ida et al., 2014) catalyzed beta-elimination from cystine and subsequent transsulfuration between CySSH and GSH. Note that these pathways exclude the direct participation of sulfide in the persulfide formation process, which is an important difference from sulfide-mediated persulfide generation reactions (see later). Based on the measured K_M and k_{cat} values, Ida et al. proposed that under biologically relevant Cys:CySSCy ratios, CySSCy could be the preferred substrate for CSE and CBS. Hence, these reactions could provide the majority of the large amounts (tens to hundreds of μM-s) of persulfides (mostly as GSSH) that were detected in cellular systems (Ida et al., 2014).

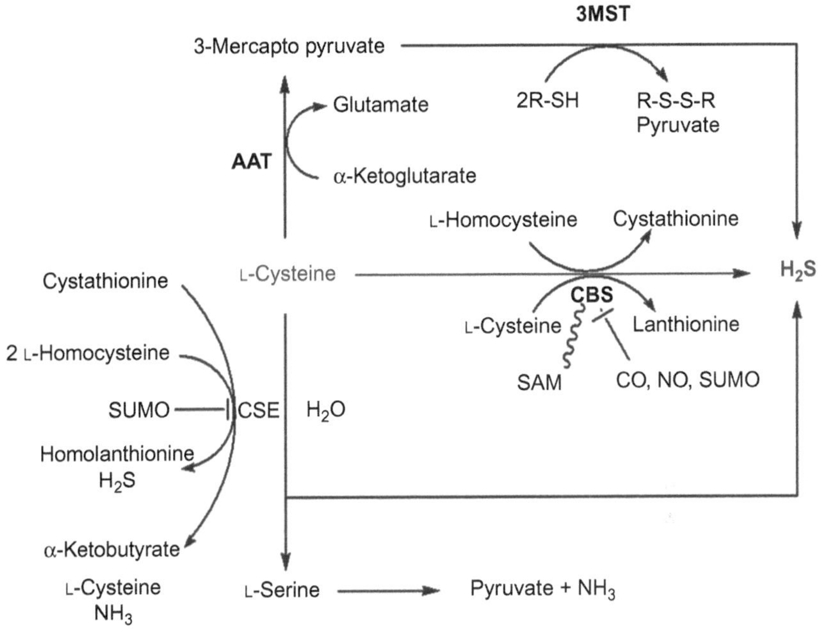

Figure 1 The main sulfide-producing enzymatic pathways via cysteine metabolism. The figure demonstrates (with examples) that beside the dominant sulfide-producing pathways CSE and CBS catalyze many other reactions using different substrates that not necessarily produce sulfide. Furthermore, both CSE and CBS are inhibited by *small ubiquitin-like modifier* (SUMO) catalyzed sumoilation. CO and NO were also shown to be inhibitory on CBS activity, but *S*-adenosylmethionine (SAM) is an allosteric activator of the enzyme (Kabil & Banerjee, 2014).

Furthermore, a newly *discovered thiosulfate:glutathione sulfurtransferase* (TST) in the transsulfuration pathway of sulfide catabolism to sulfite, sulfate, and thiosulfate (see later), could also generate substantial concentrations of GSH-persulfide (and potentially CoA-SSH and other persulfide derivatives), which may have a sulfide trafficking function particularly at sites with decreased *sulfide dioxygenase* (SDO) activities (Melideo, Jackson, & Jorns, 2014).

Importantly, the subsequent liberation of sulfide from persulfides could also be a controlling factor in sulfide production where the thermodynamic (reducing vs. oxidizing environment) and kinetic (rate of persulfide reduction) barriers should be highly dependent on the biological environment and the chemical nature of the apparent reducing reaction partner. Hence, pathways that produce sulfide via persulfide intermediates are not only orchestrated by persulfide-generating enzymatic activities, but also via persulfide reducing reactions (Kabil & Banerjee, 2014; Melideo et al., 2014; Mikami et al., 2011; Westrop, Georg, & Coombs, 2009; Yadav et al., 2013). Furthermore, a concept is emerging, that persulfides have direct biological functions rather than just being sulfide-storing molecules (Nielsen, Tachibana, Hansen, & Winther, 2011; Ono et al., 2014; Paul & Snyder, 2012; Toohey, 1989).

2.2. Sulfide catabolism

Sulfide catabolism mostly occurs in mitochondria via oxidative processes (Szabo et al., 2014) driven primarily by the *sulfide quinone reductase* (SQR) enzyme (see Fig. 2) (Hildebrandt & Grieshaber, 2008; Theissen, Hoffmeister, Grieshaber, & Martin, 2003). The molecular mechanism of this pathway involves initial reduction of an intramolecular disulfide moiety of SQR to produce an SQR-persulfide intermediate (Cherney, Zhang, Solomonson, Weiner, & James, 2010). Subsequently, this persulfide functional group is transferred catalytically on to GSH by TST to produce GSSH, which is used as a substrate by a sulfur dioxygenase to give sulfite (Melideo et al., 2014). Sulfite is than utilized by either *sulfite oxidase* (SO) or SQR to give sulfate or thiosulfate, respectively.

Despite the major mechanism of sulfide toxicity being inhibition of mitochondrial respiration via interaction with *cytochrome C oxidase* (CcO) (see later), at low concentrations sulfide can also serve as a stimulator of ATP production by the mitochondrial electron transport chain (Goubern, Andriamihaja, Nubel, Blachier, & Bouillaud, 2007; Helmy et al., 2014; Lagoutte et al., 2010; Modis, Panopoulos, Coletta, Papapetropoulos, &

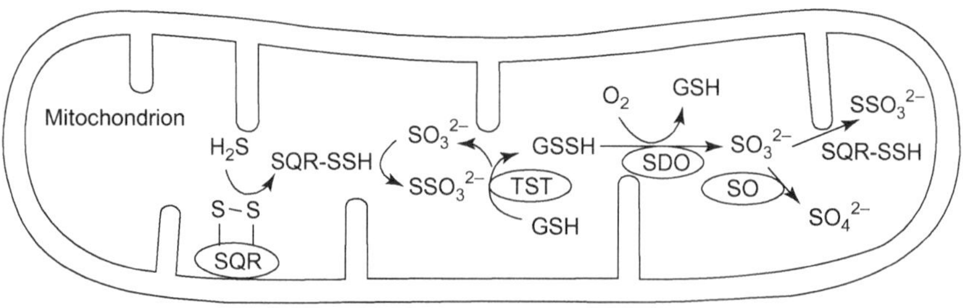

Figure 2 Proposed sulfide catabolism model in mitochondria. Reduction of an intramolecular disulfide bond in SQR gives an SQR-SSH intermediate species. TST-mediated transsulfuration from SQR-SSH to GSH gives GSSH. GSSH was proposed to engage in sulfide trafficking, but its SDO catalyzed oxidation generates sulfite. Sulfite can either be further oxidized by SO to sulfate or feed back to the catalytic cycle and react with another SQR-SSH (Melideo et al., 2014).

Szabo, 2013; Szabo et al., 2014). Hence, regulation of mitochondrial energetic functions coupled with sulfide catabolism is another prime example of a fine-tuned sulfide-mediated physiological process.

2.3. Endogenous sulfide buffers

The bioavailability of sulfide is not only governed by endogenous sulfide-producing and catabolizing reactions. It is now clear that a large amount of sulfide is stored in biomolecule-sulfide adducts. Due to the reversible nature of sulfide binding, this pool could serve as a buffer of free biological sulfide concentrations (Nagy et al., 2014). In fact, based on a large number of investigations (summarized in the SI of Nagy et al., 2014) we now know that in most cells, tissues, and biological fluids, free sulfide represents less than 1% of the potentially available sulfide, indicating that endogenous sulfide pools should have large buffering capacities. This sulfide buffer system is under strict thermodynamic control (constituted by the actual equilibrium constants describing the relationships between the dissociated and bound sulfide-biomolecule complexes and their interrelations), which is crucial for maintaining sulfide levels below the toxic threshold. In addition, the kinetics of sulfide binding and release are likely to mediate sulfide signaling, because most reported physiological sulfide functions required relatively high concentrations. *In vivo* these could only be provided by the sulfide buffer system. This notion is corroborated by the fact that many of sulfide's physiological actions are better represented *in vitro* using a slow releasing donor molecule (Whiteman, Le Trionnaire, Chopra, Fox, & Whatmore, 2011), most likely because the former could better mimic the sulfide-donating capacity of the endogenous sulfide pool. The rate of sulfide release is highly dependent on the binding partner and could show orders of magnitude differences among the different sulfide-biomolecule adducts. The large variation in on–off rates could provide selectivity and potentially serve as a mediator of time-resolved sulfide bioavailability.

3. INORGANIC POLYSULFIDES

3.1. Biological relevance

It is increasingly appreciated that some of sulfide's biological actions are due to partially oxidized sulfane-sulfur (sulfur with an oxidation state of 0) containing molecules (Greiner et al., 2013; Ida et al., 2014; Kimura, 2013, 2014; Kimura & Kimura, 2004; Kimura et al., 2013; Koike, Ogasawara, Shibuya,

Kimura, & Ishii, 2013; Nielsen et al., 2011; Ono et al., 2014; Toohey, 1989). Major players in sulfane biochemistry may be inorganic polysulfides (Greiner et al., 2013; Kimura, 2014; Nagy, Jameson, & Winterbourn, 2009; Toohey, 2011). In fact, many of the reported controversies in sulfide biology may be due to different polysulfide contents in "sulfide donor" solutions (Nagy et al., 2014). Polysulfide contamination in sulfide (and hydrosulfide) salts shows large variations. In addition, the extent and rate of air oxidation of sulfide to polysulfides in a particular experimental setup, largely depends on components of the applied medium such as oxygen, metal ions, and pH (Nagy et al., 2014). Furthermore, very small amounts of polysulfide even in the presence of a large excess of sulfide can serve as the actual regulators of enzymatic functions (Greiner et al., 2013). Based on these observations, the reported controversies and irreproducible results in sulfide biology are not at all surprising, and it is likely that a deeper insight into polysulfide-mediated processes will reconcile many conflicting observations. Moreover, polysulfides are likely to be major endogenous oxidation products of sulfide in different biological situations and participate in the cross-talk of sulfide and redox signaling events (see later) with important physiological consequences (Greiner et al., 2013; Kimura, 2014).

3.2. Speciation and redox capacity of polysulfides

From a chemical perspective, polysulfides are thermodynamically unstable sulfide oxidation products, with a general formula of H_2S_x ($x = 2$–9). Their acid dissociation constants (pK_a) decrease with an increase in chain length, but for all polysulfides it is likely to be <7, the pK_a of HS^- (Hoffmann, 1977). Therefore, at physiological pH, polysulfides are expected to be in their mono-anionic forms and it is more appropriate to use the general formula HS_x^- instead of H_2S_x. For the sake of simplicity, they can be considered as molecules that contain one sulfide and 1–8 sulfane-sulfurs, which nicely demonstrate that their oxidizing capacity increases with the number of sulfur atoms in the chain.

The speciation of polysulfides is described by a large number of interrelated equilibrium processes, exemplified by the following reactions:

$$2HS_2^- \rightleftharpoons HS_3^- + HS^- \tag{1}$$

$$HS_2^- + HS_3^- \rightleftharpoons HS_4^- + HS^- \tag{2}$$

$$HS_3^- + HS_3^- \rightleftharpoons HS_4^- + HS_2^- \tag{3}$$

$$HS_4^- + HS_2^- \rightleftharpoons HS_5^- + HS^- \tag{4}$$

This complexity together with (i) the difficulty of achieving aqueous media in which one or other species dominates and (ii) polysulfides being thermodynamically unstable and kinetically labile molecules, make the systematic investigation of polysulfide distribution extraordinary challenging. Nevertheless, Giggenbach attempted to study polysulfide speciation under different conditions based on UV-characteristics, (Giggenbach, 1972). His data indicate that polysulfide distribution is largely dependent on pH and the sulfane-sulfur–sulfide ratio. Generally, higher pH and a more reducing environment (i.e., high sulfide to polysulfide ratios) favor the formation of polysulfides with lower numbers of zerovalent sulfur atoms (i.e., shorter chain length). It has also been demonstrated that the redox potential of polysulfide solutions strictly correlates with the sulfide–sulfane-sulfur ratio and is largely influenced by the pH (Lessner, Mclarnon, Winnick, & Cairns, 1993), suggesting that the redox potential of polysulfides increases with chain length.

3.3. Polysulfide formation by sulfide oxidation

Polysulfides can be generated in both 1 and 2 electron oxidation reactions of sulfide. Importantly, some oxidants react at moderate rates (e.g., H_2O_2 $k_{\text{app}}^{\text{pH 6.8}} = 0.2 M^{-1} s^{-1}$, Hoffmann, 1977), whereas other reactions (e.g., with HOCl $k_{\text{app}}^{\text{pH 7.4}} = 2.3 \times 10^9 M^{-1} s^{-1}$, Nagy & Winterbourn, 2010a) are almost diffusion controlled. Although a number of reports suggest an antioxidant role for sulfide via neutralizing reactive oxygen species (ROS) (Fu, Liu, Geng, Fang, & Tang, 2008; Jha, Calvert, Duranski, Ramachandran, & Lefer, 2008; Kimura & Kimura, 2004; Laggner et al., 2007; Whiteman et al., 2004, 2005; Yonezawa et al., 2007), in my view, because of the orders of magnitude lower free sulfide levels compared to Cys thiol concentrations, direct scavenging of oxidants is unlikely to be responsible for the observed antioxidant properties in biological systems. Instead, antioxidant activity may be better explained by modulation of pro- and antioxidant enzyme functions via reactions with their Cys residues (such as the activation of the *nuclear factor (erythroid-derived 2)-like 2* (Nrf2) pathway via *Kelch-like ECH-associated protein 1* (Keap1) inhibition at its active site Cys residue (Hourihan, Kenna, & Hayes, 2013; Yang et al., 2013)) or metal centers (see later).

Two-electron oxidation is likely to be the dominant pathway in sulfide's reactions with closed shell oxidants such as HOCl, HOSCN, H_2O_2 and other lipid, peptide, or protein peroxides. These reactions, in analogy with Cys, are expected to produce the inorganic disulfide species via highly reactive sulfenic acid (or sulfenyl halide) intermediate species (see reactions 5 and 6).

$$HS^- + Ox \rightarrow HSX + Red \quad (X = Cl, SCN, \text{ or } OH) \tag{5}$$

$$HS^- + HSX \rightarrow HS_2^- + H^+ + X^- \tag{6}$$

In contrast to cysteine, this inorganic disulfide is thermodynamically unstable and subsequent disproportionation gives a mixture of polysulfide species (see Section 3.2). Alternatively, HS_2^- could be reduced before it has time to disproportionate, e.g., by Cys thiolates in biological systems. Hence, it could potentially serve as a sulfhydrating agent, but further kinetic information is required to establish whether this is the case.

One electron oxidation of sulfide by radicals (such as superoxide, Wedmann et al., 2014) or metal centers (such as myeloperoxidase (MPO), Palinkas et al., 2014) generate polysulfides via the intermediate production of the thiyl radical ($HS^{\bullet}$). The reaction of $HS^{\bullet}$ with another sulfide (reaction 7) is diffusion controlled and gives the disulfide anion radical species ($HSSH^{-\bullet}$) (Das, Huie, Neta, & Padmaja, 1999). $HSSH^{-\bullet}$ is then expected to be further oxidized (e.g., by oxygen) to give HS_2^- (reaction 8) and eventually polysulfides (and superoxide if the reaction partner is oxygen).

$$HS^{\bullet} + HS^- \rightarrow HSSH^{-\bullet} \tag{7}$$

$$HSSH^{-\bullet} + O_2 \rightarrow HSSH + O_2^{-\bullet} \tag{8}$$

Polysulfides generated by sulfide oxidation can be viewed as secondary reactive oxidant species. This property is best demonstrated by their reactions with Cys thiols to give persulfides (and possibly Cys-polysulfides; see later). However, sulfur can also acquire oxidation states larger than 0 and therefore polysulfides can also act as reducing molecules. Persulfides are likely to be more efficient radical reducing agents than thiols because the unpaired electron of $RSS^{\bullet-}$ is resonance stabilized by the adjacent sulfur atom (Ono et al., 2014), and this might well be true for inorganic polysulfides.

3.4. Stability of polysulfides

Polysulfides are thermodynamically unstable species in aqueous media. In the absence of reducing or oxidizing reaction partners they disproportionate to eventually give H_2S and S_8:

$$HS_{(n+1)}^- + H^+ \rightarrow n/8S_8 + H_2S \quad (n = 1 - 8) \tag{9}$$

In biological systems, polysulfides are expected to be consumed more rapidly by biomolecules before they have time to disproportionate. However, it is important to take disproportionation into consideration when working with polysulfide solutions in experimental systems. Therefore, the stability of polysulfide species under different experimental conditions needs to be systematically studied in order to avoid misinterpretation of results. Furthermore, it is important to note that it is not possible to make sulfide-free polysulfide solutions, because based on reaction (9), consumption of sulfide (e.g., by volatilization) will facilitate polysulfide decomposition and sulfur precipitation.

4. SULFIDE SIGNALING VIA PROTEIN SULFHYDRATION

The most frequently quoted mechanism for sulfide signaling is sulfhydration of protein cysteine residues. Chemically, this posttranslational modification results in an analogous functional group to those in the previously discussed CySSH and GSSH species (note that from a chemical perspective the term persulfide formation is preferred over sulfhydration). Persulfide formation on functional Cys residues can enhance or inhibit enzymatic activities either directly or indirectly as indicated by the following examples:

Examples of enzyme activation via persulfide formation: (1) Up to sevenfold greater glycolytic activity of *glyceraldehyde-3-phosphate dehydrogenase* (GAPDH) was observed when the active site Cys150 was modified to a persulfide (Mustafa et al., 2009). (2) Persulfide formation on Cys38 of the p65 subunit of *nuclear factor kappa-light-chain-enhancer of activated B cells* (NfκB) activated binding to the co-activator *ribosomal protein S3* (RPS3) and triggered antiapoptotic transcriptional activity (Sen et al., 2012). (3) ATP-sensitive potassium channels were activated via persulfide formation-induced inhibition of ATP binding (Mustafa et al., 2011). (4) Persulfide formation on Cys341 of *mitogen-activated protein kinase kinase* (MEK1) leads to *poly [ADP-ribose] polymerase 1* (PARP-1) activation and DNA damage repair via facilitation of phosphorylated *extracellular signal-regulated kinases* ½ (ERK1/2) translocation into the nucleus (Zhao et al., 2014). (5) Nrf2 was activated via Cys151-persulfide formation-induced activity loss of Keap1 (Yang et al., 2013).

Inhibition of enzymatic activities via persulfide formation: (1) The previous examples of Nrf2 or PARP-1 activation represent indirect effects, because they are facilitated via persulfide formation-mediated inactivation of their negative regulators, Keap1 and MEK1, respectively (Yang et al., 2013). (2) Polysulfide-induced persulfide generation on the active site Cys124 and/or Cys71 residues efficiently inactivated the phosphatase activity of *phosphatase and tensin homolog* (PTEN) (Greiner et al., 2013). (3) Persulfide formation on Cys215 had a similar inactivating effect on *protein-tyrosine phosphatase 1B* (PTP1B) as the well-established redox switch of the enzyme via reversible cyclic sulfenamide formation between the Cys215 thiolate and its backbone amide nitrogen (Krishnan, Fu, Pappin, & Tonks, 2011).

Persulfide formation has recently been proposed to serve as a potential protecting mechanism of thiol residues toward oxidative stress (Greiner et al., 2013; Ida et al., 2014; Ono et al., 2014). The driving force of ROS scavenging was proposed to be preferential oxidation of the persulfide moiety compared to the corresponding thiol. The protecting effect was explained by the chemical notion that the perthiosulfenic (P-CySSOH), perthiosulfinic (P-$CySSO_2H$), and perthiosulfonic acid (P-$CySSO_3H$) oxidation products of persulfides can efficiently be recycled via reduction of their -S-S- moieties, in contrast to the sulfinic ($CySO_2H$) or sulfonic acid ($CySO_3H$) products of thiol oxidation (Ono et al., 2014).

From a chemical perspective, it is important to emphasize that persulfide formation mostly inhibits the function of the modified Cys residue, with the only reported example of GAPDH, where activation is solicited by the persulfidation of the actual catalytic thiol (Mustafa et al., 2009). All the other examples of enzyme activations represent indirect effects via persulfide formation-induced inhibition of the respective negative regulator. Furthermore, the observed enhanced peroxide-scavenging potential of GSSH (Ida et al., 2014) could be simply due to a lower pK_a of GSSH versus GSH, resulting in an increase in the proportion of the more nucleophilic deprotonated form of the peroxide-scavenging sulfhydryl moiety (as in GSS^- vs. GS^-). It will be interesting to see how the reactivities of the deprotonated forms, i.e., the Cys thiolates versus the corresponding perthiolates compare to each other.

4.1. Mechanisms of persulfide formation

An early misconception that the direct reaction of sulfide with reduced Cys residues can give persulfides has been addressed (Greiner et al., 2013; Nagy & Winterbourn, 2010a; Toohey, 2012) and it is now widely appreciated that sulfide-mediated persulfide formation occurs either via the reactions of oxidized Cys derivatives with sulfide or sulfide oxidation products reacting with Cys thiols.

4.1.1 Persulfide formation via disulfide reduction

A potential mechanism for persulfide formation is the reduction of biologically abundant Cys-disulfide bonds by sulfide (Cavallini, Federici, & Barboni, 1970; Francoleon, Carrington, & Fukuto, 2011; Nielsen et al., 2011). However, the physiological relevance of this pathway has been questioned by many investigators based on either thermodynamic or kinetic considerations (Kabil & Banerjee, 2010; Paulsen & Carroll, 2013; Toohey, 2011, 2012; Zhang et al., 2014). In a comprehensive kinetic study on using model disulfides (Vasas, Doka, Fabian, & Nagy, 2014), we found that the rate of sulfide-mediated disulfide reduction is highly dependent on the reactivity of the disulfide moiety. This introduces orders of magnitude differences in second-order rate constants. On this basis, it would be predicted that activated functional disulfides, such as in SQR, *protein disulfide isomerase* (PDI), Mia40, *thioredoxin* (Trx), *ER oxidoreductin* (Ero1), *essential for respiration and vegetative growth sulfhydryl oxidase* (Erv1), *glutathione oxido-reductase* (GOR), or the *disulfide bond protein family* are likely to be rapidly reduced by sulfide. Corroborating this notion, the first proposed reaction step in SQR-mediated sulfide catabolism is the reduction of the intramolecular Cys160–Cys356 disulfide bond by sulfide to give a persulfide intermediate species (Cherney et al., 2010; Kabil & Banerjee, 2014).

Our experimental data are consistent with a kinetic model which includes a direct reaction between the disulfide and sulfide to give the corresponding persulfide and thiol derivatives (reaction 10 in Scheme 1). The persulfide can be further reduced by sulfide to give inorganic polysulfides and another thiol (reactions 12–13 in Scheme 1). However, at high cellular thiol concentrations, persulfide is likely to be reduced by another

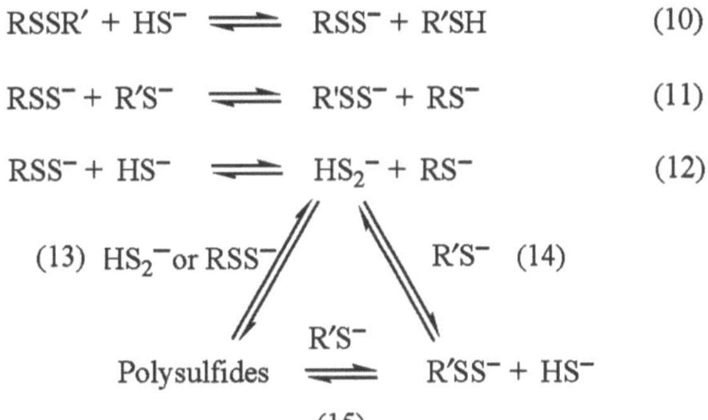

Scheme 1 Proposed pathways for the reactions of Cys-disulfides with sulfide (Vasas et al., 2014).

Cys thiol rather than sulfide (reaction 11 in Scheme 1), and even if inorganic polysulfides are generated, they are expected to eventually oxidize other thiols (reactions 14 and 15 in Scheme 1). Both pathways represent potential transsulfuration mechanisms (see Scheme 1).

These reactions are under strict thermodynamic control, with the distribution of the oxidized sulfane-sulfur moieties in Cys-persulfides and polysulfides depending on the corresponding equilibrium constants and their interrelations under the applied conditions. The two extreme scenarios in simple isolated reaction mixtures are represented by the reactions of 5,5′-dithiobis-(2-nitrobenzoic acid) (DTNB) and PTEN. Favorable reduction of DTNB by sulfide yields stoichiometric amounts of inorganic polysulfides and 5-thio-2-nitrobenzoic acid (TNB), even when sulfide is in deficit (Vasas et al., 2014). On the other hand, PTEN is efficiently oxidized by trace amounts of polysulfides even in the presence of a large excess of sulfide (Greiner et al., 2013).

Such a tight kinetic and thermodynamic control could potentiate the feasibility of intentional persulfide formation via sulfide-mediated disulfide reducing pathways in biology. We propose that in more oxidizing environments such as in the endoplasmic reticulum (ER) or mitochondria, reactive functional disulfides are likely to react with endogenous sulfide (as shown for SQR). Furthermore, the considerable steady-state concentrations of disulfides and sulfide predict that even if disulfide reducing pathways are relatively slow, they could also influence the distribution of sulfane-sulfur moieties in cytosolic environments, where redox homeostasis is mostly mediated by the glutathione system.

4.1.2 Persulfide formation via the reactions of Cys sulfenic acid species with sulfide

Recent studies mostly from the Carroll and Poole's laboratories revealed that some thiol proteins are detectable as sulfenylated derivatives in cellular systems (Gupta & Carroll, 2014; Poole, Karplus, & Claiborne, 2004). These Cys sulfenic acids (CySOH) are reactive toward thiol nucleophiles and therefore they are expected to react with sulfide to give the corresponding persulfide species.

$$\mathrm{CySOH + HS^- \rightarrow CySS^- + H_2O} \quad (16)$$

Steric factors at the active site of these proteins may stabilize their CySOH derivatives (Nagy & Winterbourn, 2010b). Therefore, the small size of sulfide should give it better access compared to larger aliphatic thiols. The rate constant for CySOH reduction by Cys at pH 7.4 is $>10^5\ M^{-1}s^{-1}$ in the absence of steric hindrance (Nagy & Ashby, 2007; Nagy, Lemma, & Ashby, 2007). However, the similar reaction of *human serum albumin*-CySOH, which is barely accessible for Cys is 4 orders of magnitude slower (Turell et al., 2008). Based on these considerations reactions of protein sulfenic acids with sulfide are likely to be rapid and considerably faster than with aliphatic thiols. In addition, the pK_a of H_2S is ~2 orders of magnitude smaller than the microscopic acid dissociation constants of GSH (Nagy, 2013; Nagy & Winterbourn, 2010b), which makes it a much better nucleophile under physiological conditions. Therefore, despite GSH being the most abundant thiol species in cells at concentrations 3–4 orders of magnitude larger than sulfide, kinetic factors could make reactions of protein sulfenic acids more favorable with sulfide than with GSH. This pathway is therefore a feasible route for protein persulfide formation. Moreover, the chemical nature, steady-state concentration and microenvironment of the CysOH moiety as well as the relative local concentrations of sulfide vs GSH should orchestrate the cross-talk between protein Cys glutathionylation (Ghezzi, 2013) and persulfide formation.

4.1.3 Persulfide formation via the reactions of oxidized sulfide species with Cys thiols

Polysulfides efficiently oxidize functional Cys residues on PTEN and *reduction–oxidation sensitive green fluorescent protein* (roGFP) (Greiner et al., 2013). Although I am not aware of a comprehensive mechanistic study on the reaction of polysulfides with Cys, some analogies can be drawn to the well-studied thiol-disulfide exchange reactions (Nagy, 2013): As I highlighted earlier in this review, the formation of polysulfides is a multistep oxidative process, which involves reactive intermediate species, such as HSOH or HSX (where X can be Cl, Br, I, or SCN) or HS_2^-. In a biological situation, these intermediates could be captured by Cys thiol residues and generate persulfides via reactions (14), (17), or (18).

$$\mathrm{CySH + HSOH \rightarrow CySSH + H_2O} \tag{17}$$

$$\mathrm{CySH + HSX \rightarrow CySSH + H^+ + X^-} \tag{18}$$

In addition, the reactions of Cys residues with polysulfides can generate both Cys-persulfides (reaction 15) and Cys-polysulfides (reaction 19).

$$\mathrm{CySH + HS_x{}^- \rightarrow CySS_yH + HS_{(x-y)}{}^-} \tag{19}$$

The relative reactivity of Cys residues with either of the above electrophilic oxidizing agents should provide kinetic selectivity for protein persulfide formation via the above pathways in biological situations. It is likely to be influenced by similar factors as in the well-studied Cys-thiol-disulfide exchange reactions, including the thiol pK_a, steric factors or charges, and coulombic or H-bonding interactions with neighboring functional groups. (Nagy, 2013).

4.1.4 Persulfide formation via radical pathways

$HS^{\bullet}$ formed by 1e^- oxidation of sulfide would be expected to react with a protein thiol to give the corresponding disulfide anion species ($RSSH^{\bullet -}$) (Das et al., 1999), which would react rapidly with oxygen to give superoxide and the corresponding persulfide:

$$HS^{\bullet} + RS^- \rightarrow RSSH^{\bullet -} \qquad (20)$$

$$RSSH^{\bullet -} + O_2 \rightarrow RSSH + O_2^{\bullet -} \qquad (21)$$

The rich radical chemistry of sulfur species predict that there are many more possibilities for persulfide formation via 1 electron-mediated pathways (Winterbourn, 2014), but in the absence of comprehensive experimental investigations it is premature to speculate on these.

5. SULFIDE SIGNALING VIA SULFIDE–HEMEPROTEIN INTERACTIONS

Sulfide has an affinity to coordinate and/or reduce metal centers in proteins. Most focus has historically been on heme proteins, which reversibly bind sulfide at their high spin Fe^{3+} centers to give the corresponding low-spin complexes. In addition, sulfide can reduce Fe^{3+} centers to Fe^{2+}, which can also form ferrous-sulfide complexes (Fig. 3).

Whether coordination or reduction occurs at an Fe^{3+} center is governed by (i) the active site environment of the protein (it has been suggested that stabilization of H_2S or HS^- at the active site promotes coordination or reduction, respectively, Pietri et al., 2009), (ii) sulfide concentration, and (iii) pH. The chemical aspects of sulfide interactions with heme proteins are discussed in detail in recent excellent review articles from Lopez-Garriga's group (Pietri, Roman-Morales, & Lopez-Garriga, 2011; Rios-Gonzalez, Roman-Morales, Pietri, & Lopez-Garriga, 2014); here the focus is on their potential role in sulfide signaling.

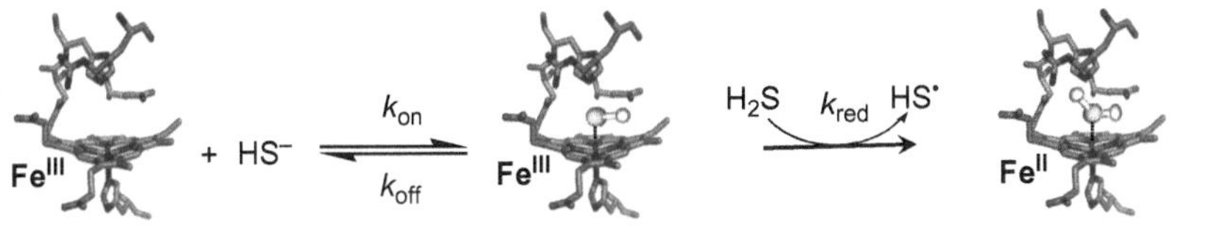

Figure 3 Reaction of sulfide with heme Fe^{3+} centers. Coordination of sulfide to a high spin Fe^{3+} center of a heme protein (k_{on}) results in a low-spin Fe^{3+}-sulfide complex. Sulfide can be liberated from this complex (k_{off}) but redox reaction with another sulfide (k_{red}) can generate the ferrous(Fe^{2+})-sulfide complex and $HS^{\bullet}$. (See the color plate.)

5.1. Sulfide mediates heme protein functions

The effect of sulfide on CcO has been extensively studied because this interaction inhibits mitochondrial respiration and is thought to be a major component of sulfide toxicity. However, recent studies have demonstrated that this interaction and the consequent biological effects are dose dependent. At high concentrations of sulfide the well-studied toxicological effect is due to inhibition of oxygen binding via a series of complex redox and coordination chemical reactions at the active site heme a_3 and Cu_B centers of CcO (Pietri et al., 2011). However, low concentrations of sulfide can fuel oxidative phosphorylation (Goubern et al., 2007; Helmy et al., 2014; Powell & Somero, 1986) by acting as an electron donor to the ferric and cupric derivatives of the enzyme and convert them to their active reduced forms (Collman, Ghosh, Dey, & Decreau, 2009). It is most likely that a reversible inhibition of CcO by moderate sulfide concentrations is responsible for the intriguing sulfide-induced hibernation-like state in mice (Blackstone & Roth, 2007; Collman et al., 2009).

Interactions of sulfide with *hemoglobins* (Hb) and *myoglobins* (Mb) have been extensively studied due to early observations of sulfhemoglobinemia (Park & Nagel, 1984). This rare medical condition is associated with a unique heme modification, where the sulfur is incorporated into a pyrrol ring of the heme porphyrin to give a so-called sulfheme derivative (Fig. 4) (Rios-Gonzalez et al., 2014).

Sulfheme production is likely to require the intermediate oxidative formation of higher valent iron centers (Michel, 1938; Roman-Morales et al., 2010; Romero, Ordonez, Arduini, & Cadenas, 1992) and appears to be irreversible. The impaired oxygen binding capacity of the metal centers in these sulfheme derivatives is thought to be solely responsible for their toxicological effect, and low concentrations of sulfhemoglobin or sulfmyoglobin are

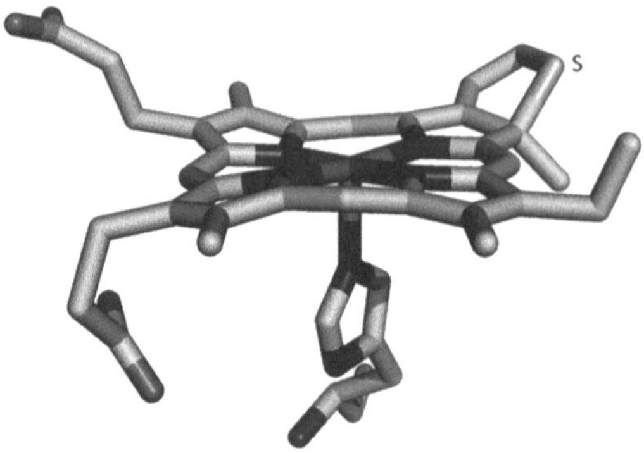

Figure 4 Sulfheme moiety from the crystal structure of sulfmyoglobin using PDB file 1YMC. (See the color plate.)

well tolerated (Nicholls, 1961). The irreversible nature of sulfheme formation suggests that this chemistry is unlikely to have a role in sulfide signaling. However, reduction of the strong oxidant ferryl forms to nontoxic species could have a protective antioxidant function (see later). Reversible sulfide binding to Hb and Mb (Pietri et al., 2011) and consequent buffering free steady-state sulfide levels could have an indirect signaling role via sulfide trafficking and preventing sulfide toxicity (Nagy et al., 2014).

Sulfide interacts with the heme peroxidases, MPO (Palinkas et al., 2014) *lactoperoxidase* (LPO) (Nakamura, Nakamura, Yamazaki, & Morrison, 1984), and *thyroid peroxidase* (TPO) (Ohtaki, Nakagawa, Nakamura, & Yamazaki, 1982) and with *catalase* (Nicholls, 1961). These enzymes have important roles in the production, destruction, and interconversion of ROS and therefore the modulation of their activities by sulfide could represent a cross-talk between redox signaling and sulfide signaling. We recently published a comprehensive mechanistic analysis of the interaction of sulfide with MPO (Palinkas et al., 2014). Similar reactions were proposed for LPO, TPO, and *catalase*. With MPO, sulfide readily coordinates to the active site Fe^{3+} center of the native enzyme to give a low-spin complex (see Fig. 5). Subsequent reduction of the Fe^{3+} center occurs at higher sulfide levels albeit at a slower rate. These interactions translate to highly efficient enzyme inhibition under physiological conditions. Inhibition is fully reversible and a drop in sulfide levels quickly restores enzymatic activity, thus providing the potential for a fine-tuned regulatory function. Sulfide is also a good substrate for the redox intermediates of MPO, reducing Compound I and Compound II in consecutive rapid 1 and 2 electron processes, respectively. The resulting ferrous MPO intermediate also binds sulfide (our study

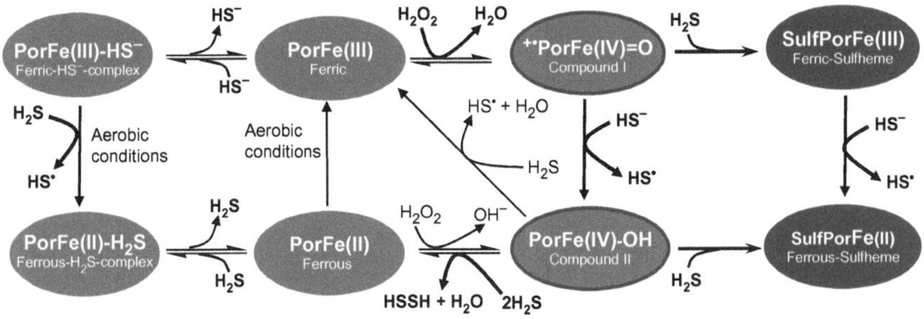

Figure 5 Proposed model for the interactions of heme protein derivatives with sulfide. The reactions of the green (light gray in the print version) enzyme forms represent the proposed model for the interactions of sulfide with the catalytic cycle of MPO. The red (dark gray in the print version) enzyme derivatives depict the proposed reduction pathways of Hb and Mb ferryl species by sulfide.

presented the first experimental evidence for the formation of a ferrous heme-sulfide complex), which is slowly recycled and therefore serves as a sink in the catalytic cycle.

MPO has numerous activities including the conversion of H_2O_2 to hypohalous acids (Klebanoff, Kettle, Rosen, Winterbourn, & Nauseef, 2013), oxidation of Tyr residues and other phenols in radical-mediated processes (Heinecke, Li, Daehnke, & Goldstein, 1993; Klebanoff, 2005; Nagy, Kettle, & Winterbourn, 2010), and removal of H_2O_2 (Kettle & Winterbourn, 2001) and superoxide (Kettle, Anderson, Hampton, & Winterbourn, 2007; Kettle, Maroz, Woodroffe, Winterbourn, & Anderson, 2011). Most other heme peroxidases exhibit similar ROS producing/destroying/regulating functions (Dunford, 1999). Therefore, the reversible tight regulation of peroxidase activities could serve as a prime example for sulfide's mediatory role in redox biology.

5.2. Heme proteins generate sulfide oxidation products

As noted above, sulfide not only coordinates but also engages in redox reactions with the metal centers of heme proteins. These can either be 1 or 2 electron reactions to produce the thiyl radical, polysulfides, thiosulfate, and other sulfide oxidation products (Collman et al., 2009; Palinkas et al., 2014; Pietri et al., 2009; Wedmann et al., 2014). Reduction of the Fe^{3+} center to Fe^{2+} appears to be a common reaction for heme proteins (k_{red} in Fig. 3); it was shown for MPO, LPO, catalase CcO, Hb, or Mb. These reactions generate the highly reactive $HS^{\bullet}$. As noted previously, $HS^{\bullet}$ can eventually produce protein persulfides or inorganic polysulfides. It may also interact with endogenous radical species such as NO. The interaction of NO with sulfide and its importance in the cross-talk between NO and sulfide signaling is gaining ground (Ali et al., 2006; Altaany, Ju, Yang, & Wang, 2014; Cortese-Krott et al., 2014; Filipovic et al., 2012; King et al., 2014; Lo Faro, Fox, Whatmore, Winyard, & Whiteman, 2014; Whiteman et al., 2006), but these studies are still in their infancy and therefore I would like to avoid any speculation or discussion of the current literature data.

In general, most sulfide oxidation products are bioactive species with distinct chemical and biological properties; thereby sulfide oxidation could potentially represent a modulating function for heme proteins in sulfide signaling.

5.3. Antioxidant properties of sulfide via reduction of metal centers with higher oxidation states

Reduction of oxidized heme centers could be a potential pathway of sulfide-mediated antioxidant effects. We have recently shown that the rapid ($k = 1.1 \times 10^6\ M^{-1}s^{-1}$) $1e^-$ reduction of MPO Compound I (the first redox intermediate in the catalytic cycle of the enzyme which contains an oxoiron(IV) heme center, i.e., Fe(IV)=O, and a porphyrin π-cation radical) by sulfide gives Compound II (an oxoiron(IV) heme). Compound II, which represents the resting state of the enzyme during endogenous catalysis, is also favorably reduced by sulfide ($k = 2 \times 10^5\ M^{-1}s^{-1}$), albeit in a $2e^-$ process to give the ferrous MPO species. In contrast, Compound I and/or II forms of Hb and Mb were proposed to be reduced by sulfide to give the corresponding sulfheme derivatives (see Fig. 5). Ferryl to ferrous peroxidase reduction (as in MPO) maintains catalytic activity and generates polysulfides and/or other bioactive sulfide oxidation products, but sulfheme formation (as in Hb or Mb) could potentially represent an antioxidant mechanism (Smith & Gosselin, 1964), where both sulfide and the ferryl heme are eliminated from the redox cycle. Therefore, the partitioning of these pathways will likely be protein specific with intentional biological functions. It was suggested that the presence of a distal His residue in a hemeprotein is an important driving force of sulfheme formation (Roman-Morales et al., 2010). However, we have not observed evidence for substantial amounts of sulfheme formation in human MPO, which apparently contains a distal His residue. Hence a detailed comprehensive mechanistic study is required for all individual proteins to further elucidate the factors that drive the mechanisms of ferryl heme reduction by sulfide.

6. CONCLUSIONS

Major molecular pathways for the signaling and antioxidant actions of sulfide include redox and/or coordination chemical interactions with heme proteins (or other metalloproteins) and redox chemistry with functional cysteine residues. These interactions can promote the apparent biological effects via modulation of enzymatic activities, production of bioactive sulfide oxidation products, or sulfide trafficking.

Regulation of thiol proteins mostly occur via Cys-persulfide formation for which the major proposed pathways are (1) polysulfide-mediated oxidation of Cys, (2) reduction of disulfides, and (3) condensation of sulfide with CySOH. Persulfide formation mostly inhibits Cys thiol functions, but it may also activate cellular pathways via diminishing the activities of negative regulators. For example, persulfide formation on Keap1 or MEK1 can activate Nrf2 or PARP-1 (respectively), which are important elements of the cellular antioxidant stratagem.

The rich chemistry of sulfide with heme proteins includes coordination, reduction of iron centers, and formation of sulfheme derivatives. These interactions could orchestrate the bioavailability of ROS, regulate heme-enzyme functions, produce bioactive sulfide metabolites, and protect against heme-induced oxidative stress.

Coordination to heme proteins and covalent binding in persulfides contribute to sulfide storage in biological systems, and therefore play an important role in sulfide bioavailability. Large biological sulfide pools serve as buffers of steady-state free sulfide levels. The mechanism of sulfide release depends on the binding partner: e.g., with persulfides a reducing partner is required, whereas release from heme coordination requires only a shift in chemical equilibria. Importantly, thermodynamic control of the sulfide buffer system keeps free sulfide levels below the toxic threshold, whereas kinetic control is most likely to have a regulating function in signaling processes. The latter is best exemplifies by observations that slow releasing and fast releasing sulfide donors trigger different biological events.

Acceptance of the physiological significance of polysulfides is gaining ground. An increasing number of publications propose that they could be responsible for many of the biological functions that have been attributed to sulfide. Polysulfides are produced endogenously, by direct reactions of sulfide with ROS and by enzymatic processes. Their complex chemistry is not yet well understood and rigorous investigations are needed to better understand how to handle these chemicals and learn more about their biological actions.

ACKNOWLEDGMENTS

Financial support from FP7-PEOPLE-2010-RG (Marie Curie International Reintegration Grant; grant No.: PIRG08-GA-2010-277006) and The Hungarian National Science Foundation (OTKA; grant No.: K 109843) are greatly acknowledged. I thank Christine and Michael Winterbourn for critical reading of the chapter and Zsuzsanna Lénárt for her help in formatting.

REFERENCES

Abe, K., & Kimura, H. (1996). The possible role of hydrogen sulfide as an endogenous neuromodulator. *Journal of Neuroscience*, *16*(3), 1066–1071.

Ali, M. Y., Ping, C. Y., Mok, Y. Y., Ling, L., Whiteman, M., Bhatia, M., et al. (2006). Regulation of vascular nitric oxide in vitro and in vivo; a new role for endogenous hydrogen sulphide? *British Journal of Pharmacology*, *149*(6), 625–634.

Altaany, Z., Ju, Y., Yang, G., & Wang, R. (2014). The coordination of S-sulfhydration, S-nitrosylation, and phosphorylation of endothelial nitric oxide synthase by hydrogen sulfide. *Science Signaling*, 7(342), ra87.

Blackstone, E., Morrison, M., & Roth, M. B. (2005). H_2S induces a suspended animation-like state in mice. *Science*, *308*(5721), 518.

Blackstone, E., & Roth, M. B. (2007). Suspended animation-like state protects mice from lethal hypoxia. *Shock*, *27*(4), 370–372.

Cavallini, D., Federici, G., & Barboni, E. (1970). Interaction of proteins with sulfide. *European Journal of Biochemistry*, *14*(1), 169–174.

Cherney, M. M., Zhang, Y., Solomonson, M., Weiner, J. H., & James, M. N. (2010). Crystal structure of sulfide:quinone oxidoreductase from acidithiobacillus ferrooxidans: Insights into sulfidotrophic respiration and detoxification. *Journal of Molecular Biology*, *398*(2), 292–305.

Chiku, T., Padovani, D., Zhu, W., Singh, S., Vitvitsky, V., & Banerjee, R. (2009). H_2S biogenesis by human cystathionine gamma-lyase leads to the novel sulfur metabolites lanthionine and homolanthionine and is responsive to the grade of hyperhomocysteinemia. *Journal of Biological Chemistry*, *284*(17), 11601–11612.

Coletta, C., Papapetropoulos, A., Erdelyi, K., Olah, G., Modis, K., Panopoulos, P., et al. (2012). Hydrogen sulfide and nitric oxide are mutually dependent in the regulation of angiogenesis and endothelium-dependent vasorelaxation. *Proceedings of the National Academy of Sciences of the United States of America*, *109*(23), 9161–9166.

Collman, J. P., Ghosh, S., Dey, A., & Decreau, R. A. (2009). Using a functional enzyme model to understand the chemistry behind hydrogen sulfide induced hibernation. *Proceedings of the National Academy of Sciences of the United States of America*, *106*(52), 22090–22095.

Cortese-Krott, M. M., Fernandez, B. O., Santos, J. L., Mergia, E., Grman, M., Nagy, P., et al. (2014). Nitrosopersulfide (SSNO(−)) accounts for sustained NO bioactivity of S-nitrosothiols following reaction with sulfide. *Redox Biology*, *2*, 234–244.

Das, T. N., Huie, R. E., Neta, P., & Padmaja, S. (1999). Reduction potential of the sulfhydryl radical: Pulse radiolysis and laser flash photolysis studies of the formation and reactions of center dot SH and HSSH center dot(−) in aqueous solutions. *Journal of Physical Chemistry A*, *103*(27), 5221–5226.

Dunford, H. B. (1999). *Heme peroxidases*. New York: John Wiley, xiii + 475 pp.

Elrod, J. W., Calvert, J. W., Morrison, J., Doeller, J. E., Kraus, D. W., Tao, L., et al. (2007). Hydrogen sulfide attenuates myocardial ischemia-reperfusion injury by preservation of mitochondrial function. *Proceedings of the National Academy of Sciences of the United States of America*, *104*(39), 15560–15565.

Filipovic, M. R., Miljkovic, J., Nauser, T., Royzen, M., Klos, K., Shubina, T., et al. (2012). Chemical characterization of the smallest S-nitrosothiol, HSNO; cellular cross-talk of H_2S and S-nitrosothiols. *Journal of the American Chemical Society*, *134*(29), 12016–12027.

Flannigan, K. L., Agbor, T. A., Blackler, R. W., Kim, J. J., Khan, W. I., Verdu, E. F., et al. (2014). Impaired hydrogen sulfide synthesis and IL-10 signaling underlie hyperhomocysteinemia-associated exacerbation of colitis. *Proceedings of the National Academy of Sciences of the United States of America*, *111*(37), 13559–13564.

Francoleon, N. E., Carrington, S. J., & Fukuto, J. M. (2011). The reaction of H_2S with oxidized thiols: Generation of persulfides and implications to H_2S biology. *Archives of Biochemistry and Biophysics*, *516*(2), 146–153.

Fu, Z., Liu, X., Geng, B., Fang, L., & Tang, C. (2008). Hydrogen sulfide protects rat lung from ischemia-reperfusion injury. *Life Sciences*, *82*(23–24), 1196–1202.

Ghezzi, P. (2013). Protein glutathionylation in health and disease. *Biochimica et Biophysica Acta—Reviews on Cancer*, *1830*(5), 3165–3172.

Giggenbach, W. (1972). Optical-spectra and equilibrium distribution of polysulfide ions in aqueous-solution at 20°. *Inorganic Chemistry*, *11*(6), 1201–1207.

Goubern, M., Andriamihaja, M., Nubel, T., Blachier, F., & Bouillaud, F. (2007). Sulfide, the first inorganic substrate for human cells. *FASEB Journal*, *21*(8), 1699–1706.

Greiner, R., Palinkas, Z., Basell, K., Becher, D., Antelmann, H., Nagy, P., et al. (2013). Polysulfides link H_2S to protein thiol oxidation. *Antioxidants & Redox Signaling*, *19*(15), 1749–1765.

Gupta, V., & Carroll, K. S. (2014). Sulfenic acid chemistry, detection and cellular lifetime. *Biochimica et Biophysica Acta—Reviews on Cancer*, *1840*(2), 847–875.

Heinecke, J. W., Li, W., Daehnke, H. L., 3rd., & Goldstein, J. A. (1993). Dityrosine, a specific marker of oxidation, is synthesized by the myeloperoxidase-hydrogen peroxide system of human neutrophils and macrophages. *Journal of Biological Chemistry*, *268*(6), 4069–4077.

Helmy, N., Prip-Buus, C., Vons, C., Lenoir, V., Abou-Hamdan, A., Guedouari-Bounihi, H., et al. (2014). Oxidation of hydrogen sulfide by human liver mitochondria. *Nitric Oxide*, *41*, 105–112.

Hildebrandt, T. M., & Grieshaber, M. K. (2008). Three enzymatic activities catalyze the oxidation of sulfide to thiosulfate in mammalian and invertebrate mitochondria. *FEBS Journal*, *275*(13), 3352–3361.

Hoffmann, M. R. (1977). Kinetics and mechanism of oxidation of hydrogen-sulfide by hydrogen-peroxide in acidic solution. *Environmental Science & Technology*, *11*(1), 61–66.

Hourihan, J. M., Kenna, J. G., & Hayes, J. D. (2013). The gasotransmitter hydrogen sulfide induces nrf2-target genes by inactivating the keap1 ubiquitin ligase substrate adaptor through formation of a disulfide bond between cys-226 and cys-613. *Antioxidants & Redox Signaling*, *19*(5), 465–481.

Ida, T., Sawa, T., Ihara, H., Tsuchiya, Y., Watanabe, Y., Kumagai, Y., et al. (2014). Reactive cysteine persulfides and S-polythiolation regulate oxidative stress and redox signaling. *Proceedings of the National Academy of Sciences of the United States of America*, *111*(21), 7606–7611.

Jha, S., Calvert, J. W., Duranski, M. R., Ramachandran, A., & Lefer, D. J. (2008). Hydrogen sulfide attenuates hepatic ischemia-reperfusion injury: Role of antioxidant and anti-apoptotic signaling. *American Journal of Physiology. Heart and Circulatory Physiology*, *295*(2), H801–H806.

Kabil, O., & Banerjee, R. (2010). Redox biochemistry of hydrogen sulfide. *Journal of Biological Chemistry*, *285*(29), 21903–21907.

Kabil, O., & Banerjee, R. (2014). Enzymology of H_2S biogenesis, decay and signaling. *Antioxidants & Redox Signaling*, *20*(5), 770–782.

Kettle, A. J., Anderson, R. F., Hampton, M. B., & Winterbourn, C. C. (2007). Reactions of superoxide with myeloperoxidase. *Biochemistry*, *46*(16), 4888–4897.

Kettle, A. J., Maroz, A., Woodroffe, G., Winterbourn, C. C., & Anderson, R. F. (2011). Spectral and kinetic evidence for reaction of superoxide with compound I of myeloperoxidase. *Free Radical Biology and Medicine*, *51*(12), 2190–2194.

Kettle, A. J., & Winterbourn, C. C. (2001). A kinetic analysis of the catalase activity of myeloperoxidase. *Biochemistry*, *40*(34), 10204–10212.

Kimura, H. (2013). Physiological role of hydrogen sulfide and polysulfide in the central nervous system. *Neurochemistry International*, *63*(5), 492–497.

Kimura, H. (2014). Signaling molecules: Hydrogen sulfide and polysulfide. *Antioxidants & Redox Signaling*. http://dx.doi.org/10.1089/ars.2014.5869.

Kimura, Y., & Kimura, H. (2004). Hydrogen sulfide protects neurons from oxidative stress. *FASEB Journal*, *18*(10), 1165–1167.

Kimura, Y., Mikami, Y., Osumi, K., Tsugane, M., Oka, J., & Kimura, H. (2013). Polysulfides are possible H_2S-derived signaling molecules in rat brain. *FASEB Journal*, *27*(6), 2451–2457.

King, A. L., Polhemus, D. J., Bhushan, S., Otsuka, H., Kondo, K., Nicholson, C. K., et al. (2014). Hydrogen sulfide cytoprotective signaling is endothelial nitric oxide synthase-nitric oxide dependent. *Proceedings of the National Academy of Sciences of the United States of America*, *111*(8), 3182–3187.

Klebanoff, S. J. (2005). Myeloperoxidase: Friend and foe. *Journal of Leukocyte Biology*, *77*(5), 598–625.

Klebanoff, S. J., Kettle, A. J., Rosen, H., Winterbourn, C. C., & Nauseef, W. M. (2013). Myeloperoxidase: A front-line defender against phagocytosed microorganisms. *Journal of Leukocyte Biology*, *93*(2), 185–198.

Koike, S., Ogasawara, Y., Shibuya, N., Kimura, H., & Ishii, K. (2013). Polysulfide exerts a protective effect against cytotoxicity caused by t-butylhydroperoxide through Nrf2 signaling in neuroblastoma cells. *FEBS Letters*, *587*(21), 3548–3555.

Krishnan, N., Fu, C., Pappin, D. J., & Tonks, N. K. (2011). H_2S-Induced sulfhydration of the phosphatase PTP1B and its role in the endoplasmic reticulum stress response. *Science Signaling*, *4*(203), ra86.

Laggner, H., Muellner, M. K., Schreier, S., Sturm, B., Hermann, M., Exner, M., et al. (2007). Hydrogen sulphide: A novel physiological inhibitor of LDL atherogenic modification by HOCl. *Free Radical Research*, *41*(7), 741–747.

Lagoutte, E., Mimoun, S., Andriamihaja, M., Chaumontet, C., Blachier, F., & Bouillaud, F. (2010). Oxidation of hydrogen sulfide remains a priority in mammalian cells and causes reverse electron transfer in colonocytes. *Biochimica et Biophysica Acta—Reviews on Cancer*, *1797*(8), 1500–1511.

Lessner, P. M., Mclarnon, F. R., Winnick, J., & Cairns, E. J. (1993). The dependence of aqueous sulfur-polysulfide redox potential on electrolyte-composition and temperature. *Journal of The Electrochemical Society*, *140*(7), 1847–1849.

Lo Faro, M. L., Fox, B., Whatmore, J. L., Winyard, P. G., & Whiteman, M. (2014). Hydrogen sulfide and nitric oxide interactions in inflammation. *Nitric Oxide*, *41C*, 38–47.

Melideo, S. L., Jackson, M. R., & Jorns, M. S. (2014). Biosynthesis of a central intermediate in hydrogen sulfide metabolism by a novel human sulfurtransferase and its yeast ortholog. *Biochemistry*, *53*(28), 4739–4753.

Michel, H. O. (1938). A study of sulfhemoglobin. *Journal of Biological Chemistry*, *126*(1), 323–348.

Mikami, Y., Shibuya, N., Kimura, Y., Nagahara, N., Ogasawara, Y., & Kimura, H. (2011). Thioredoxin and dihydrolipoic acid are required for 3-mercaptopyruvate sulfurtransferase to produce hydrogen sulfide. *Biochemical Journal*, *439*(3), 479–485.

Modis, K., Panopoulos, P., Coletta, C., Papapetropoulos, A., & Szabo, C. (2013). Hydrogen sulfide-mediated stimulation of mitochondrial electron transport involves inhibition of the mitochondrial phosphodiesterase 2A, elevation of cAMP and activation of protein kinase A. *Biochemical Pharmacology, 86*(9), 1311–1319.

Mustafa, A. K., Gadalla, M. M., Sen, N., Kim, S., Mu, W., Gazi, S. K., et al. (2009). H_2S signals through protein S-sulfhydration. *Science Signaling, 2*(96), ra72.

Mustafa, A. K., Sikka, G., Gazi, S. K., Steppan, J., Jung, S. M., Bhunia, A. K., et al. (2011). Hydrogen sulfide as endothelium-derived hyperpolarizing factor sulfhydrates potassium channels. *Circulation Research, 109*(11), 1259–1268.

Nagahara, N. (2013). Regulation of mercaptopyruvate sulfurtransferase activity via intrasubunit and intersubunit redox-sensing switches. *Antioxidants & Redox Signaling, 19*(15), 1792–1802.

Nagy, P. (2013). Kinetics and mechanisms of thiol-disulfide exchange covering direct substitution and thiol oxidation-mediated pathways. *Antioxidants & Redox Signaling, 18*(13), 1623–1641.

Nagy, P., & Ashby, M. T. (2007). Reactive sulfur species: Kinetics and mechanisms of the oxidation of cysteine by hypohalous acid to give cysteine sulfenic acid. *Journal of the American Chemical Society, 129*(45), 14082–14091.

Nagy, P., Jameson, G. N., & Winterbourn, C. C. (2009). Kinetics and mechanisms of the reaction of hypothiocyanous acid with 5-thio-2-nitrobenzoic acid and reduced glutathione. *Chemical Research in Toxicology, 22*(11), 1833–1840.

Nagy, P., Kettle, A. J., & Winterbourn, C. C. (2010). Neutrophil-mediated oxidation of enkephalins via myeloperoxidase-dependent addition of superoxide. *Free Radical Biology and Medicine, 49*(5), 792–799.

Nagy, P., Lemma, K., & Ashby, M. T. (2007). Reactive sulfur species: Kinetics and mechanisms of the reaction of cysteine thiosulfinate ester with cysteine to give cysteine sulfenic acid. *Journal of Organic Chemistry, 72*(23), 8838–8846.

Nagy, P., Palinkas, Z., Nagy, A., Budai, B., Toth, I., & Vasas, A. (2014). Chemical aspects of hydrogen sulfide measurements in physiological samples. *Biochimica et Biophysica Acta—General Subjects, 1840*(2), 876–891.

Nagy, P., & Winterbourn, C. C. (2010a). Rapid reaction of hydrogen sulfide with the neutrophil oxidant hypochlorous acid to generate polysulfides. *Chemical Research in Toxicology, 23*(10), 1541–1543.

Nagy, P., & Winterbourn, C. C. (2010b). Redox chemistry of biological thiols. (vol. 4) *Advances in molecular toxicology*. 183–222.

Nakamura, S., Nakamura, M., Yamazaki, I., & Morrison, M. (1984). Reactions of ferryl lactoperoxidase (compound II) with sulfide and sulfhydryl compounds. *Journal of Biological Chemistry, 259*(11), 7080–7085.

Nicholls, P. (1961). The formation and properties of sulphmyoglobin and sulphcatalase. *Biochemical Journal, 81*, 374–383.

Nielsen, R. W., Tachibana, C., Hansen, N. E., & Winther, J. R. (2011). Trisulfides in proteins. *Antioxidants & Redox Signaling, 15*(1), 67–75.

Ohtaki, S., Nakagawa, H., Nakamura, M., & Yamazaki, I. (1982). Reactions of purified hog thyroid peroxidase with H_2O_2, tyrosine, and methylmercaptoimidazole (goitrogen) in comparison with bovine lactoperoxidase. *Journal of Biological Chemistry, 257*(2), 761–766.

Ono, K., Akaike, T., Sawa, T., Kumagai, Y., Wink, D. A., Tantillo, D. J., et al. (2014). The redox chemistry and chemical biology of H_2S, hydropersulfides and derived species: Implications to their possible biological activity and utility. *Free Radical Biology and Medicine, 77*, 82–94.

Palinkas, Z., Furtmuller, P. G., Nagy, A., Jakopitsch, C., Pirker, K. F., Magierowski, M., et al. (2014). *Interactions of hydrogen sulfide with myeloperoxidase.* British Journal of Pharmacology. http://dx.doi.org/10.1111/bph.12769.

Papapetropoulos, A., Pyriochou, A., Altaany, Z., Yang, G., Marazioti, A., Zhou, Z., et al. (2009). Hydrogen sulfide is an endogenous stimulator of angiogenesis. *Proceedings of the National Academy of Sciences of the United States of America, 106*(51), 21972–21977.
Park, C. M., & Nagel, R. L. (1984). Sulfhemoglobinemia. Clinical and molecular aspects. *New England Journal of Medicine, 310*(24), 1579–1584.
Paul, B. D., & Snyder, S. H. (2012). H_2S signalling through protein sulfhydration and beyond. *Nature Reviews Molecular Cell Biology, 13*(8), 499–507.
Paulsen, C. E., & Carroll, K. S. (2013). Cysteine-mediated redox signaling: Chemistry, biology, and tools for discovery. *Chemical Reviews, 113*(7), 4633–4679.
Pietri, R., Lewis, A., Leon, R. G., Casabona, G., Kiger, L., Yeh, S. R., et al. (2009). Factors controlling the reactivity of hydrogen sulfide with hemeproteins. *Biochemistry, 48*(22), 4881–4894.
Pietri, R., Roman-Morales, E., & Lopez-Garriga, J. (2011). Hydrogen sulfide and hemeproteins: Knowledge and mysteries. *Antioxidants & Redox Signaling, 15*(2), 393–404.
Poole, L. B., Karplus, P. A., & Claiborne, A. (2004). Protein sulfenic acids in redox signaling. *Annual Review of Pharmacology and Toxicology, 44*, 325–347.
Powell, M. A., & Somero, G. N. (1986). Hydrogen sulfide oxidation is coupled to oxidative phosphorylation in mitochondria of Solemya reidi. *Science, 233*(4763), 563–566.
Rios-Gonzalez, B. B., Roman-Morales, E. M., Pietri, R., & Lopez-Garriga, J. (2014). Hydrogen sulfide activation in hemeproteins: The sulfheme scenario. *Journal of Inorganic Biochemistry, 133*, 78–86.
Roman-Morales, E., Pietri, R., Ramos-Santana, B., Vinogradov, S. N., Lewis-Ballester, A., & Lopez-Garriga, J. (2010). Structural determinants for the formation of sulfhemeprotein complexes. *Biochemical and Biophysical Research Communications, 400*(4), 489–492.
Romero, F. J., Ordonez, I., Arduini, A., & Cadenas, E. (1992). The reactivity of thiols and disulfides with different redox states of myoglobin. Redox and addition reactions and formation of thiyl radical intermediates. *Journal of Biological Chemistry, 267*(3), 1680–1688.
Sen, N., Paul, B. D., Gadalla, M. M., Mustafa, A. K., Sen, T., Xu, R., et al. (2012). Hydrogen sulfide-linked sulfhydration of NF-kappaB mediates its antiapoptotic actions. *Molecular Cell, 45*(1), 13–24.
Smith, R. P., & Gosselin, R. E. (1964). The influence of methemoglobinemia on the lethality of some toxic anions. Ii. Sulfide. *Toxicology and Applied Pharmacology, 6*, 584–592.
Suzuki, K., Olah, G., Modis, K., Coletta, C., Kulp, G., Gero, D., et al. (2011). Hydrogen sulfide replacement therapy protects the vascular endothelium in hyperglycemia by preserving mitochondrial function. *Proceedings of the National Academy of Sciences of the United States of America, 108*(33), 13829–13834.
Szabo, C., Coletta, C., Chao, C., Modis, K., Szczesny, B., Papapetropoulos, A., et al. (2013). Tumor-derived hydrogen sulfide, produced by cystathionine-beta-synthase, stimulates bioenergetics, cell proliferation, and angiogenesis in colon cancer. *Proceedings of the National Academy of Sciences of the United States of America, 110*(30), 12474–12479.
Szabo, C., Ransy, C., Modis, K., Andriamihaja, M., Murghes, B., Coletta, C., et al. (2014). Regulation of mitochondrial bioenergetic function by hydrogen sulfide. Part I. Biochemical and physiological mechanisms. *British Journal of Pharmacology, 171*(8), 2099–2122.
Theissen, U., Hoffmeister, M., Grieshaber, M., & Martin, W. (2003). Single eubacterial origin of eukaryotic sulfide:quinone oxidoreductase, a mitochondrial enzyme conserved from the early evolution of eukaryotes during anoxic and sulfidic times. *Molecular Biology and Evolution, 20*(9), 1564–1574.
Toohey, J. I. (1989). Sulphane sulphur in biological systems: A possible regulatory role. *Biochemical Journal, 264*(3), 625–632.

Toohey, J. I. (2011). Sulfur signaling: Is the agent sulfide or sulfane? *Analytical Biochemistry, 413*(1), 1–7.
Toohey, J. I. (2012). The conversion of H_2S to sulfane sulfur. *Nature Reviews Molecular Cell Biology, 13*(12), 803, author reply p 803.
Turell, L., Botti, H., Carballal, S., Ferrer-Sueta, G., Souza, J. M., Duran, R., et al. (2008). Reactivity of sulfenic acid in human serum albumin. *Biochemistry, 47*(1), 358–367.
Vasas, A., Doka, E., Fabian, I., & Nagy, P. (2014). Kinetic and thermodynamic studies on the disulfide-bond reducing potential of hydrogen sulfide. *Accepted for publication.*
Vitvitsky, V., Kabil, O., & Banerjee, R. (2012). High turnover rates for hydrogen sulfide allow for rapid regulation of its tissue concentrations. *Antioxidants & Redox Signaling, 17*(1), 22–31.
Wedmann, R., Bertlein, S., Macinkovic, I., Boltz, S., Miljkovic, J., Munoz, L. E., et al. (2014). Working with "H_2S": Facts and apparent artifacts. *Nitric Oxide, 41*, 85–96.
Westrop, G. D., Georg, I., & Coombs, G. H. (2009). The mercaptopyruvate sulfurtransferase of Trichomonas vaginalis links cysteine catabolism to the production of thioredoxin persulfide. *Journal of Biological Chemistry, 284*(48), 33485–33494.
Whiteman, M., Armstrong, J. S., Chu, S. H., Jia-Ling, S., Wong, B. S., Cheung, N. S., et al. (2004). The novel neuromodulator hydrogen sulfide: An endogenous peroxynitrite 'scavenger'? *Journal of Neurochemistry, 90*(3), 765–768.
Whiteman, M., Cheung, N. S., Zhu, Y. Z., Chu, S. H., Siau, J. L., Wong, B. S., et al. (2005). Hydrogen sulphide: A novel inhibitor of hypochlorous acid-mediated oxidative damage in the brain? *Biochemical and Biophysical Research Communications, 326*(4), 794–798.
Whiteman, M., Le Trionnaire, S., Chopra, M., Fox, B., & Whatmore, J. (2011). Emerging role of hydrogen sulfide in health and disease: Critical appraisal of biomarkers and pharmacological tools. *Clinical Science (London, England: 1979), 121*(11), 459–488.
Whiteman, M., Li, L., Kostetski, I., Chu, S. H., Siau, J. L., Bhatia, M., et al. (2006). Evidence for the formation of a novel nitrosothiol from the gaseous mediators nitric oxide and hydrogen sulphide. *Biochemical and Biophysical Research Communications, 343*(1), 303–310.
Winterbourn, C. C. (2014). Are free radicals involved in thiol-based redox signaling? *Free Radical Biology and Medicine.* http://dx.doi.org/10.1016/j.freeradbiomed.2014.08.017.
Yadav, P. K., Yamada, K., Chiku, T., Koutmos, M., & Banerjee, R. (2013). Structure and kinetic analysis of H_2S production by human mercaptopyruvate sulfurtransferase. *Journal of Biological Chemistry, 288*(27), 20002–20013.
Yamanishi, T., & Tuboi, S. (1981). The mechanism of the L-cystine cleavage reaction catalyzed by rat liver gamma-cystathionase. *Journal of Biochemistry, 89*(6), 1913–1921.
Yang, G., Wu, L., Jiang, B., Yang, W., Qi, J., Cao, K., et al. (2008). H_2S as a physiologic vasorelaxant: Hypertension in mice with deletion of cystathionine gamma-lyase. *Science, 322*(5901), 587–590.
Yang, G., Zhao, K., Ju, Y., Mani, S., Cao, Q., Puukila, S., et al. (2013). Hydrogen sulfide protects against cellular senescence via S-sulfhydration of Keap1 and activation of Nrf2. *Antioxidants & Redox Signaling, 18*(15), 1906–1919.
Yonezawa, D., Sekiguchi, F., Miyamoto, M., Taniguchi, E., Honjo, M., Masuko, T., et al. (2007). A protective role of hydrogen sulfide against oxidative stress in rat gastric mucosal epithelium. *Toxicology, 241*(1–2), 11–18.
Zhang, D., Macinkovic, I., Devarie-Baez, N. O., Pan, J., Park, C. M., Carroll, K. S., et al. (2014). Detection of protein S-sulfhydration by a tag-switch technique. *Angewandte Chemie International Edition (English), 53*(2), 575–581.
Zhao, K., Ju, Y., Li, S., Altaany, Z., Wang, R., & Yang, G. (2014). S-sulfhydration of MEK1 leads to PARP-1 activation and DNA damage repair. *EMBO Reports, 15*(7), 792–800.

CHAPTER TWO

Measurement of H_2S *In Vivo* and *In Vitro* by the Monobromobimane Method

Xinggui Shen, Gopi K. Kolluru, Shuai Yuan, Christopher G. Kevil[1]
Department of Pathology, Louisiana State University Health Sciences Center–Shreveport, Shreveport, Louisiana, USA
[1]Corresponding author: e-mail address: ckevil@LSUHSC.edu

Contents

Abstract

The gasotransmitter hydrogen sulfide (H_2S) is known as an important regulator in several physiological and pathological responses. Among the challenges facing the field is the accurate and reliable measurement of hydrogen sulfide bioavailability. We have reported an approach to discretely measure sulfide and sulfide pools using the monobromobimane (MBB) method coupled with reversed phase high-performance liquid chromatography (RP-HPLC). The method involves the derivatization of sulfide with excess MBB under precise reaction conditions at room temperature to form sulfide dibimane (SDB). The resultant fluorescent SDB is analyzed by RP-HPLC using fluorescence detection with the limit of detection for SDB (2 n*M*). Care must be taken to avoid conditions that may confound H_2S measurement with this method. Overall, RP-HPLC with fluorescence detection of SDB is a useful and powerful tool to measure biological sulfide levels.

Methods in Enzymology, Volume 554
ISSN 0076-6879
http://dx.doi.org/10.1016/bs.mie.2014.11.039

1. INTRODUCTION

1.1. Properties of hydrogen sulfide

Hydrogen sulfide (H_2S) is a colorless gas with the odor of rotten eggs and can be oxidized to form sulfur dioxide, sulfates, sulfite, and elemental sulfur. Based on its lipophilic property, hydrogen sulfide easily penetrates the lipid bilayer of cell membranes (Wang, 2012); however, it is less membrane permeable than nitric oxide (NO) and carbon monoxide (CO). The difference in membrane permeability between NO, CO, and hydrogen sulfide is also reflected by their dipole moments, which have values of 0.16, 0.13, and 0.97, respectively. Hydrogen sulfide is slightly soluble in water and acts as a weak acid with an acid dissociation constant (pK_{a1}) of 7.04 and pK_{a2} of 19 at 37 °C (Hughes, Centelles, & Moore, 2009). It can dissociate into H^+ and hydrosulfide anion (HS^-), which in turn may dissociate into H^+ and sulfide anion (S^{2-}) in the following reaction:

$$H_2S \leftrightarrow H^+ + HS^- \leftrightarrow 2H^+ + S^{2-}$$

At physiological pH and 37 °C, ~20% of sulfide is present as H_2S, whereas at physiological pH and 25 °C, ~40% of sulfide is present as H_2S conversely at pH 9.5, hydrogen sulfide mainly exists as HS^- (Hughes et al., 2009; Shen et al., 2011). *In vivo*, pH favors sulfide existence primarily as H_2S and its highly reactive anion, HS^-.

1.2. Hydrogen sulfide pools

Hydrogen sulfide is produced from a variety of sources, including chemical reactions (e.g., hydrogen gas and elemental sulfur, ferrous sulfide and HCl, aluminum sulfide and water), sulfate-reducing bacteria, and in mammalian tissues. During hydrogen sulfide production in mammalian tissues, there are three tissue-specific enzymes involved, viz., cystathionine-β-synthase, cystathionine-γ-lyase, and 3-mercaptosulfurtransferase (Moore, Bhatia, & Moochhala, 2003).

Acid-labile sulfide and bound sulfane sulfur are two main forms of hydrogen sulfide stored in mammalian cells. They can release hydrogen sulfide under acidic and on reducing conditions, respectively (Shen, Peter, Bir, Wang, & Kevil, 2012). Examples of bound sulfane sulfur include thiosulfate, persulfide, thiosulfonate, polysulfides, polythionates, and elemental sulfur. Overall, these different biochemical forms are important for regulating the amount of bioavailable hydrogen sulfide (Ishigami et al., 2009; Wintner et al., 2010).

1.3. Physiological and pathophysiological roles of hydrogen sulfide

Hydrogen sulfide is best known for its toxicity. Indeed, H_2S at high concentrations irreversibly inhibits the respiratory chain by binding to the ferric heme a3 center and the CuB center of cytochrome c oxidase. Similarly, H_2S reacts with oxygenated ferrous hemoglobin and myoglobin and converts them to sulfhemoglobin or sulfmyoglobin, which are unable to carry O_2. However, mounting evidence implicates H_2S as an endogenous signaling molecule that plays important roles in physiological and pathological processes (Kolluru, Shen, Bir, & Kevil, 2013; Kolluru, Shen, & Kevil, 2013; Wang, 2011).

To date, a large spectrum of proteins has been shown to be targeted by H_2S. H_2S post-translationally modifies the regulatory sulfonylurea receptor of the ATP-sensitive potassium (K_{ATP}) channel in vascular smooth muscle cells (SMCs), resulting in an increased potassium flow, resultant hyperpolarization, and vasodilation (Tang, Wu, Liang, & Wang, 2005). The wide distribution of K_{ATP} and its isoforms makes H_2S important for the regulation of heart contractility and rate (cardiomyocytes), sensation (neurons), insulin secretion (β-islet cells), and mitochondrial functions (mitoK_{ATP}). Other ion channels are also shown to be the target of H_2S, including intermediate and small conductance potassium channels (IK_{Ca}/SK_{Ca}), L-type calcium channels, and the transient receptor potential cation channel A1 (Avanzato et al., 2014; Mustafa et al., 2011; Streng et al., 2008; Tang, Zhang, Yang, Wu, & Wang, 2013).

Moreover, recent evidence shows vascular endothelial growth factor receptor 2 is modified by H_2S to facilitate its activation after ligand binding (Tao et al., 2013). Meanwhile, SMC proliferation and survival have been shown to be inhibited by H_2S involving ERK activation, which indicates critical regulation of vascular remodeling by H_2S (Baskar, Sparatore, Del Soldato, & Moore, 2008). H_2S also targets and inhibits PTP1B and regulates endoplasmic reticulum stress (Krishnan, Fu, Pappin, & Tonks, 2011). Additionally, persulfidation of p65 by H_2S promotes its nucleus translocation and increases transcription of antiapoptotic proteins (Sen et al., 2012). Last but not least, H_2S can serve as an anti-oxidant that counter balances with reactive oxygen species, including superoxide, hypochlorous acid, peroxynitrite, and lipid peroxidation. Interestingly, H_2S is also thought to interact with the other gasotransmitters NO and CO on the levels of enzymatic synthesis, oxidative stress, and small adducts, such as HNO, HSNO, and GSNO (Nagy et al., 2014).

1.4. Measurement of hydrogen sulfide bioavailability

Accurate and reliable measurement of biological hydrogen sulfide can provide critical information associated with various pathophysiological functions. However, significant uncertainty exists regarding levels of H_2S associated with health, disease, and therapeutics. While there are several reasons for the current uncertainties it is now clear that a wide range of values for hydrogen sulfide have been reported (Levitt, Abdel-Rehim, & Furne, 2011; Nagy et al., 2014; Shen et al., 2011, 2012; Zheng et al., 2012).

The methylene blue method is the most commonly reported method used in the literature to measure hydrogen sulfide in biological samples (Zhu et al., 2007). The method is based on spectrophotometry of methylene blue dye after the reaction of sulfide and *N,N*-dimethyl-p-phenylenediamine. However, this method can be highly problematic, making it inappropriate for measuring biological levels of hydrogen sulfide (Shen et al., 2011). Key problems include: (1) interference of other colored substances, (2) methylene blue dimer and trimer formation, (3) strong acid chemical pretreatment, and (4) low sensitivity.

Other analytical methods have been reported but are limited for various reasons. Gas chromatography is sensitive enough to measure physiological sulfide levels, but it potentially liberates loosely-bound sulfide because of irreversible sulfide binding or shifts in phase transition equilibria (Levitt et al., 2011; Ubuka, Abe, Kajikawa, & Morino, 2001). Sulfide-specific ion-selective electrodes have also been in use to detect H_2S levels in biological samples, with a detection range of 1–10 μ*M* but are prone to fouling and limited sensitivity detection. Lastly, fluorescent probes for intracellular measurement of hydrogen sulfide have greatly evolved in the last couple of years. Yet, a major challenge exists with the regard to interference by other thiol species (Nagy et al., 2014).

The purpose of this chapter is to provide detailed techniques to perform the MBB derivatized method for detecting hydrogen sulfide in various biological matrices. The main methodology used is RP-HPLC with fluorescence detection or in combination with mass spectrometry. Additionally, different sample treatment workflows allow for the separation and quantification of free sulfide, acid-labile sulfide, and bound sulfane sulfur (Shen et al., 2012). With a 2.0 n*M* limit of detection, this method is sensitive and reliable enough for use with most biological samples.

2. EXPERIMENTAL METHODS

2.1. Derivatization reaction of H_2S with monobromobimane

The fluorescent reagent monobromobimane (MBB) has been widely used to measure various thiols by alkylation (Newton, Dorian, & Fahey, 1981). As shown in Fig. 1A, S-alkylation occurs twice with sulfide, forming sulfide dibimane (SDB). According to our previous report (Shen et al., 2011), free sulfide samples are prepared after sulfide derivatization with MBB.

2.1.1 Procedure

The following steps are performed in a hypoxic chamber (1% O_2) at room temperature.

(1) 30 μl of sample was added to a PCR tube containing 70 μl of 100 m*M* Tris–HCl buffer (pH 9.5, 0.1 m*M* DTPA), followed by addition of 50 μl of 10 m*M* MBB (dissolved in deoxygenated acetonitrile).

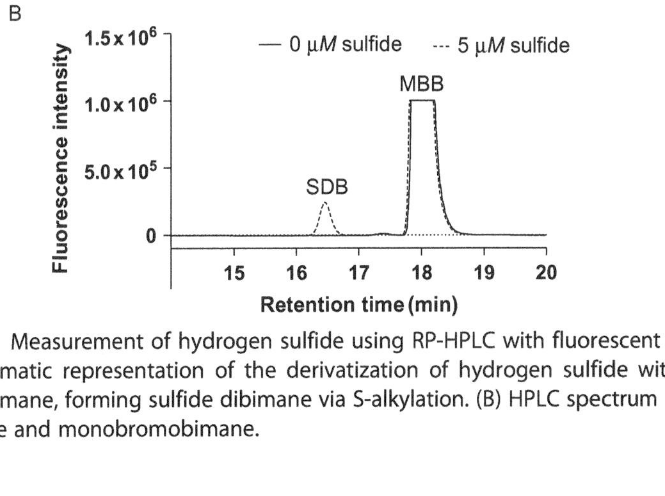

Figure 1 Measurement of hydrogen sulfide using RP-HPLC with fluorescent detector. (A) Schematic representation of the derivatization of hydrogen sulfide with monobromobimane, forming sulfide dibimane via S-alkylation. (B) HPLC spectrum of sulfide dibimane and monobromobimane.

(2) The reaction was stopped by adding 50 μl of 200 m*M* 5-sulfosalicylic acid after a 30-min incubation.
(3) All derivatized samples are removed from the hypoxic chamber and stored at 4 °C until analyzed by RP-HPLC.

2.1.2 Comment and limitations

(1) The reaction buffer should be deoxygenated by using nitrogen gas and kept in a hypoxic chamber (1% O_2, room temperature).
(2) Samples should be protected from light.
(3) The samples should be kept at 4 °C after derivatization of sulfide.
(4) For biological samples, the sample lysis buffer cannot include *N*-ethylmaleimide (NEM), 2-mercaptoethanol, etc., because they can interface with the H_2S/MBB reaction and directly impact SDB formation in an inversely proportional manner.

2.2. H_2S detection in biological samples: Effects of sample preparation

Hydrogen sulfide is easily lost from biological samples through volatilization and oxidation, especially at micromolar or lower levels of sulfide. However, using RP-HPLC, we are able to detect changes in sulfide levels in mice after sulfide treatment (Bir et al., 2012; Shen et al., 2011). For the measurement of sulfide in biological samples, it is very important that sample preparation be carefully and consistently performed for accurate measurement.

Currently, there are a number of research groups using pH 8.0 buffer for derivative reaction of sulfide with MBB. Figure 2 demonstrates that the remaining sulfide after the derivative reaction of sulfide with MBB under different pH conditions. During the first sulfide derivative reaction, 98.6% and 84.6% of sulfide were converted to SDB at pH 9.5 and 8.0, respectively. After the second sulfide derivative reaction, 1.4% and 11.4% of sulfide were converted to SDB at pH 9.5 and 8.0, respectively. These data demonstrate that hydrogen sulfide cannot be completely converted to SDB under reaction conditions at pH 8.0 unlike pH 9.5.

The deoxygenated Tris–HCl buffer (100 m*M*, pH 9.5, 0.1 m*M* DTPA) is not only the best derivatization buffer to use for MBB/sulfide reaction but also serves as an ideal stabilization buffer for sulfide samples in that plasma sulfide levels can be stabilized at room temperature for at least 15 min, and in liquid nitrogen for up to 4 weeks in this solution (Peter et al., 2013; Shen et al., 2011).

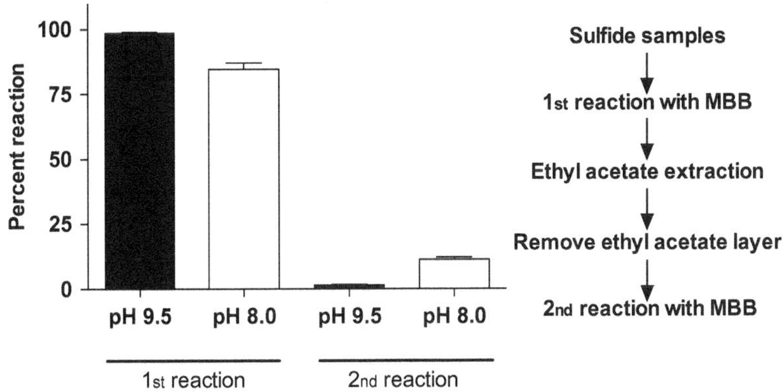

Figure 2 Comparison of remaining sulfide after the derivatization of hydrogen sulfide with MBB under different pH. Amount of ten micromolar sulfide solutions were reacted with MBB for 30 min at pH 9.5 and 8.0, respectively. After ethyl acetate extraction, the remaining sulfide was reacted with MBB again at pH 9.5 and 8.0, respectively.

Photolysis of MBB will result in the formation of fluorescent bimane (Kosower & Kosower, 1987), thus it is important to use dark amber vials and dim room lighting. In addition, sulfide derivatization is affected by the pH of the reaction buffer, oxygen concentration, trace metals, and the reaction vessel. As our previous optimization of sulfide derivatization (Shen et al., 2011) shows the highest reaction percentage that was achieved in PCR tubes with reaction buffer (pH 9.5, 0.1 m*M* DTPA) in a 1% O_2 hypoxic chamber. NEM is a thiol-reactive compound, which was widely used to protect thiols from oxidation. However, we have shown that the use of NEM to block free thiols may interfere with the follow-up measurements of hydrogen sulfide by MBB (Shen et al., 2011). In the laboratory, ethylenediaminetetraacetic acid (EDTA) is widely used for scavenging metal ions, which is the same role as DTPA. However, we have shown that red cells may lyse in EDTA tubes, resulting in increased levels of sulfide (Peter et al., 2013).

2.3. RP-HPLC with fluorescence detection

In our laboratory, SDB is detected using Shimadzu Prominence HPLC (LC-20), which consists of a binary, high-pressure gradient solvent delivery pump (model LC 20AB), an autosampler equipped with a cooling sample device (model SIL-20AC HT), a fluorescence detector (model RF10-AXL), and data processing software (LCsolution Version 1.23).

2.3.1 Procedure

(1) 10 μl of sample is injected into the RP-HPLC with an XDB-C18 column (4.6 × 250 mm, 80 Å), a guard column (4.6 × 12.5 mm, 80 Å) using the fluorescence detector setting of (ex/em, 390 nm/475 nm).

(2) The mobile phase for elution of SDB was composed of water (A, 0.1% trifluoroacetic acid (TFA), v/v) and 99.9% pure acetonitrile (B, 0.1% TFA, v/v), which is applied in the following gradient elution at 0.6 ml/min flow rate (Table 1).

(3) With this method, the retention time of SDB and MBB are 16.4 and 17.8 min, respectively (Fig. 1B).

2.3.2 Preparation of SDB standard

The calibration standard of SDB is prepared following our previous report (Shen et al., 2011). Briefly, 4 ml of 6 m*M* sodium sulfide was added to a 50 ml tube with 10 ml of 100 m*M* deoxygenated Tris–HCl buffer (pH 9.5, 0.1 m*M* DTPA) followed by addition of 5 ml of a 10 m*M* MBB solution. After 30 min of incubation in a 1% O_2 hypoxic chamber at room temperature, 1 ml of 2-mercaptoethanol was added to quench excess MBB. SDB was extracted by 10 ml of ethyl acetate, evaporated by nitrogen stream, and then purified by an Alltech Prevail SPE cartridge. SDB should be eluted using a 50% methanol solution (v/v). After evaporation of the solvent in the fractions containing pure SDB, dry powder of purified SDB can be weighed and then configured to standard solution at different concentrations. The concentration of purified SDB is verified by spectrophotometry, using the extinction coefficient values of 4883.257 and 4694.125 $dm^3\ mol^{-1}\ cm^{-1}$ at 370 nm (methanol solution) and 380 nm (HCl solution), respectively, to calculate an average SDB concentration.

Table 1 Mobile phase gradient table

Time (min)	% Phase A	% Phase B
0	85	15
5	65	35
16	45	55
23	30	70
24	10	90
26	10	90
28	85	15

2.3.3 Comment and limitations

(1) Though the fluorescent nature of SDB allows low concentrations of hydrogen sulfide to be easily measured, the prepared samples should be kept in the dark and 4 °C in the HPLC autosampler.
(2) Prepared samples should be centrifuged before being injected into the HPLC.
(3) Increasing the acetonitrile level in the mobile phase will decrease the elution time of SDB, and conversely decreasing the acetonitrile level will lengthen the retention time of SDB. It should also be noted that other thiol derivatives (e.g., cysteine, glutathione, homocysteine, and thiosulfate) can be simultaneously detected with this fluorescent system.
(4) All phase solutions should be filtered or be HPLC grade solvents.
(5) The column should be equilibrated and never allowed to dry.
(6) Sample injection cannot be overloaded into the column.
(7) Though sulfide solution is unstable, a known concentration of sulfide solution can also be used as a calibration standard.

2.4. H_2S and sulfide pool detection in biological samples

Hydrogen sulfide may exist in three biological forums: free hydrogen sulfide, acid-labile sulfide, and bound sulfane sulfur (Ishigami et al., 2009; Johnson, Dean, Smith, & Johnson, 2005; Shen et al., 2012; Ubuka, 2002). At pH 7.4 and 37 °C, 18.5% of free hydrogen sulfide exists as H_2S gas and the remainder is almost all hydrosulfide anion (HS^-) with a negligible contribution of sulfide anion (S^{2-}) (Hughes et al., 2009). Acid-labile sulfide consists of sulfur present in the iron–sulfur clusters contained in iron–sulfur proteins (Johnson et al., 2005). Bound sulfane sulfur includes thiosulfate, persulfide, thiosulfonates, polysulfides, polythionates, and elemental sulfur (Ubuka, 2002). A previous limitation of the sulfide field has been the lack of precise methodology for accurate and reproducible measurement of hydrogen sulfide and sulfide pools in biological samples. We developed an approach to measure them (Shen et al., 2012) which involves selective liberation, trapping, and derivatization of hydrogen sulfide, which is illustrated in Fig. 3.

2.4.1 Procedure

The following steps are performed in a hypoxic chamber (1% O_2) at room temperature.

(1) 50 μl of sample is transferred into two vacutainer tubes, which are labeled A and B.

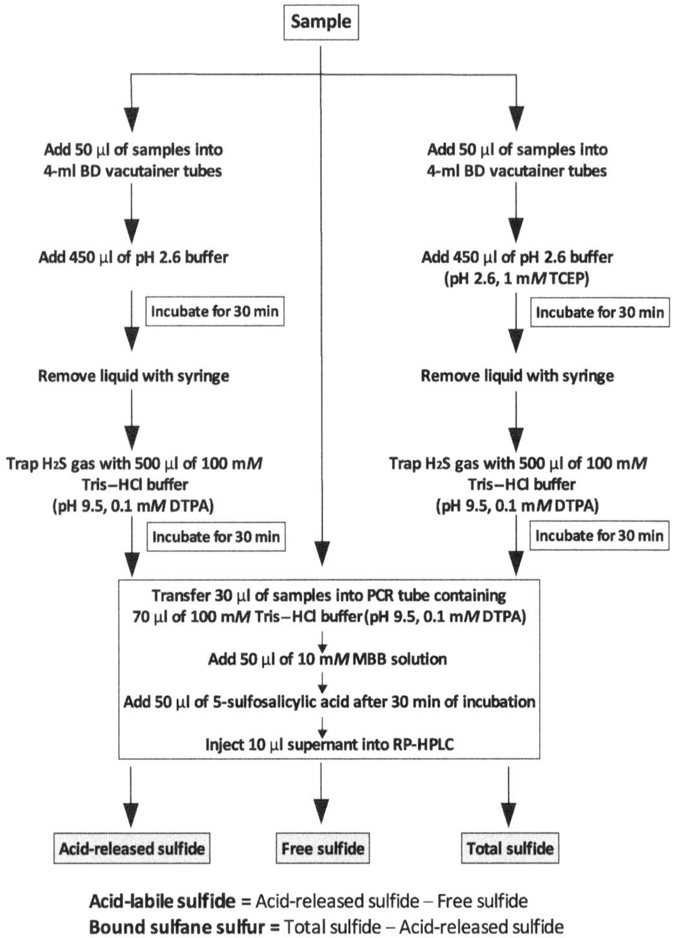

Figure 3 Schematic illustration of the workflow for sulfide measurement.

(2) For measurement of acid-labile sulfide, 450 μl of 100 m*M* phosphate buffer (pH 2.6, 0.1 m*M* DTPA) is added into tube A. For measurement of bound sulfane sulfur, 450 μl of 100 m*M* phosphate buffer (pH 2.6, 0.1 m*M* DTPA and 1 m*M* TCEP) is added into tube B.

(3) All vacutainer tubes are put onto a rocker and rocked for 30 min.

(4) All of the solutions in the vacutainer tubes are removed by a 1 ml syringe with a spinal needle so that the solution may be removed without inverting the tube.

(5) Volatilized hydrogen sulfide is then trapped by adding 500 μl of 100 m*M* Tris–HCl buffer (pH 9.5, 0.1 m*M* DTPA).

(6) All vacutainer tubes are then put onto the rocker and rocked for 30 min.
(7) Derivatization reactions of sulfide with MBB are performed following the procedure in Section 2.1.
(8) After HPLC assay, the acid-labile sulfide level is calculated using the A value minus the free sulfide value. The bound sulfane sulfur level is calculated using the B value minus the A value.

2.4.2 Comment and limitations

(1) TCEP solution should be freshly made.
(2) Volatilized hydrogen sulfide should be immediately trapped.

2.5. Confirmation of HPLC and SDB by mass spectrometer

Hydrogen sulfide and sulfide pools were detected using MBB in biological samples using RP-HPLC with a fluorescence detector (Shen et al., 2011, 2012; Wintner et al., 2010). A LCQ-Deca ion trap mass spectrometer was used to identify the SDB peak measured by RP-HPLC (Shen, Chakraborty, Dugas, & Kevil, 2014). Additionally, nonradioactive isotopic ^{34}S can be used to generate sodium sulfide to detect changes in sulfide distribution in biological samples (Shen et al., 2014). The mass spectra of SDB and ^{34}S-labeled SDB produced by ESI (+) MS are shown in Fig. 4A and B, respectively, with the major abundant ions identified. The m/z of 415.17 and 417.16 correspond to the molecular ion of SDB and ^{34}S-labeled SDB, respectively. For SDB and ^{34}S-labeled SDB, a similar MS2 fragment is produced from the 415.17 → 192.94 or 417.16 → 192.91 transitions. ^{34}S-labeled SDB can also be used as an internal standard to sensitively and accurately measure biological sulfide levels by selected reaction monitoring (SRM).

2.5.1 Procedure

(1) 3 μl of the 1.0 mg/ml IS stock solution (^{34}S-labeled SDB), 30 μl of SDB samples, and 267 μl of 50% acetonitrile were added into the empty tube, which was then vortexed for 30 s and injected into the mass spectrometer.
(2) Electrospray data were acquired using a Finnigan LCQ Deca mass spectrometer. The following conditions were used for MS: the ionization mode was positive electrospray ionization (ESI); sheath gas is 10 units; sweep gas is 5 units; maximum inject time is 400 ms; isolation width is 1.0; capillary temperature is 250 °C; spray voltage is 5.0 kV; and the ion gauge pressure was 2.41×10^{-5} Torr; a SRM mode was used, in which

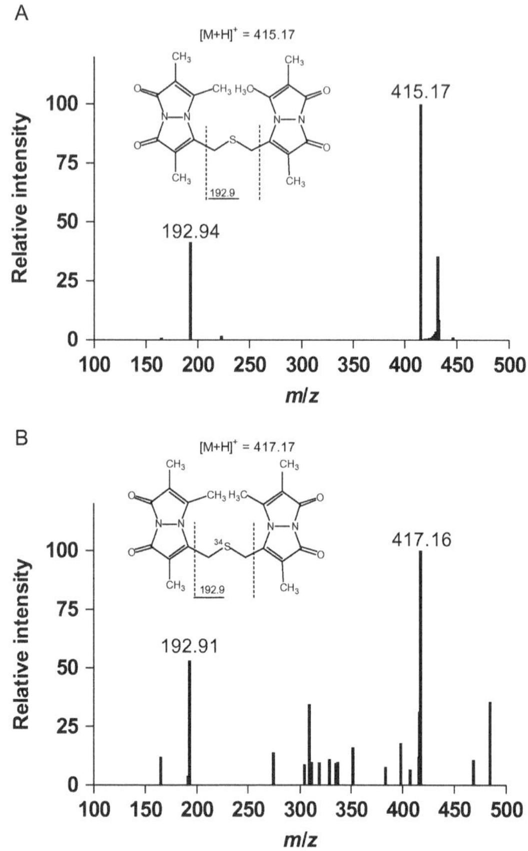

Figure 4 ESI (+) product ion mass spectra of SDB (A) and ^{34}S-labled SDB (B) with identification of the abundant ions. Molecular ion of SDB and ^{34}S-labled SDB $[M+H]^+$ correspond to 415.17 and 417.16, respectively. Insets, structure of SDB (A) and ^{34}S-labled SDB (B) with abundant fragments indicated.

ions of SDB and ^{34}S-labeled SDB were selected at *m/z* 415.1 and 417.1 as the precursor ion, and *m/z* 192.9 and 192.9 as product ions, respectively, with an analysis time of 20 min. Full scan mass spectra were acquired from 150 to 500 amu at unit mass resolution. Collision energy was 35 V, isolation width was 1.0, and the activation Q was 0.25. The automatic gain control (AGC) was activated and the value was set to 5×10^6.

(3) All acquired data were analyzed with Xcalibur software (Thermo Fisher Scientific, Inc.).

2.5.2 Comment and limitations

(1) Proper quantitative MS methods are developed employing SRM along with defined internal standards to account for sample and run fluctuations.

3. SUMMARY

The MBB method for measuring H_2S by RP-HPLC with fluorescence detection is a useful and sensitive quantitative method to measure sulfide levels in biological samples. This method is also advantageous in that workflow modification can be used to measure all biochemical forms of sulfide, including free sulfide, acid-labile sulfide, and bound sulfane sulfur from the same sample. While the MBB method has many advantages, important facts must be kept in mind, including:

(1) MBB is a light-sensitive reagent so the derivatization of the sulfide and MBB solution should be performed in the dark.
(2) All solutions must be deoxygenated before preparation of sulfide samples.
(3) Hydrogen sulfide readily binds to glass, therefore, high-quality polypropylene plastic tubes should be used for preparation of samples.
(4) The chemical derivatization reaction of H_2S with MBB is heavily influenced by pH, time, oxygen tension, and volatilization, which all affect the amount of sulfide measured.
(5) In the acidic environment the SDB formed from sulfide derivatization is stable at 4 °C.

ACKNOWLEDGMENT

This work was supported by NIH grant HL113303 to C. G. K.

REFERENCES

Avanzato, D., Merlino, A., Porrera, S., Wang, R., Munaron, L., & Mancardi, D. (2014). Role of calcium channels in the protective effect of hydrogen sulfide in rat cardiomyoblasts. *Cellular Physiology and Biochemistry: International Journal of Experimental Cellular Physiology, Biochemistry, and Pharmacology, 33*, 1205–1214.

Baskar, R., Sparatore, A., Del Soldato, P., & Moore, P. K. (2008). Effect of S-diclofenac, a novel hydrogen sulfide releasing derivative inhibit rat vascular smooth muscle cell proliferation. *European Journal of Pharmacology, 594*, 1–8.

Bir, S. C., Kolluru, G. K., McCarthy, P., Shen, X., Pardue, S., Pattillo, C. B., et al. (2012). Hydrogen sulfide stimulates ischemic vascular remodeling through nitric oxide synthase and nitrite reduction activity regulating hypoxia-inducible factor-1alpha and vascular endothelial growth factor-dependent angiogenesis. *Journal of the American Heart Association [Electronic Resource], 1*(5), e004093. http://dx.doi.org/10.1161/JAHA.112.004093.

Hughes, M. N., Centelles, M. N., & Moore, K. P. (2009). Making and working with hydrogen sulfide: The chemistry and generation of hydrogen sulfide in vitro and its measurement in vivo: A review. *Free Radical Biology & Medicine*, *47*, 1346–1353.

Ishigami, M., Hiraki, K., Umemura, K., Ogasawara, Y., Ishii, K., & Kimura, H. (2009). A source of hydrogen sulfide and a mechanism of its release in the brain. *Antioxidants & Redox Signaling*, *11*, 205–214.

Johnson, D. C., Dean, D. R., Smith, A. D., & Johnson, M. K. (2005). Structure, function, and formation of biological iron-sulfur clusters. *Annual Review of Biochemistry*, *74*, 247–281.

Kolluru, G. K., Shen, X., Bir, S. C., & Kevil, C. G. (2013). Hydrogen sulfide chemical biology: Pathophysiological roles and detection. *Nitric Oxide: Biology and Chemistry/Official Journal of the Nitric Oxide Society*, *35*, 5–20.

Kolluru, G. K., Shen, X., & Kevil, C. G. (2013). A tale of two gases: NO and H_2S, foes or friends for life? *Redox Biology*, *1*, 313–318.

Kosower, N. S., & Kosower, E. M. (1987). Thiol labeling with bromobimanes. *Methods in Enzymology*, *143*, 76–84.

Krishnan, N., Fu, C., Pappin, D. J., & Tonks, N. K. (2011). H_2S-induced sulfhydration of the phosphatase PTP1B and its role in the endoplasmic reticulum stress response. *Science Signaling*, *4*, ra86.

Levitt, M. D., Abdel-Rehim, M. S., & Furne, J. (2011). Free and acid-labile hydrogen sulfide concentrations in mouse tissues: Anomalously high free hydrogen sulfide in aortic tissue. *Antioxidants & Redox Signaling*, *15*, 373–378.

Moore, P. K., Bhatia, M., & Moochhala, S. (2003). Hydrogen sulfide: From the smell of the past to the mediator of the future? *Trends in Pharmacological Sciences*, *24*, 609–611.

Mustafa, A. K., Sikka, G., Gazi, S. K., Steppan, J., Jung, S. M., Bhunia, A. K., et al. (2011). Hydrogen sulfide as endothelium-derived hyperpolarizing factor sulfhydrates potassium channels. *Circulation Research*, *109*, 1259–1268.

Nagy, P., Palinkas, Z., Nagy, A., Budai, B., Toth, I., & Vasas, A. (2014). Chemical aspects of hydrogen sulfide measurements in physiological samples. *Biochimica et Biophysica Acta*, *1840*, 876–891.

Newton, G. L., Dorian, R., & Fahey, R. C. (1981). Analysis of biological thiols: Derivatization with monobromobimane and separation by reverse-phase high-performance liquid chromatography. *Analytical Biochemistry*, *114*, 383–387.

Peter, E. A., Shen, X., Shah, S. H., Pardue, S., Glawe, J. D., Zhang, W. W., et al. (2013). Plasma free H_2S levels are elevated in patients with cardiovascular disease. *Journal of the American Heart Association [Electronic Resource]*, *2*(5), e000387. http://dx.doi.org/10.1161/JAHA.113.000387.

Sen, N., Paul, B. D., Gadalla, M. M., Mustafa, A. K., Sen, T., Xu, R., et al. (2012). Hydrogen sulfide-linked sulfhydration of NF-kappaB mediates its antiapoptotic actions. *Molecular Cell*, *45*, 13–24.

Shen, X., Chakraborty, S., Dugas, T. R., & Kevil, C. G. (2014). Hydrogen sulfide measurement using sulfide dibimane: Critical evaluation with electrospray ion trap mass spectrometry. *Nitric Oxide: Biology and Chemistry/Official Journal of the Nitric Oxide Society*, *41*, 97–104.

Shen, X., Pattillo, C. B., Pardue, S., Bir, S. C., Wang, R., & Kevil, C. G. (2011). Measurement of plasma hydrogen sulfide in vivo and in vitro. *Free Radical Biology & Medicine*, *50*, 1021–1031.

Shen, X., Peter, E. A., Bir, S., Wang, R., & Kevil, C. G. (2012). Analytical measurement of discrete hydrogen sulfide pools in biological specimens. *Free Radical Biology & Medicine*, *52*, 2276–2283.

Streng, T., Axelsson, H. E., Hedlund, P., Andersson, D. A., Jordt, S. E., Bevan, S., et al. (2008). Distribution and function of the hydrogen sulfide-sensitive TRPA1 ion channel in rat urinary bladder. *European Urology, 53*, 391–399.

Tang, G., Wu, L., Liang, W., & Wang, R. (2005). Direct stimulation of K(ATP) channels by exogenous and endogenous hydrogen sulfide in vascular smooth muscle cells. *Molecular Pharmacology, 68*, 1757–1764.

Tang, G., Zhang, L., Yang, G., Wu, L., & Wang, R. (2013). Hydrogen sulfide-induced inhibition of L-type Ca^{2+} channels and insulin secretion in mouse pancreatic beta cells. *Diabetologia, 56*, 533–541.

Tao, B. B., Liu, S. Y., Zhang, C. C., Fu, W., Cai, W. J., Wang, Y., et al. (2013). VEGFR2 functions as an H_2S-targeting receptor protein kinase with its novel Cys1045-Cys1024 disulfide bond serving as a specific molecular switch for hydrogen sulfide actions in vascular endothelial cells. *Antioxidants & Redox Signaling, 19*, 448–464.

Ubuka, T. (2002). Assay methods and biological roles of labile sulfur in animal tissues. *Journal of Chromatography. B, Analytical Technologies in the Biomedical and Life Sciences, 781*, 227–249.

Ubuka, T., Abe, T., Kajikawa, R., & Morino, K. (2001). Determination of hydrogen sulfide and acid-labile sulfur in animal tissues by gas chromatography and ion chromatography. *Journal of Chromatography. B, Biomedical Sciences and Applications, 757*, 31–37.

Wang, R. (2011). Signaling pathways for the vascular effects of hydrogen sulfide. *Current Opinion in Nephrology and Hypertension, 20*, 107–112.

Wang, R. (2012). Physiological implications of hydrogen sulfide: A whiff exploration that blossomed. *Physiological Reviews, 92*, 791–896.

Wintner, E. A., Deckwerth, T. L., Langston, W., Bengtsson, A., Leviten, D., Hill, P., et al. (2010). A monobromobimane-based assay to measure the pharmacokinetic profile of reactive sulphide species in blood. *British Journal of Pharmacology, 160*, 941–957.

Zheng, Y., Liao, F., Du, J. B., Tang, C. S., Xu, G. H., & Geng, B. (2012). Modified methylene blue method for measurement of hydrogen sulfide level in plasma. *Sheng li xue bao: [Acta physiologica Sinica], 64*, 681–686.

Zhu, Y. Z., Wang, Z. J., Ho, P., Loke, Y. Y., Zhu, Y. C., Huang, S. H., et al. (2007). Hydrogen sulfide and its possible roles in myocardial ischemia in experimental rats. *Journal of Applied Physiology, 102*, 261–268.

CHAPTER THREE

Hydrogen Sulfide Detection Using Nucleophilic Substitution–Cyclization-Based Fluorescent Probes

Bo Peng, Ming Xian[1]
Department of Chemistry, Washington State University, Pullman, Washington, USA
[1]Corresponding author: e-mail address: mxian@wsu.edu

Contents

Abstract

While previously known as a toxic pollutant, hydrogen sulfide (H_2S) has been recently classified as an important signaling molecule with potent cytoprotective actions. The study on H_2S is a rapidly growing field in biomedical research. Nevertheless, the detection of H_2S in biological samples is still a challenge. Fluorescent assays can be very useful in this field owing to their high sensitivity and convenience. Based on the dual nucleophilicity of hydrogen sulfide, a series of reaction-based fluorescence "turn-on" probes have been developed by our lab. In this chapter, we provide an overview of the reactivity/property of these probes and detailed protocols for the use of them in experiments.

Methods in Enzymology, Volume 554
ISSN 0076-6879
http://dx.doi.org/10.1016/bs.mie.2014.11.030

1. INTRODUCTION

Hydrogen sulfide (H_2S) has been known as a toxic gas for years. However, recent studies have revealed that H_2S is generated endogenously by enzymes including cystathionine β-synthase, cystathionine γ-lyase, and cysteine aminotransferase in combination with 3-mercaptopyruvate sulfurtransferase (Kimura, 2011; Kimura, Shibuya, & Kimura, 2012; Li, Rose, & Moore, 2011; Olson, 2012; Predmore, Lefer, & Gojon, 2012). It has been proved that the endogenous formation of H_2S or exogenous administration of H_2S can exert protective effects in many pathologies, for example, regulation of blood pressure, inhibition of leukocyte adherence, and neuromodulation (Olson, 2011; Szabo, 2007; Tang, Wu, & Wang, 2010; Vandiver & Snyder, 2012; Wang, 2012). As a highly reactive molecule, H_2S can undergo reactions with a number of biological substrates such as reactive oxygen/nitrogen species (superoxide, hydrogen peroxide, peroxynitrite, etc.), cysteine derivatives (cysteine *S*-nitrosothiols, cysteine sulfenic acids, etc.), and metal-containing proteins (methemoglobin, hemoglobin, cytochrome *c*, etc.) (Fukuto et al., 2012; Kabil, Motl, & Banerjee, 2014; Kashfi & Olson, 2013; Li & Lancaster, 2013; Nagy et al., 2014; Pietri, Roman-Morales, & Lopez-Garriga, 2011). These reactions may be responsible for the biological functions of H_2S. The complexity of these reactions and high reactivity of H_2S also make the detection of H_2S in biological systems a difficult task. It has become important, therefore, to study the fundamental chemistry of H_2S and to develop effective and convenient methods for H_2S detection.

Because H_2S has been known as a pollutant for years, many detection techniques have been developed, which include colorimetric measurements, electrochemical assays, gas chromatography, and polarographic sensors (Olson, 2009; Tangerman, 2009). However, these conventional measurements can hardly be used to monitor H_2S in biological samples because they usually need complicated sample destruction which may cause inconsistent results due to the high reactivity and volatility of H_2S (Doeller et al., 2005; Furne, Saeed, & Levitt, 2008; Kabil & Banerjee, 2010; Kolluru, Shen, Bir, & Kevil, 2013; Ubuka, 2002). In this regard, fluorescence-based measurements have great advantages as they are suitable for nondestructive detections and can be used for imaging in live cells or tissues with readily available instruments.

In the past several years, a number of reaction-based fluorescent probes for H_2S detection have been reported (Lin & Chang, 2012; Peng et al., 2012;

Peng & Xian, 2014). These probes are usually designed using three strategies: (1) H_2S-mediated reductions (often use azide-containing substrates); (2) nucleophilic reactions of H_2S; and (3) metal-sulfide precipitations. Based on the dual nucleophilicity of H_2S, a series of disulfide-containing probes (WSP1–5) were developed by our lab (Liu et al., 2011; Peng et al., 2014). These probes showed high sensitivity and selectivity toward H_2S. Currently, WSP1 is the only commercially available H_2S fluorescent probe (by Cayman Chemicals Co.), and it has been used in monitoring both endogenously and exogenously produced H_2S in many studies (Devarie-Baez et al., 2013; Li et al., 2014; Markand et al., 2013; Trionnaire et al., 2014; Zhao et al., 2013). In this chapter, we provide a brief summary about the development of this type of probes. We also provide some detailed experimental protocols about using these probes in H_2S detection and imaging.

2. DESIGN AND SYNTHESIS OF THE PROBES

The design of these probes (WSP1–5) was based on the dual nucleophilicity of H_2S. The pK_a of H_2S is 6.8 and it is a potent nucleophile under physiological conditions. Moreover, H_2S can undergo nucleophilic reaction twice, which is different from other biothiols such as cysteine and glutathione. This property can be used to differentiate H_2S from biothiols. We expected that compounds containing dual-electrophilic centers were useful reagents for H_2S detection. As shown in Scheme 1, H_2S should react with the disulfide bond, which was the most electrophilic component of the probe **A**, to form a –SH containing intermediate **B**. With an ester group presented at the *ortho*- position, the SH group should undergo a spontaneous cyclization to release the fluorophore and form product **D**. This design could not only capture H_2S as a stable and analyzable product **D** but also visualize H_2S by convenient and sensitive fluorescence measurements.

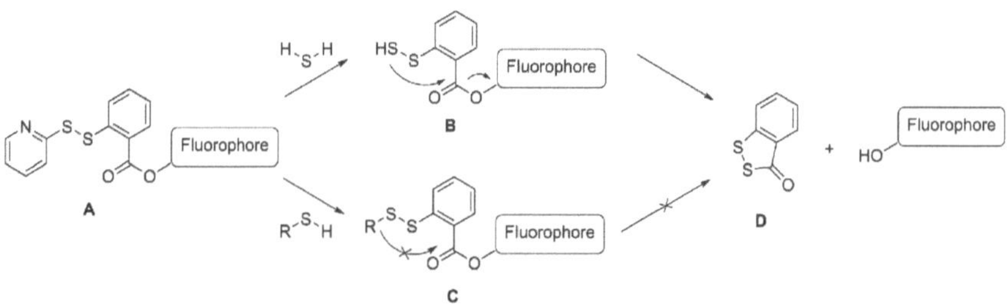

Scheme 1 General design of the probes.

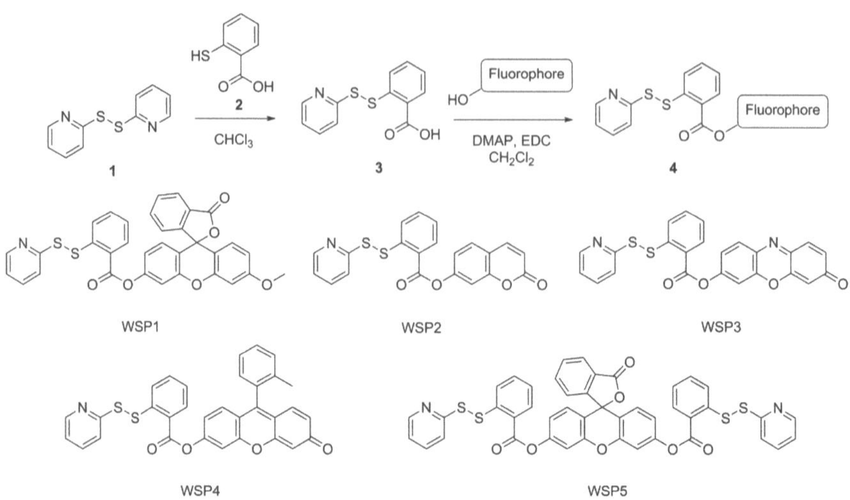

Scheme 2 Synthesis and structures of the probes.

It should be noted that probe **A** might also react with biothiols. However, the product **C** should not proceed the cyclization; therefore, fluorescence could not be turned on.

Based on this idea, a series of probes were synthesized (Scheme 2). The key intermediate **3** was prepared from thiosalicylic acid **2** and 2,2′-dipyridyl disulfide **1**. Compound **3** was then conjugated with several hydroxyl-containing fluorophores to provide the probes (WSP1–5) in good yields. Pyridyl disulfide was expected to be an effective electrophile for trapping H_2S and the benzene ring should be an appropriate linker to promote the cyclization. Fluorescein, 7-hydroxycoumarin, resorufin, and TokyoGreen were used as the fluorophores because of their readily availability and excellent fluorescence properties. It is also known that fluorescence of these fluorophores can be quenched by blocking their hydroxy groups.

3. CHEMISTRY AND PROPERTIES OF THE PROBES

These probes exhibited different spectroscopic properties with varied exciting and emission wavelengths. For example, WSP1 had an excitation wavelength of 476 nm and an emission wavelength of 516 nm, while the excitation and emission wavelengths of WSP3 were 550 and 586 nm, respectively. As expected, these probes showed very weak fluorescence with low quantum yields (from 0.003 to 0.02) due to esterification of the hydroxy groups of the fluorophores. This low background fluorescence is important for highly sensitive detection of H_2S. Their fluorescence intensities increased by 20–275-folds when they were treated with H_2S, thus demonstrating high sensitivity. Here, we describe detailed experimental protocols that were used to evaluate these probes.

3.1. Materials

(1) The probes are stable in dry, solid form and can be stored for several months at 4 °C. The probes can dissolve easily in most organic solvents such as dichloromethane, acetonitrile, and methanol. The probes may be sensitive to light, so both solid and stock solutions should be kept away from light by storing in dark or being covered by aluminum foil.

To prepare the stock solutions, the probes were first dissolved in DMSO (Reagent grade, MP Biomedicals) to prepare 2.5 m*M* stock solutions. These stock solutions were then diluted in PBS or DMEM and mixed thoroughly by pipetting or vortexing to give final concentrations of 10–50 μ*M*. The stock solutions should be freshly prepared before use.

(2) PBS buffer (10 m*M*, pH 7.4) was prepared by dissolving NaCl (4 g), KCl (0.1 g), Na_2HPO_4 (0.72 g), and KH_2PO_4 (0.12 g) in 500 mL double deionized water. Then pH was adjusted to 7.4 by 1 *M* HCl or 1 *M* NaOH. All aqueous solutions were prepared using double deionized water from a Nanopure Ultrapure Water System.

(3) NaHS stock solution was prepared freshly by dissolving 7.4 mg sodium hydrosulfide hydrate (ACROS ORGANICS, 95%) in 10 mL PBS buffer (10 m*M*, pH 7.4), concentration: 10 m*M*.

(4) CTAB (hexadecyltrimethylammonium bromide/cetyltrimethylammonium bromide, ACROS, 99+%) stock solution was prepared freshly by dissolving 109.2 mg CTAB in 3 mL ethanol (200 proof), concentration: 100 m*M*.

(5) Spectroscopy was carried out in acetonitrile/PBS (10 m*M*, pH 7.4, 1:1 v/v) or in PBS buffer (10 m*M*, pH 7.4) containing 1 m*M* CTAB. Fluorescence spectra were recorded on a Cary Eclipse Fluorescence Spectrophotometer. Samples for emission measurements were contained in a quartz cuvette (1 cm length).

(6) Cysteine (Cys) stock solution was prepared freshly by dissolving 12.1 mg L-cysteine (ACROS, 99%) in 2 mL PBS, concentration: 100 m*M*.

(7) Glutathione (GSH) stock solution was prepared freshly by dissolving 30.5 mg glutathione (ACROS, 98%) in 2 mL PBS, concentration: 100 m*M*.

(8) Homocysteine (Hcy) stock solution was prepared freshly by dissolving 13.5 mg homocysteine (Sigma-Aldrich, 95%) in 2 mL PBS, concentration: 100 m*M*.

(9) Sodium sulfite (Na_2SO_3) stock solution was prepared freshly by dissolving 12.6 mg sodium sulfite (J. T. Baker, 99%) in 2 mL PBS, concentration: 100 m*M*.

(10) Sodium thiosulfate ($Na_2S_2O_3$) stock solution was prepared freshly by dissolving 24.8 mg sodium thiosulfate 5-hydrate (J. T. Baker, 99.5%) in 2 mL PBS, concentration: 100 m*M*.

3.2. Test the reaction between the probes and H_2S

To test the reactivity of the probes toward H_2S, we carried out a reaction using WSP2 and NaHS (the equivalent of H_2S) in a mixed PBS–acetonitrile solvent system (Scheme 3). This reaction resulted in the desired products benzodithiolone, 2-mercaptopyridine, and 7-hydroxycoumarin in high yields (94%), which supported the reaction mechanism we proposed.

The reaction was performed as following:

(1) WSP2 (41 mg) was dissolved in 4 mL acetonitrile. NaHS (74 mg) was dissolved in 4 mL PBS (10 m*M*, pH 7.4).

(2) These two solutions were mixed and stirred at room temperature for 1 h.

(3) The reaction mixture was concentrated by a rota-evaporator to remove acetonitrile. The remaining aqueous mixture was extracted by CH_2Cl_2 (10 mL × 3).

(4) The combined organic layers were dried by $MgSO_4$, filtered, and concentrated. The crude products were purified by flash chromatography (5% methanol in CH_2Cl_2) to afford the three products shown in Scheme 3. The structures of the products were confirmed by ^{1}H-NMR and MS.

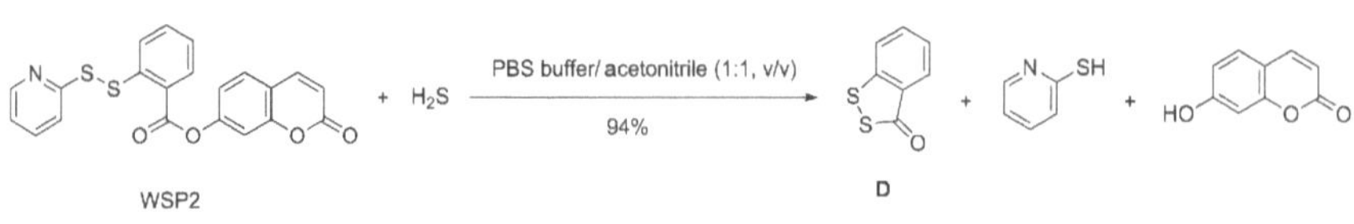

Scheme 3 The reaction between WSP2 and H_2S.

3.3. Fluorescence turn-on properties by H_2S

We next tested the probes' fluorescence responses in the presence of H_2S. All of them showed strong fluorescence enhancements by H_2S. One example using WSP2 is shown in Fig. 1. This experiment was carried out using the following procedure:

(1) In a 4-mL vial, 16 μL WSP2 stock solution was added and followed by the addition of 3974 μL acetonitrile/PBS (10 m*M*, pH 7.4, 1:1 v/v) mixed solvent. The concentration of WSP2 was 10 μ*M*.

(2) 2 mL of the solution was transferred to a cuvette and the cuvette was placed in the spectrophotometer.

(3) Fluorescence signal was measured (λ_{ex} 385 nm). This result was recorded as the background.

(4) To the cuvette was added 10 μL NaHS stock solution. The concentration of NaHS was 50 μ*M*.

(5) The fluorescence intensities were recorded every 3 min. Results are shown in Fig. 1.

In this solvent system (PBS/acetonitrile, 1/1), fluorescence turn-on rate was somewhat slow, probably due to poor water solubility of the probe. To solve this problem, other solvent systems were tested. Eventually we found that the fluorescence turn-on rates could be significantly enhanced when hexadecyltrimethylammonium bromide (CTAB) was added into the detection system. Under this condition, fluorescence signals could reach the maximum within 3 min. In addition, fluorescence intensities were also increased

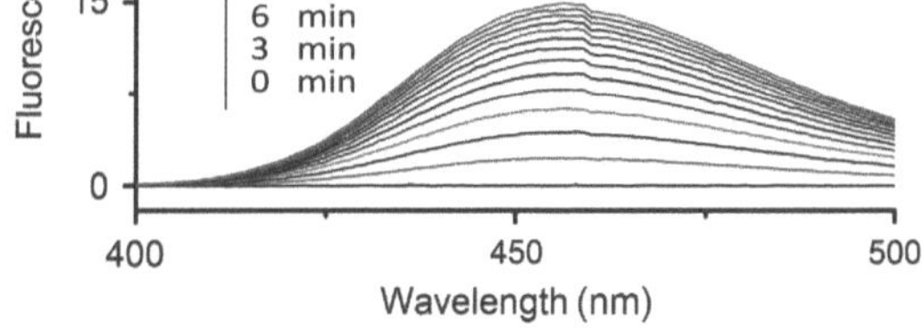

Figure 1 Time-dependent fluorescence changes of 10 μ*M* WSP2 in the presence of 50 μ*M* NaHS. Data were acquired at room temperature in acetonitrile/PBS (10 m*M*, pH 7.4, 1:1 v/v) with excitation at 385 nm.

dramatically. The effects of CTAB may be attributed to (a) CTAB can increase probes' solubility in aqueous buffers; and (b) as a cationic surfactant, CTAB may absorb sulfide anion (HS^-) and facilitate the reaction between sulfide anion and the probes. The procedure for this experiment is as following:

(1) In a 4-mL vial, 16 μL WSP2 stock solution was added and followed by the addition of 3934 μL PBS (10 m*M*, pH 7.4) and 40 μL CTAB stock solution. The concentration of the probe was 10 μ*M*.

(2) 2 mL of the solution was transferred to a cuvette and the cuvette was placed in the spectrophotometer.

(3) Fluorescence signal was measured (λ_{ex} 385 nm). This result was recorded as the background.

(4) To the cuvette was added 10 μL NaHS stock solution. The concentration of NaHS was 50 μ*M*.

(5) Fluorescence signals were measured every 3 min. Results are shown in Fig. 2.

3.4. Test the probes' selectivity for H_2S

In order to prove the probes were selective for H_2S, we also tested their fluorescence responses to other reactive sulfur species, including Cys, GSH, Hcy, sulfite (SO_3^{2-}), and thiosulfate ($S_2O_3^{2-}$). Experimental procedure is as following:

(1) Nine empty vials (4 mL) were numbered #1 to #9. To each vial was added 3944 μL of PBS.

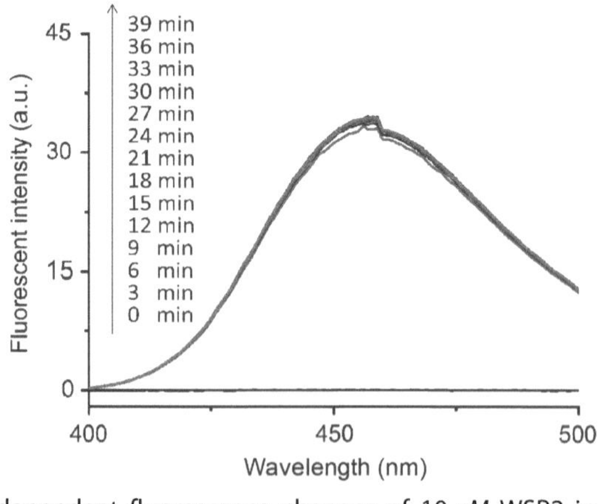

Figure 2 Time-dependent fluorescence changes of 10 μ*M* WSP2 in the presence of 50 μ*M* NaHS. Data were acquired at room temperature in PBS (10 m*M*, pH 7.4) containing 1 m*M* CTAB.

(2a) 16 μL probe's stock solution was added to vials #1 to #7. The concentration of the probe was 10 μ*M*.

(2b) 20 μL NaHS stock solution and 8 μL Cys stock solution were added to vial #8. 20 μL NaHS stock solution and 8 μL GSH stock solution were added to vial #9. Final concentration of NaHS was 50 μ*M*. Final concentration of RSH was 200 μ*M*.

(3) 40 μL CTAB stock solution was added to each vial (#1 to #9). Final concentration of CTAB was 1 m*M*.

(4a) 20 μL DI water, 20 μL NaHS stock solution, and 8 μL other reactive sulfur species' stock solution were added to vials #1 to #7, respectively. In vial #2, the concentration of NaHS was 50 μ*M*. In vials #3 to #7, the concentration of reactive sulfur species was 200 μ*M*. Vial #1 was the control (blank sample).

(4b) 16 μL probe's stock solution was added to vials #8 and #9. The concentration of the probe was 10 μ*M*.

(5) In each vial, after incubation for 5 min, 2 mL solution was transferred to a cuvette and the cuvette was placed in the spectrophotometer.

(6) Fluorescence signals were measured at the corresponding emission wavelength.

The results obtained from WSP1, WSP4, and WSP5 are shown in Fig. 3.

As shown in Fig. 3, only NaHS gave strong fluorescence responses to these probes. Other sulfur-containing species (vials #3 to #7) did not cause any fluorescence enhancements. These results demonstrated good selectivity of the probes. As the probes were expected to react with biothiols as well, we wondered if the probes could still give effective fluorescence signals when biothiols and H_2S coexist. Experiments in vials #8 and #9 were designed to address this question. As can be seen in Fig. 3, significant fluorescence enhancements were observed for each probe (10 μ*M*) toward NaHS (50 μ*M*) and biothiols (200 μ*M*), although at decreased levels. These results suggested that the probes were quite sensitive for H_2S.

4. APPLICATIONS OF THE PROBES IN H_2S IMAGING IN CELL-BASED EXPERIMENTS

4.1. Materials

Probes' stock solutions, PBS buffer, NaHS stock solution, and CTAB stock solution were prepared using the same protocols described in Section 3.1.

(1) Growth media: Dulbecco's modified Eagle's medium (DMEM, Cellgro) supplemented with 10% fetal bovine serum (FBS), 4 m*M* glutamine, 100 IU/mL penicillin, and 100 μg/mL streptomycin.

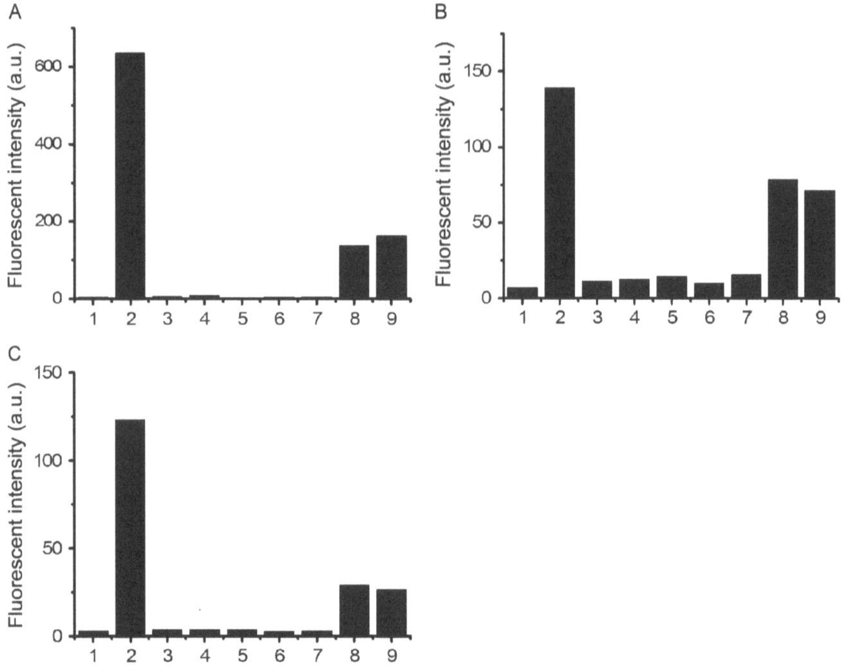

Figure 3 Fluorescence intensity of the probe (10 μ*M*) in the presence of various reactive sulfur species: (1) control, (2) 50 μ*M* NaHS, (3) 200 μ*M* Cys, (4) 200 μ*M* GSH, (5) 200 μ*M* Hcy, (6) 200 μ*M* Na_2SO_3, (7) 200 μ*M* $Na_2S_2O_3$, (8) 50 μ*M* NaHS+200 μ*M* Cys, and (9) 50 μ*M* NaHS+200 μ*M* GSH. WSP1 (A), WSP4 (B), and WSP5 (C).

(2) Washing media: DMEM (Cellgro) supplemented without glutamine and phenol red, Dulbecco's phosphate-buffered saline (DPBS, Hyclone).

(3) HeLa cells were cultured in DMEM with high glucose supplemented with 10% FBS at 37 °C in an incubator under an atmosphere of 5% CO_2 and 95% air. Cells were passaged every 3–4 days at nearly 100% confluence.

(4) Fluorescence images were taken on an AMG EVOS fl LED fluorescence microscope. Samples were excited at 470 nm and emission was collected at 525 nm by a GFP light cube.

(5) H_2S donor (YZ-4-074) (Zhao et al., 2013) stock solution was freshly prepared by dissolving 3.82 mg donor in 1 mL DMSO, concentration: 10 m*M*.

(6) UVGL-15 handheld UV lamps 4-W with 365 nm wavelength were used as UV source.

(7) H_2S donor (NOD-1d) (Devarie-Baez et al., 2013) stock solution was freshly prepared by dissolving 8.72 mg donor in 1 mL DMSO, concentration: 20 m*M*.

4.2. Fluorescence imaging of exogenous H_2S in HeLa cells

To study the capability of the probes in monitoring exogenous H_2S in cells, the following experiments were carried out (WSP5 was used as the example):

(1) One day before imaging, HeLa cells were passaged to a 24-well plate and cultured overnight to give around 80% confluence.
(2) Medium was removed from the plate and the cells were washed with 500 μL DMEM (without FBS) once.
(3) A solution of 50 μ*M* WSP5 was prepared by adding 10 μL probe stock solution to 490 μL medium and vortexing thoroughly. The cell media were exchanged with this WSP5 solution and the cells were incubated for 30 min at 37 °C.
(4) Extracellular solution was removed and cells were washed by 500 μL DPBS once.
(5) DPBS (containing 100 μ*M* CTAB) was added to cells. At the same time, 5 μL H_2O (as control) or NaHS stock solution was added.
(6) The cells were incubated for another 30 min at 37 °C.
(7) The cells were washed by 500 μL DPBS twice before images were recorded.
(8) Images were obtained on an AMG EVOS fl LED fluorescence microscope in green channel.

As can be seen in Fig. 4, NaHS triggered significant fluorescence enhancements while the control experiment did not show any fluorescence.

4.3. Fluorescence imaging of H_2S generated by persulfide-based H_2S donors

Persulfide-based H_2S donors were developed by our laboratory and have been demonstrated to release H_2S in cells (Zhao et al., 2013). Probe WSP4 was used to monitor H_2S generation from these donors. The procedure is shown below:

(1) One day before imaging, HeLa cells were passaged to a 24-well plate and cultured overnight to give around 80% confluence.
(2) Medium was removed from the cell plate and the cells were washed with 500 μL DMEM (without FBS) one time.
(3) 5 μL DMSO (control) or YZ-4-074 stock solution (10 m*M*) was added to 495 μL DPBS buffer and the cells were then incubated in these DPBS solutions (with or without donor) for 30 min at 37 °C. The structure of YZ-4-074 is shown in Scheme 4.

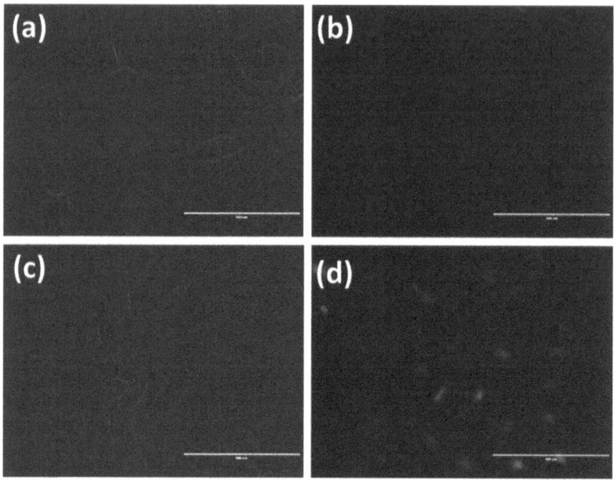

Figure 4 Fluorescence images of H_2S in HeLa cells using WSP5. Cells on a 24-well plate were incubated with WSP5 (50 μ*M*) for 30 min and then washed and subjected to different treatments: (A, B) Control (no NaHS was added) and (C, D) treated with 100 μ*M* NaHS. Scale bar: 100 nm. *Reprinted with permission from Peng et al. (2014). Copyright 2014 John Wiley & Sons, Inc.* (See the color plate.)

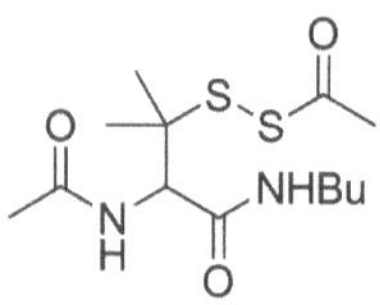

YZ-4-074

Scheme 4 Structure of H_2S donor YZ-4-074.

(4) After washed by 500 μL DMEM (without FBS) once, the cells were incubated in the solution of H_2S probe (WSP4, 30 μ*M*) for 30 min at 37 °C. This solution was prepared as follows: 6 μL WSP4's stock solution and 0.5 μL CTAB's stock solution were added to 494 μL DMEM (without FBS) and vortexed thoroughly.

(5) The cells were washed by 500 μL DPBS twice before images were recorded.

(6) Images were recorded by an AMG EVOS fl LED fluorescence microscope in green channel.

As shown in Fig. 5, only the cells treated with YZ-4-074 showed strong fluorescence signals.

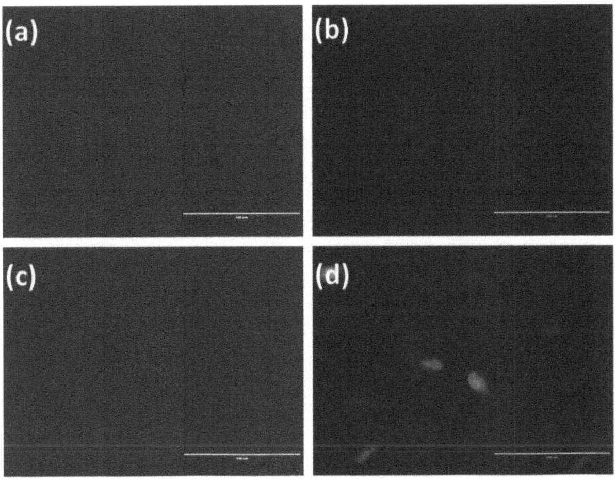

Figure 5 Fluorescence images of H_2S production from a H_2S donor YZ-4-074 in HeLa cells. Cells on a 24-well plate were incubated with or without YZ-4-074 for 30 min and then washed and incubated with 30 μ*M* WSP4 for 30 min. (A, B) Controls (no donor was added) and (C, D) treated with 100 μ*M* YZ-4-074. Scale bar: 100 nm. *Reprinted with permission from Peng et al. (2014). Copyright 2014 John Wiley & Sons, Inc.* (See the color plate.)

4.4. Fluorescence imaging of H_2S generated from photo-sensitive H_2S donors

A series of photo-activated H_2S donors were recently developed in our laboratory (Devarie-Baez et al., 2013). Probe WSP1 was used to measure H_2S generation from these donors in cell-based experiments. The procedure is shown below:

(1) One day before imaging, HeLa cells were passaged to a 6-well plate and cultured overnight to give around 80% confluence.

(2) Medium was removed from the cell plate and the cells were washed with 2 mL DMEM (without FBS).

(3) Cells were incubated in a solution of H_2S donor (NOD-1d, 200 μ*M*) for 30 min at 37 °C. This solution was prepared by adding 20 μL donor's stock solution to 1980 μL medium and vortexing thoroughly. The structure of NOD-1d is shown in Scheme 5.

(4) The cells were irradiated for 15 min with three handheld UV lamps that were placed 1.5 in. above the cells. For the control experiments, the cells were not exposed to UV light.

(5) The solution of donor was removed and the cells were washed quickly by 2 mL DPBS.

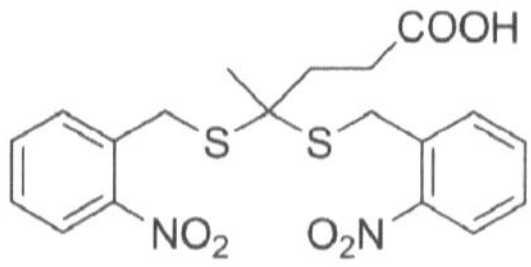

NOD-1d

Scheme 5 Structure of H_2S donor NOD-1d.

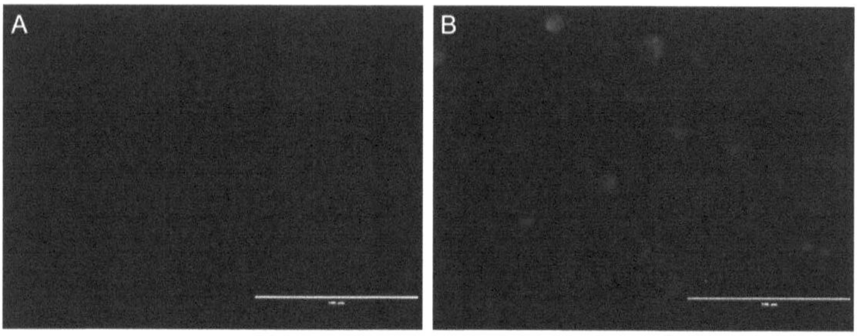

Figure 6 H_2S release from NOD-1d in HeLa cells: (A) cells treated with donor (200 μ*M*) and WSP-1 (50 μ*M*), no UV irradiation; (B) cells treated with donor (200 μ*M*) and WSP1 (50 μ*M*), under UV irradiation. *Reprinted with permission from Devarie-Baez et al. (2013). Copyright 2013 American Chemical Society.* (See the color plate.)

(6) A solution of 50 μ*M* WSP1 was prepared by adding 10 μL probe's stock solution and 0.5 μL CTAB stock solution to 490 μL DMEM (without FBS) and vortexing thoroughly. Then the cells were incubated with this probe solution for 30 min at 37 °C.

(7) The cells were washed by 2 mL DPBS twice before images were taken.

(8) Images were obtained with an AMG EVOS fl LED fluorescence microscope in green channel.

As shown in Fig. 6, only the cells treated with NOD-1d showed strong fluorescence signals.

5. CONCLUSIONS

In this chapter, we described the design, synthesis, and evaluation of a series of nucleophilic substitution/cyclization-based H_2S fluorescent probes. These probes showed good sensitivity and selectivity for H_2S over other reactive sulfur species, and have been used in imaging H_2S in cells. These probes can be useful research tools in exploring biological functions of H_2S.

ACKNOWLEDGMENTS

This work was supported by NIH (R01HL116571) and the ACS-Teva USA Scholar Award. We thank Prof. Viveka Vadyvaloo for providing the fluorescence microscope for this work.

REFERENCES

Devarie-Baez, N. O., Bagdon, P. E., Peng, B., Zhao, Y., Park, C.-M., & Xian, M. (2013). Light-induced hydrogen sulfide release from "caged" *gem*-dithiols. *Organic Letters*, *15*, 2786–2789.

Doeller, J. E., Isbell, T. S., Benavudes, G., Koenitzer, J., Patel, H., Patel, R. P., et al. (2005). Polarographic measurement of hydrogen sulfide production and consumption by mammalian tissues. *Analytical Biochemistry*, *341*, 40–51.

Fukuto, J. M., Carrington, S. J., Tantillo, D. J., Harrison, J. G., Ignarro, L. J., Freeman, B. A., et al. (2012). Small molecule signaling agents: The integrated chemistry and biochemistry of nitrogen oxides, oxides of carbon, dioxygen, hydrogen sulfide, and their derived species. *Chemical Research in Toxicology*, *25*, 769–793.

Furne, J., Saeed, A., & Levitt, M. D. (2008). Whole tissue hydrogen sulfide concentrations are orders of magnitude lower than presently accepted values. *American Journal of Physiology—Regulatory, Integrative and Comparative Physiology*, *295*, R1479–R1485.

Kabil, O., & Banerjee, R. (2010). Redox biochemistry of hydrogen sulfide. *The Journal of Biological Chemistry*, *285*, 21903–21907.

Kabil, O., Motl, N., & Banerjee, R. (2014). H_2S and its role in redox signaling. *Biochimica et Biophysica Acta*, *1844*, 1355–1366.

Kashfi, K., & Olson, K. R. (2013). Biology and therapeutic potential of hydrogen sulfide and hydrogen sulfide-releasing chimeras. *Biochemical Pharmacology*, *85*, 689–703.

Kimura, H. (2011). Hydrogen sulfide: Its production, release and functions. *Amino Acids*, *41*, 113–121.

Kimura, H., Shibuya, N., & Kimura, Y. (2012). Hydrogen sulfide is a signaling molecule and a cytoprotectant. *Antioxidants & Redox Signaling*, *17*, 45–57.

Kolluru, G. K., Shen, X., Bir, S. C., & Kevil, C. G. (2013). Hydrogen sulfide chemical biology: Pathophysiological roles and detection. *Nitric Oxide*, *35*, 5–20.

Li, Y.-J., Chen, J., Xian, M., Zhou, L.-G., Han, F. X., Gan, L.-J., et al. (2014). In site bioimaging of hydrogen sulfide uncovers its pivotal role in regulating nitric oxide-induced lateral root formation. *PLoS One*, *9*, e90340.

Li, Q., & Lancaster, J. R., Jr. (2013). Chemical foundations of hydrogen sulfide biology. *Nitric Oxide*, *35*, 21–34.

Li, L., Rose, P., & Moore, P. K. (2011). Hydrogen sulfide and cell signaling. *Annual Review of Pharmacology and Toxicology*, *51*, 169–187.

Lin, V. S., & Chang, C. J. (2012). Fluorescent probes for sensing and imaging biological hydrogen sulfide. *Current Opinion in Chemical Biology*, *16*, 595–601.

Liu, C., Pan, J., Li, S., Zhao, Y., Wu, L. Y., Berkman, C. E., et al. (2011). Capture and visualization of hydrogen sulfide by a fluorescent probe. *Angewandte Chemie International Edition*, *50*, 10327–10329.

Markand, S., Tawfik, A., Ha, Y., Gnana-Prakasam, J., Sonne, S., Ganapathy, V., et al. (2013). Cystathionine beta synthase expression in mouse retina. *Current Eye Research*, *38*, 597–604.

Nagy, P., Palinkas, Z., Nagy, A., Budai, B., Toth, I., & Vasas, A. (2014). Chemical aspects of hydrogen sulfide measurements in physiological samples. *Biochimica et Biophysica Acta*, *1840*, 876–891.

Olson, K. R. (2009). Is hydrogen sulfide a circulating "gasotransmitter" in vertebrate blood? *Biochimica et Biophysica Acta*, *1787*, 856–886.

Olson, K. R. (2011). The therapeutic potential of hydrogen sulfide: Separating hype from hope. *American Journal of Physiology—Regulatory, Integrative and Comparative Physiology, 301*, R297–R312.

Olson, K. R. (2012). A practical look at the chemistry and biology of hydrogen sulfide. *Antioxidants & Redox Signaling, 17*, 32–44.

Peng, H., Chen, W., Cheng, Y., Hakuna, L., Strongin, R., & Wang, B. (2012). Thiol reactive probes and chemosensors. *Sensors, 12*, 15907–15946.

Peng, B., Chen, W., Liu, C., Rosser, E. W., Pacheco, A., Zhao, Y., et al. (2014). Fluorescent probes on nucleophilic substitution-cyclization for hydrogen sulfide detection and bioimaging. *Chemistry—A European Journal, 20*, 1010–1016.

Peng, B., & Xian, M. (2014). Fluorescent probes for hydrogen sulfide detection. *Asian Journal of Organic Chemistry, 3*, 914–924.

Pietri, R., Roman-Morales, E., & Lopez-Garriga, J. (2011). Hydrogen sulfide and hemeproteins: Knowledge and mysteries. *Antioxidants & Redox Signaling, 15*, 393–404.

Predmore, B. L., Lefer, D. J., & Gojon, G. (2012). Hydrogen sulfide in biochemistry and medicine. *Antioxidants & Redox Signaling, 17*, 119–140.

Szabo, C. (2007). Hydrogen sulphide and its therapeutic potential. *Nature Reviews. Drug Discovery, 6*, 917–935.

Tang, G., Wu, L., & Wang, R. (2010). Interaction of hydrogen sulfide with ion channels. *Clinical and Experimental Pharmacology and Physiology, 37*, 753–763.

Tangerman, A. (2009). Measurement and biological significance of the volatile sulfur compounds hydrogen sulfide, methanethiol and dimethyl sulfide in various biological matrices. *Journal of Chromatography B, 877*, 3366–3377.

Trionnaire, S. L., Perry, A., Szczesny, B., Szabo, C., Winyard, P. G., Whatmore, J. L., et al. (2014). The synthesis and functional evaluation of a mitochondria-targeted hydrogen sulfide donor, (10-oxo-10-(4-(3-thioxo-3H-1,2-dithiol-5-yl)phenoxy)decyl) triphenylphosphonium bromide (AP39). *Medicinal Chemistry Communications, 5*, 728–736.

Ubuka, T. (2002). Assay methods and biological roles of labile sulfur in animal tissues. *Journal of Chromatography B, 781*, 227–249.

Vandiver, M. S., & Snyder, S. H. (2012). Sulfhydration mediates neuroprotective actions of parkin. *Journal of Molecular Medicine, 90*, 255–263.

Wang, R. (2012). Physiological implications of hydrogen sulfide: A whiff exploration that blossomed. *Physiological Reviews, 92*, 791–896.

Zhao, Y., Bhushan, S., Yang, C., Otsuka, H., Stein, J. D., Pacheco, A., et al. (2013). Controllable hydrogen sulfide donors and the activity against myocardial ischemia-reperfusion injury. *ACS Chemical Biology, 8*, 1283–1290.

CHAPTER FOUR

Azide-Based Fluorescent Probes: Imaging Hydrogen Sulfide in Living Systems

Vivian S. Lin*, Alexander R. Lippert[†,‡,1], Christopher J. Chang[*,§,¶,1]

*Department of Chemistry, University of California, Berkeley, California, USA
[†]Department of Chemistry, Southern Methodist University, Dallas, Texas, USA
[‡]Center for Drug Discovery, Design, and Delivery (CD4), Southern Methodist University, Dallas, Texas, USA
[§]Department of Molecular and Cell Biology, Howard Hughes Medical Institute, University of California, Berkeley, California, USA
[¶]Howard Hughes Medical Institute, Chevy Chase, Maryland, USA
[1]Corresponding authors: e-mail address: alippert@smu.edu; chrischang@berkeley.edu

Contents

Abstract

Hydrogen sulfide is a redox active sulfur species that is endogenously generated in mammalian systems as an antioxidant and signaling molecule to support cellular function. The fundamental and ubiquitous actions of hydrogen sulfide demand sensitive and specific methods to track this biomolecule as it is produced within living organisms with temporal and spatial regulation. In this context, the hydrogen sulfide-mediated reduction of an azide to an amine is a useful method for organic synthesis, and this

Methods in Enzymology, Volume 554
ISSN 0076-6879
http://dx.doi.org/10.1016/bs.mie.2014.11.011

reaction has successfully been exploited to yield biocompatible fluorescent probes for hydrogen sulfide detection *in vitro* and in cells. This chapter provides protocols and guidelines for applying azide-based fluorescence probes to detecting hydrogen sulfide in living systems, including a protocol that was used to detect endogenous hydrogen sulfide in living single cells using a confocal microscope.

1. INTRODUCTION

Hydrogen sulfide (H_2S) has been intimately involved in the history of life, contributing to the origin of biomolecules (Parker et al., 2011) and orchestrating the cataclysmic Permian-Triassic mass extinction 250 million years ago (Grice et al., 2005). This dichotomy of action is reflected in its biological effects. While displaying rapid toxicity at high concentrations (Guidotti, 2010), lower physiological levels of H_2S can enhance neuronal signaling (Abe & Kimura, 1996), induce vasorelaxation (Zhao & Wang, 2002), and even increase lifespan in model organisms (Miller & Roth, 2007). Intriguingly, nature has evolved a diverse array of enzymes to utilize H_2S as a signaling molecule (Li, Rose, & Moore, 2011), antioxidant (Kimura & Kimura, 2004; Lee, Schwab, Yu, McGeer, & McGeer, 2009), and supporter of cellular energetics (Fu et al., 2012). In mammals, H_2S is generated from cystathionine γ-lyase (CSE) (Stipnauk & Beck, 1982), cystathionine β-synthase (CBS) (Singh, Padovani, Leslie, Chiku, & Banerjee, 2009), and the coordinated action of cysteine amino transferase (CAT) and 3-mercaptopyruvate sulfurtransferase (3-MST) (Shibuya et al., 2009). These enzymes are ubiquitous in human tissues (Kimura, 2011) where H_2S helps mediate cardiovascular health, respiration, and neuronal signaling (Kimura, 2010; Vandiver & Snyder, 2012).

Given these broad, fundamental roles in physiology, it is not surprising that misregulation of H_2S is often associated with disease. Aberrantly low levels of H_2S lead to hypertension in the cardiovascular system (Yang et al., 2008) and accompany neurodegeneration in Huntington's disease (Paul et al., 2014). On the other hand, high levels of H_2S are observed in Down syndrome (Kamoun, Belardinelli, Chabli, Lallouchi, & Chadefaux-Vekemans, 2003), Zucker diabetic rats (Wu et al., 2009), and the urine of patients with cancer (Huang et al., 2013). In colon cancer cells, over production of CBS-derived H_2S has a pro-tumorigenic effect by both promoting angiogenesis and acting as an energy source for cancer cell metabolism (Szabó et al., 2013). Recently, persulfides have been identified and hypothesized as mediators of H_2S signaling (Ida et al., 2014). These transient molecules may play a fundamentally important role, particularly for intracellular H_2S signaling (Miranda & Wink, 2014), and new tools are emerging for their detection (Liu et al., 2014).

In this context, the broad, concentration-dependent effects of H_2S in health and disease demand precise methods to track the production of this important signaling molecule in living organisms. The most common classical methods for detecting H_2S include the methylene blue method, monobromobimane, gas chromatography, and amperometric sensors (Olson, DeLeon, & Liu, 2014). Such traditional techniques have provided bulk measurements of H_2S concentrations in serum, cells, and tissue samples, but they are incompatible with detection of H_2S levels in living, intact cells. Fluorescence imaging probes offer a powerful approach for H_2S imaging using fluorescence microscopy. Key advantages include high sensitivity, spatiotemporal resolution, and compatibility with living cells. Several criteria need to be met for a fluorescent probe to be useful for imaging H_2S in living cells: (1) biological compatibility, (2) cell permeability, (3) high sensitivity and selectivity, particularly with regards to intracellular thiols, and (4) optical properties that are compatible with common microscopy filter sets.

Several families of reaction-based fluorescent probes for H_2S have been developed in just the past few years. Due to space limitations, we cite review articles covering the field here (Lin & Chang, 2012; Lippert, 2014; Peng & Xian, 2014; Yu, Han, & Chen, 2014). Among these, azide-based probes have been widely adopted due to their sensitivity, selectivity, and ease of preparation. The H_2S-mediated reduction of azides has been used synthetically for a number of years, and starting in 2011 our laboratory utilized this functionality in the Sulfidefluor (SF) series of probes and Wang's lab concomitantly utilized sulfonyl azide versions for this same purpose (Peng et al., 2011). The first-generation SF probes were found to be compatible in live cells and showed good selectivity for H_2S over other thiol molecules, including glutathione (Lippert, New, & Chang, 2011). Optimization of the design to enable cellular trappability provided fluorescent H_2S probes that accumulate in cells and are capable of imaging endogenous H_2S generated for cellular purposes (Lin, Lippert, & Chang, 2013). These probes are based on rhodamine dye scaffolds that display low photobleaching, high quantum yields, and significant cellular compatibility.

This chapter is outlined as follows. Section 2 provides an overview of the design and reactivity of the probes. In Section 3, we describe the *in vitro* characterization of these probes using fluorescence spectrophotometry.

Section 4 details methods for using these probes in confocal microscopy applications and ends with an example of imaging endogenous H_2S production in human endothelial cells.

2. FLUORESCENT AZIDE-BASED H_2S PROBES

2.1. Probe design

Azide-based sulfide probes have been widely adopted in the field due to good chemical properties and simple methods for their preparation. This latter attribute has enabled the development of a wide variety of probes with tunable excitation and emission wavelengths, targeting to specific subcellular locales (Bae et al., 2013; Liu, Xu, Spring, & Cui, 2013), ratiometric readouts (Bae et al., 2013; Liu, Sun, et al., 2013; Wan, Song, Li, Gao, & Ma, 2013; Yu et al., 2012), two-photon excitation (Bae et al., 2013), and chemiluminescent emission (Bailey & Pluth, 2013).

We have focused on developing fluorescent sulfide probes to visualize endogenous H_2S produced during cellular signaling (Fig. 1). Our general strategy involves manipulating the "closed" and "open" forms of a rhodamine scaffold. The anilinic nitrogens on the rhodamine dye are masked as electron withdrawing azides, enforcing a "closed" lactone form. This structure displays very low fluorescence emission due to a break in the conjugation of the aromatic rings. Reduction of the azide with H_2S produces an aniline, which subsequently shifts the equilibrium toward the "open" form. Conjugation of the aromatic rings results in a highly fluorescent rhodamine dye.

The key to obtaining dyes that can image endogenous H_2S with scanning laser confocal microscopy was engineering probe molecules that accumulate within cells, accomplished by appending multiple acetoxymethyl (AM) esters (Minta, Kao, & Tsien, 1989). The AM ester dyes bear neutral charge and are highly permeable to the cellular membrane. Once inside the cell, however, these esters are rapidly cleaved leaving the dye molecule in an ionic form. Probe molecules will accumulate intracellularly providing both high concentrations of dye and a greatly reduced background signal. These factors provide a drastic increase in sensitivity that, under carefully controlled conditions, are amenable to imaging endogenous levels of H_2S.

2.2. Reactivity

Azides have long been used as "bioorthogonal" functional groups (Kolb, Finn, & Sharpless, 2001; Prescher & Bertozzi, 2005) defined as "nonnative,

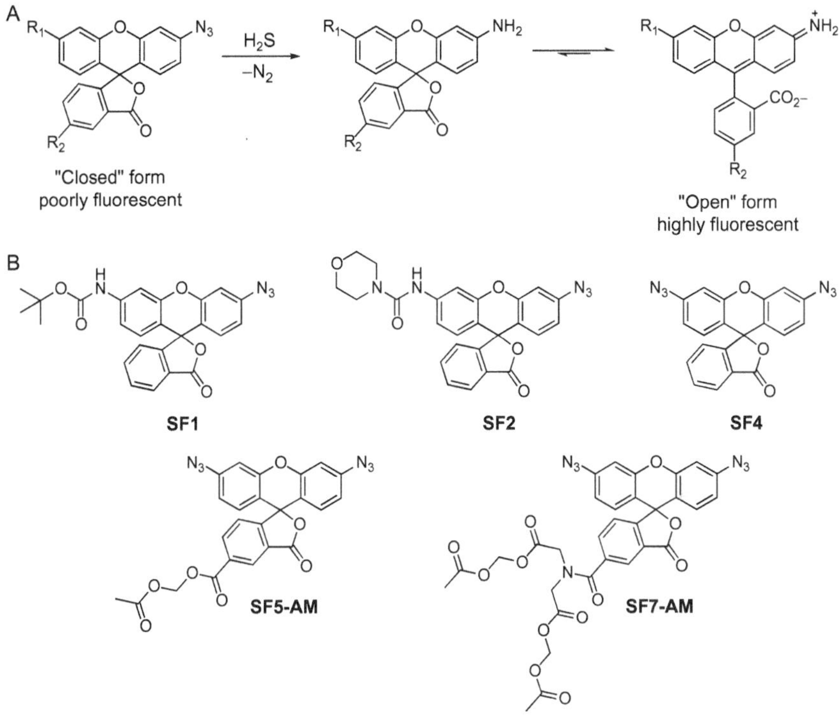

Figure 1 (A) Azide-based reduction strategy for fluorescence detection of H_2S. (B) Molecular structures of Sulfidefluor probes.

nonperturbing chemical handles that can be modified in living systems through highly selective reactions with exogenously delivered probes." We have taken advantage of this documented insensitivity to most biological molecules to design probes that are selective for H_2S inside living systems. *In vitro*, these H_2S probes display relatively low reactivity toward common biological thiols such as glutathione and cysteine. The reaction of an azide with H_2S is irreversible and these probes are properly classified as chemodosimeters. We also note caveats for the proper use of these and related azide-based fluorescent reagents for H_2S detection. The field of fluorescent probe design is in many ways in its relative infancy, and developing technologies will continue to push the frontiers of this area. First, it should be noted here that the azide functionality is potentially sensitive toward dithiothreitol, so this reducing agent should be avoided in the cellular media. The SF series probes may exhibit some similar cross-reactivity toward other highly reducing sulfur species, so appropriate controls should always be performed to confirm changes in fluorescent signal are a product of H_2S or reactive sulfur species production. Some aryl azides may also display photoactivation when exposed to high power and high energy light. As such, fluorescent azide-based probes that can be excited in the visible region are therefore preferred over probes that require excitation with UV light. Excessive exposure to light should be avoided during experiments and low light power should be used when possible. In this context, confocal microscopy is preferred over epifluorescence microscopy to minimize undesirable photoreactions.

2.3. Use and storage of probes

To prepare the probes for imaging use, an aliquot of dry probe is dissolved in the designated amount of DMF or DMSO to give a stock solution of 1–10 m*M*. This DMF or DMSO stock is then diluted in 1 mL of an aqueous buffer such as PBS, HEPES, or Tris and mixed thoroughly by pipetting or vortexing to give working concentrations of 1–20 μ*M*. Relatively dry DMF or DMSO should be used to dissolve the probe, as high water content may prevent full dissolution of the solid material and interfere with the preparation of a homogenous stock solution.

These probes are most stable in dry, solid form and can be stored for up to 6 months at −80 °C, protected from light and moisture. As with many other fluorescent dyes, freeze–thaw cycles may result in degradation of the probe, so single-use aliquots of probe are preferred. Once dissolved in dry DMF or DMSO, stocks can be stored for up to 1–2 weeks at −80 °C if necessary, although they should be carefully monitored for any changes in appearance that might indicate degradation, e.g., color change. NMR spectroscopy can be employed to monitor quality.

Any aqueous solutions of SF probes should be freshly prepared and used the same day, as hydrolysis of the ester functionalities reduces permeability of the probe. Azides are potentially unstable to high heat and light exposure, so all SF probe stocks and prepared solutions should be stored in the dark or covered with aluminum foil when not in use.

Low laser powers should be used to avoid photoactivation of the dye, since azides can be cleaved upon irradiation with UV or high-intensity light. To reduce the risk of photoactivation for a given field of cells, image collection at low wavelengths should be performed after collecting images for the SF probe. Furthermore, broadband excitation sources such as those commonly used in epifluorescence microscopes and microplate readers may not be compatible with azide-based fluorescent probes. In experiments with a Zeiss 710 confocal microscopy setup, we observed very minimal photoexcitation when using 488 nm light with low laser power. When using 405 nm light, however, significant increases in fluorescence intensity can be seen due to photoactivation.

3. *IN VITRO* CHARACTERIZATION OF PROBES

For all *in vitro* fluorometry experiments, a positive control using H_2S and a negative control using buffer should be performed on each day of experiments using the same probe aliquot between samples to account for potential variation in environmental factors, aliquot quality, and reagent preparation. Reactions with 100 μM H_2S are incomplete at 60 min. Excessive irradiation should be avoided to prevent photoactivation of the azide groups.

The following protocols yield samples with a final DMF concentration of 0.1%. In experiments where a minimal volume of organic cosolvent is desired, reduced amounts of stock solution may be used or the concentration of the stock solution may be increased. However, care must be taken to ensure that sufficient organic cosolvent is present to prevent the probe from precipitating out of solution upon addition to aqueous buffer, especially for the more hydrophobic SF4 probe. All solutions should appear optically clear and without particulates.

3.1. Safety precautions

Experimentalists should familiarize themselves with the hazards associated with H_2S, including inhalation risks, eye irritation, and flammability. Experiments should always be performed with sufficient ventilation, away from any ignition sources, using the appropriate personal protective equipment. While the concentrations and volumes of H_2S produced in the experiments detailed in this chapter are relatively small, even low levels of exposure to H_2S can result in discomfort. Thus, all practical precautions should be taken to limit exposure, such as preparing solutions of H_2S donors in a fume hood whenever possible.

3.2. Instrumentation and materials

Millipore water was used to prepare all aqueous solutions. All spectroscopic measurements were performed in 20 mM HEPES buffer, pH 7.4. Fluorescence spectra were recorded on a Photon Technology International Quanta Master 4L-format scanning spectrofluorometer (Lawrenceville, NJ) equipped with an LPS-220B 75-W xenon lamp and power supply,

A-1010B lamp housing with an integrated igniter, switchable 814 photon-counting/analog photomultiplier detection unit, and MD5020 motor driver. Samples for emission measurements were contained in 1 cm × 0.1 cm quartz cuvettes (1.5 mL volume, Starna, Atascadero, CA). Due to the toxicity of H_2S, screw-cap cuvettes (Cat no. 9 F-Q-10-GL14-S, Starna) were utilized to limit exposure to any gas that might escape from each reaction during the course of these experiments. Screw-cap cuvettes were also used for any other air-sensitive and gaseous reactive species.

Probe. SF7-AM stock solution (5 m*M*) in anhydrous DMF.

H_2S. A 25-m*M* aqueous stock solution made from solid Na_2S nonahydrate (Sigma-Aldrich) or NaHS (Cayman) in H_2O.

3.3. Time-course assays

Before performing any biological experiment, the integrity of the fluorescent probe can be examined *in vitro* by monitoring fluorescence response to H_2S over time. A 10-μ*M* solution of SF7-AM was prepared in 3 mL buffered aqueous solution. Vortexing was done to assure full dissolution of the dye. Two 1 mL aliquots of this probe solution were placed into different microcentrifuge tubes. To one tube, 4 μL of the 25 m*M* Na_2S solution was added, the solution was vortexed, and then transferred to a quartz cuvette. Absorption and fluorescence spectra were then acquired every 10 min. To the second tube, 4 μL of H_2O was added as a control, and spectra were acquired as described above (Fig. 2).

3.4. Selectivity experiments

Selectivity experiments are conducted to confirm the probe has higher reactivity toward H_2S over other reactive species that may be present in a system of interest. Generally, reactive sulfur, oxygen, and nitrogen species are tested at a final concentration of 100–200 μ*M*. Any highly abundant species should be tested at their most biologically relevant concentration, e.g., 5–10 m*M* for glutathione. The concentration of dye or probe being tested typically ranges from 1 to 10 μ*M*. Dye concentration may need adjustment if the dye fluorescence falls outside of the linear range for the detection system during selectivity tests. Readings are typically taken at 5–15 min intervals, and fluorescence response is typically represented as integrated emission intensity over a given range (Fig. 2B). For each day of data collection, an untreated sample of probe should be tested as a negative control.

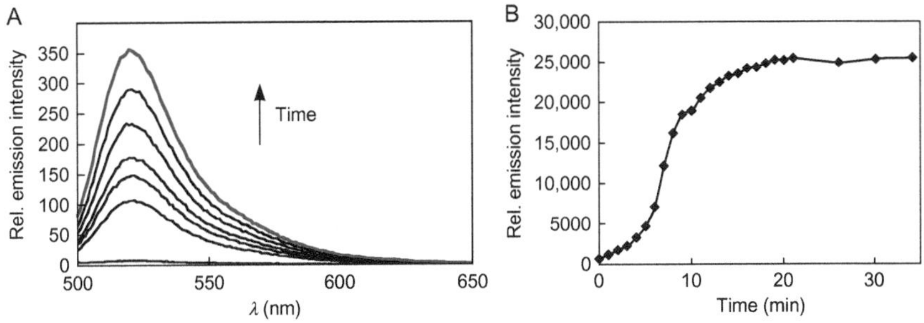

Figure 2 *In vitro* characterization of SF7-AM. (A) Fluorescence response of SF7-AM. Time points represent 0, 10, 20, 30, 40, 50, and 60 min (red trace (gray in the print version)) after addition of 100 μ*M* NaHS to 10 μ*M* SF7-AM. (B) Fluorescence response of 5 μ*M* SF7-AM in the presence of a 200-fold excess of NaHS. Fluorescence response was monitored at $\lambda_{em}=525$ nm until the reaction was complete. *Reprinted with permission from Lin et al. (2013). Copyright 2013 Proceedings of the National Academy of Sciences.*

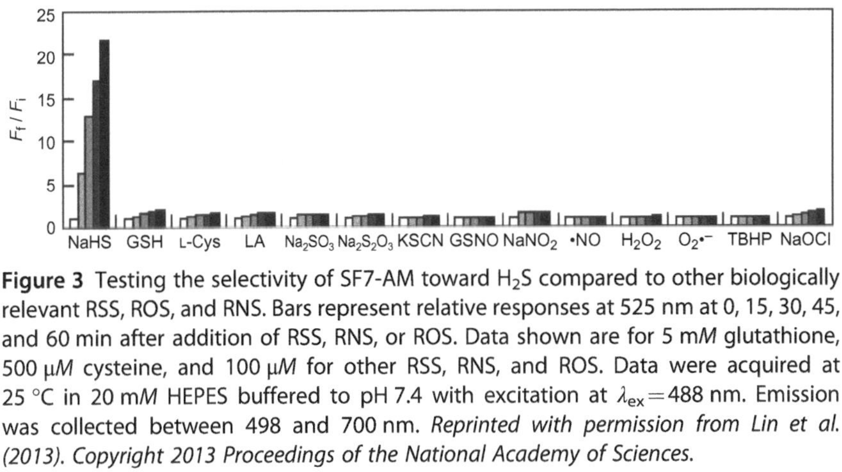

Figure 3 Testing the selectivity of SF7-AM toward H_2S compared to other biologically relevant RSS, ROS, and RNS. Bars represent relative responses at 525 nm at 0, 15, 30, 45, and 60 min after addition of RSS, RNS, or ROS. Data shown are for 5 m*M* glutathione, 500 μ*M* cysteine, and 100 μ*M* for other RSS, RNS, and ROS. Data were acquired at 25 °C in 20 m*M* HEPES buffered to pH 7.4 with excitation at $\lambda_{ex}=488$ nm. Emission was collected between 498 and 700 nm. *Reprinted with permission from Lin et al. (2013). Copyright 2013 Proceedings of the National Academy of Sciences.*

A positive control should also be performed, typically using 100 μ*M* Na_2S or NaHS (Fig. 3).

3.5. Data processing and analysis

The integrated fluorescence intensity—the area under the fluorescence emission curve—was plotted against time. Fluorescence intensity at a single point, typically the emission maximum, may also be used. For selectivity tests, integrated fluorescence intensity is typically expressed as the ratio of the integrated fluorescence intensity at a given time point to the integrated fluorescence intensity before addition of the reactive species.

4. DETECTION OF H_2S IN LIVE CELLS USING FLUORESCENT PROBES

This section will provide detailed protocols for imaging H_2S produced during a cellular model of angiogenesis. Section 4.1 describes control experiments imaging H_2S generated from an exogenous donor such as Na_2S or NaHS. This validation should be performed in any new experimental system to confirm dye efficacy. Section 4.2 details the experimental parameters needed to provide reliable measurements of endogenous H_2S generated for signaling purposes. Lastly, Section 4.3 provides methods using these fluorescent probes to reveal factors that are needed for H_2S production.

4.1. Imaging exogenous H_2S using confocal microscopy

The SF H_2S probes each possess different properties, particularly in terms of cell permeability and trappability. These attributes should be considered when selecting a dye to use for live cell imaging. SF4 is permeable to live cells and displays cytosolic staining while being excluded from the cell nucleus. Notably, SF4 is not a trappable dye, so exchanging the cellular medium typically leads to loss of fluorescence signal as the dye is washed out of the cells. SF7-AM is highly trappable and well retained within live cells after multiple media exchanges, displaying cytosolic and nuclear distribution. Owing to the mechanism of trapping for AM ester-based dyes, fixation of cells is generally incompatible with these probes. SF7-AM relies upon charged carboxylate groups to enhance retention in live cells, so permeabilization of the cell membrane results in leakage of the dye from the cells and loss of fluorescence signal. If using serum-containing media, SF7-AM must be loaded into cells shortly after dilution in media due to hydrolysis of the ester groups. Once the ester groups have been hydrolyzed, the probe is no longer cell permeable. This premature hydrolysis is not as readily observed in serum-free medium or buffer.

4.1.1 Materials and instrumentation

Media. Dulbecco's Modified Eagle Medium (Gibco) with 10% fetal bovine serum (FBS) or other standard medium appropriate for the cell line of interest.

Probes and reagents. A 5-m*M* SF7-AM stock solution in DMF made the day of the experiment. A 2.5-μ*M* solution of SF7-AM in complete media should be prepared from the DMF stock immediately before the experiment. A 10-m*M* stock solution of Na_2S in water made the day of the experiment.

Cell culture supplies. Standard supplies for mammalian cell culture.

Confocal microscope. Fluorescence microscopy studies were performed with a LSM 710 laser scanning confocal microscope (Carl Zeiss) using a 40 × water objective lens with Immersol W 2010 immersion medium and Zen 2010 software (Carl Zeiss). SF7-AM was excited using a 488-nm Ar laser, and emission was collected using a META detector between 500 and 650 nm. Hoechst 33342 was excited with a 405-nm diode laser, and emission was collected using a META detector between 450 and 500 nm. The cells were imaged at 37 °C and 5% CO_2 throughout the course of the experiment. Data processing and analysis were conducted using ImageJ (National Institutes of Health) or Zen 2010 software (Carl Zeiss).

4.1.2 Cell culture and dye loading

Two days before imaging. Cells were passaged in media without phenol red and plated in four-well Lab-Tek II glass chamberslides (Thermo Scientific, Cat no. 155382 or 155409) with an appropriate coating, if needed. At the time of the experiment, confluence should be between 70% and 80%.

60 min before imaging. The culture media in each well was replaced with 500 μL of a 2.5-μ*M* solution of SF7-AM prepared by dissolving 2 μL of the 5 m*M* stock solution into 4 mL of media. The cells were then incubated at 37 °C and 5% CO_2 for 30 min.

30 min before imaging. The media containing dye was replaced with fresh media. Five microliter of the Na_2S stock solution was added to positive control wells for a final concentration of 100 μ*M* Na_2S, and 5 μL of water was added as a vehicle control to negative control wells. The wells were capped and incubated for 30 min at 37 °C and 5% CO_2. The cells were then placed on the microscope stage, which was prewarmed and equilibrated to 37 °C and 5% CO_2.

4.1.3 Imaging and results

The cells were imaged on the microscope stage at 37 °C and 5% CO_2. Images of three to four fields per well were acquired using low laser power (1.0–1.2%). Images were quantified using the Zen 2010 Software

(Carl Zeiss) by setting common lower and upper thresholds for all images and evaluating the mean pixel intensity for each image.

4.2. Imaging endogenous H_2S production in HUVECs

4.2.1 Materials and instrumentation

Media. 1% Roswell Park Memorial Institute (RPMI) medium with 20% FBS (HyClone), 0.05 mg/mL endothelial growth supplement (ECGS, BD Biosciences), and 28 μg/mL heparin. Media should be used within 1 month of preparation.

Probes and reagents. A 5-m*M* SF7-AM stock solution in DMF made the day of the experiment. A 2.5-μ*M* solution of SF7-AM in complete media should be prepared from the DMF stock immediately before the experiment. A 25-m*M* stock solution of D,L-propargyl glycine (PAG) in water made the day of the experiment. A 10-μg/mL stock solution of VEGF in 0.1% BSA thawed the day of the experiment.

Cell culture supplies. Standard supplies for mammalian cell culture.

Confocal microscope. Fluorescence microscopy studies were performed with a LSM 710 laser scanning confocal microscope (Carl Zeiss) with a 40 × oil objective lens, atmosphere control incubator, motorized stage, and Zen 2010 software (Carl Zeiss). SF7-AM was excited using a 488-nm Ar laser, and emission was collected using a META detector between 500 and 650 nm. Hoechst 33342 was excited with a 405-nm diode laser, and emission was collected using a META detector between 450 and 500 nm. The cells were imaged at 37 °C and 5% CO_2 throughout the course of the experiment. Data processing and analysis were conducted using ImageJ (National Institutes of Health) or Zen 2010 software (Carl Zeiss).

4.2.2 Cell culture and dye loading

Human umbilical vein endothelial cells (HUVECs) were obtained as a gift from the Netherlands and cultured using RPMI supplemented with 20% FBS, 0.05 mg/mL ECGS, and 28 μg/mL heparin. Cells were passaged every 2–3 days and experiments were performed between passage numbers of 2–17.

Two days before imaging. Cells were passaged and plated in four-well Lab-Tek II glass chamberslides (Thermo Scientific, Cat no. 155382 or 155409) coated with 0.2% gelatin for 1 h before seeding. At the time of the experiment, confluence should be between 70% and 80%.

30 min before imaging. The culture media in each well was replaced with a 2.5-μ*M* solution of SF7-AM in complete media. The cells were then incubated at 37 °C and 5% CO_2.

20 min before imaging. PAG is added to well 2. The cells are further incubated for 20 min, and the cellular media was then replaced with complete RPMI media (+FBS, +ECGS, −phenol red). The cells were placed on the microscope stage, which was prewarmed and equilibrated to 37 °C and 5% CO_2.

4.2.3 Imaging and results

SF7-AM can image small increases in endogenous H_2S levels using confocal microscopy if care is taken to: (1) observe the same cells over time and (2) acquire *z*-stack images that capture fluorescence response from the entire cell thickness. Without careful attention to these experimental details, we found that cell-to-cell variability and user biases introduced by focusing on different *z*-stacks diminished the statistical significance of our results. All experiments should be accompanied by vehicle-treated controls and a method to decrease endogenous H_2S production, such as the use of pharmacological inhibitors, scavengers, or genetic knockdown.

After cells have been incubated with probe and are on the microscope stage, *z*-stack images (8 × 2 μm) are acquired of three to four fields per well. The *x*,*y* coordinates of each imaged field are saved so that the same cells can be reimaged. Two microliter of the 10 μg/mL VEGF solution was added on stage to wells 1 and 2, and 2 μL of a 0.1% BSA vehicle control was added to well 3. Each well contained 500 μL of RPMI media (+FBS, +ECGS, −phenol red). The cells were incubated on the microscope stage at 37 °C and 5% CO_2. After 30 min, *z*-stack images (8 × 2 μm) were once again acquired of the same three to four fields per well. Images were quantified using the Zen 2010 Software (Carl Zeiss) by taking the maximum intensity projection of the *z*-stacks, setting a common lower threshold for all images, ensuring that the cells are selected in the vehicle control images, and evaluating the mean pixel intensity for each image. Under these conditions, we observe a statistically significant 26% increase in the mean pixel intensity from the cell images before and after VEGF stimulation versus a 7% increase in cell images before and after stimulated with a vehicle control (Fig. 4).

4.3. Interrogating pathways involved in H_2S production using confocal microscopy

The real-time visualization of H_2S with SF7-AM provides an opportunity to investigate the cellular signaling cascades that lead to its production. The general strategy entails pharmacological inhibition of enzyme activity

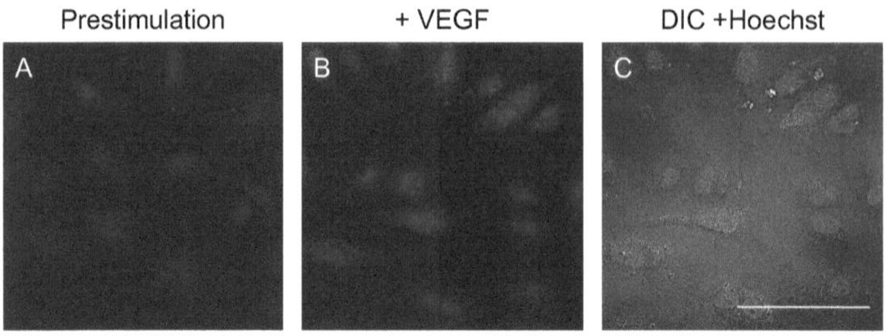

Figure 4 (A) HUVECs incubated with 5 μ*M* SF7-AM for 30 min at 37 °C, washed, and then imaged. (B) The same field of HUVECs in (A) was treated on stage with 2 μL of 10 μg/mL VEGF for 30 min at 37 °C, and then imaged. (C) Brightfield images of the same field of cells in (B) overlaid with Hoechst 33342 stain (1 μ*M*) at 37 °C. Images in (A) and (B) are the maximum intensity projections of 8 × 2 μm *z*-stacks. Scale bar represents 100 μm. *Adapted with permission from Lin et al. (2013). Copyright 2013 Proceedings of the National Academy of Sciences.* (See the color plate.)

and/or direct scavenging of molecular mediators while monitoring the optical signal from SF7-AM. In this example, we observed an increase in fluorescent signal from SF7-AM upon treatment of HUVECs with VEGF, in agreement with previous observations that VEGF stimulation increases H_2S production in this cell type (Papapetropoulos et al., 2009). Exposing cells to the pharmacological inhibitors PAG and AAL-993 which affect this H_2S production pathway correspondingly modulated the response from SF7-AM.

PAG. Incubate cells with 100 μ*M* PAG (Sigma-Aldrich) from a 25-m*M* stock in water for 10 min prior to VEGF stimulation (directly after media exchange).

AAL-993. Incubate cells with 30 μ*M* AAL-993 (Millipore) for 40 min prior to VEGF stimulation and again added at 30 μ*M* after media exchange.

Acquisition of *z*-stack images of the same cells over time as described in Section 4.2.3 revealed upstream molecular mediators of VEGF-dependent H_2S production in HUVECs. Inhibition of VEGFR2 with AAL-993 and CSE with PAG attenuated signal, establishing the involvement of these proteins at the beginning and end of the cascade leading to H_2S production. The observation that PAG fails to completely abolish signal combined with the observation that HUVECs express CBS (Lin et al., 2013) may point toward alternative production pathways (Fig. 5).

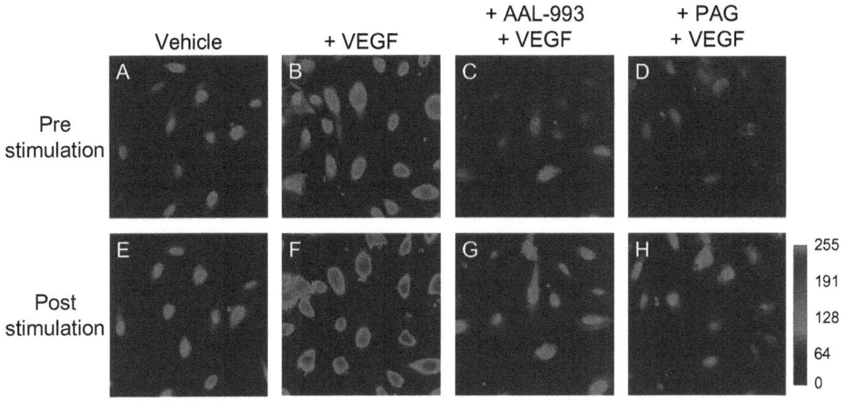

Figure 5 Representative confocal fluorescence images of H_2S signaling in live HUVECs. HUVECs were incubated with 2.5 μ*M* SF7-AM, washed, and imaged before (A) and after (E) treatment with 0.1% BSA in H_2O as a vehicle control. HUVECs were incubated with 2.5 μ*M* SF7-AM, washed, and imaged before (B) and after (F) treatment with VEGF. SF7-AM-labeled HUVECs pretreated with 30 μ*M* AAL-993 (AAL) for 40 min before (C) and after (G) treatment with VEGF. SF7-AM-labeled HUVECs pretreated with 100 μ*M* PAG for 10 min before (D) and after (H) treatment with VEGF. *Adapted with permission from Lin et al. (2013). Copyright 2013 Proceedings of the National Academy of Sciences.* (See the color plate.)

5. CONCLUSIONS

The ready installation of azides into molecular scaffolds has led to a rapid development of a large family of optical probes for H_2S with a variety of excitation and emission characteristics, luminescence mechanisms, and cellular localization. Despite the considerable growth of the field, examples of endogenous H_2S detection with a fluorescent probe remain rare, and typically require advanced microscopy techniques like live cell scanning laser confocal microscopy (Lin et al., 2013) or two-photon imaging (Bae et al., 2013). The method described here provides researchers with a detailed protocol to enable the real-time detection of H_2S generated for signaling purposes in living cells. In order to address the needs of the field (Papapetropoulos, Whiteman, & Giuseppe, 2014), we and other groups are continually working to develop H_2S probes with improved sensitivity, ease of use, and compatibility with widely available instrumentation. This work provides another successful illustration of how the reaction-based approach to sensing offers a general and versatile way to study molecules of biological interest (Chan, Dodani, & Chang, 2012; Lippert, Van de Bittner, & Chang, 2011).

ACKNOWLEDGMENTS

We thank the University of California, Berkeley, Southern Methodist University the Packard Foundation, the National Institute of General Medical Sciences (NIH GM 79465), as well as Amgen, Astra Zeneca, and Novartis for funding our work on redox imaging probes. C. J. C. is an investigator with the Howard Hughes Medical Institute.

REFERENCES

Abe, K., & Kimura, H. (1996). The possible role of hydrogen sulfide as an endogenous neuromodulator. *The Journal of Neuroscience*, *16*, 1066–1071.

Bae, S. K., Heo, C. H., Choi, D. J., Sen, D., Joe, E. H., Cho, B. R., et al. (2013). A ratiometric two-photon fluorescent probe reveals reduction in mitochondrial H_2S production in Parkinson's disease gene knockout astrocytes. *Journal of the American Chemical Society*, *135*, 9915–9923.

Bailey, T. S., & Pluth, M. P. (2013). Chemiluminescent detection of enzymatically produced hydrogen sulfide: Substrate hydrogen bonding influences selectivity for H_2S over biological thiols. *Journal of the American Chemical Society*, *135*, 16697–16704.

Chan, J., Dodani, S. C., & Chang, C. J. (2012). Reaction-based small-molecule fluorescent probes for chemoselective bioimaging. *Nature Chemistry*, *4*, 973–984.

Fu, M., Zhang, W., Wu, L., Yang, G., Li, H., & Wang, R. (2012). Hydrogen sulfide (H_2S) metabolism and its regulatory role in energy production. *Proceedings of the National Academy of Sciences of the United States of America*, *109*, 2943–2948.

Grice, K., Cao, C., Love, G. D., Böttcher, M. E., Twitchett, R. J., Grosjean, E., et al. (2005). Photic zone euxinia during the Permian-triassic superanoxic event. *Science*, *307*, 706–709.

Guidotti, T. L. (2010). Hydrogen sulfide: Advances in understanding human toxicity. *International Journal of Toxicology*, *29*, 569–581.

Huang, J., Kumar, S., Abbassi-Ghadi, N., Španěl, P., Smith, D., & Hanna, G. B. (2013). Selected ion flow tube mass spectrometry analysis of volatile metabolites in urine headspace for the profiling of gastro-esophageal cancer. *Analytical Chemistry*, *85*, 3409–3416.

Ida, T., Sawa, T., Ihara, H., Tsuchiya, Y., Watanabe, Y., Kumagai, Y., et al. (2014). Reactive cysteine persulfides and S-polythiolation regulate oxidative stress and redox signaling. In: *Proceedings of the National Academy of Sciences of the United States of America*, *111*, 7606–7611. http://dx.doi.org/10.1073/pnas.1321232111.

Kamoun, P., Belardinelli, M. C., Chabli, A., Lallouchi, B., & Chadefaux-Vekemans, B. (2003). Endogenous hydrogen sulfide overproduction in Down syndrome. *American Journal of Medical Genetics. Part A*, *116A*, 310–311.

Kimura, H. (2010). Hydrogen sulfide: From brain to gut. *Antioxidants & Redox Signaling*, *12*, 1111–1123.

Kimura, H. (2011). Hydrogen sulfide: Its production and functions. *Experimental Physiology*, *96*, 833–835.

Kimura, Y., & Kimura, H. (2004). Hydrogen sulfide protects neurons from oxidative stress. *The FASEB Journal*, *18*, 1165–1167.

Kolb, H. C., Finn, M. G., & Sharpless, K. B. (2001). Click chemistry: Diverse chemical function from a few good reactions. *Angewandte Chemie International Edition in English*, *40*, 2004–2021.

Lee, M., Schwab, C., Yu, S., McGeer, E., & McGeer, P. L. (2009). Astrocytes produce the antiinflammatory and neuroprotective agent hydrogen sulfide. *Neurobiology of Aging*, *30*, 1523–1534.

Li, L., Rose, P., & Moore, P. K. (2011). Hydrogen sulfide and cell signaling. *Annual Review of Pharmacology and Toxicology*, *51*, 169–187.

Lin, V. S., & Chang, C. J. (2012). Fluorescent probes for sensing and imaging biological hydrogen sulfide. *Current Opinion in Chemical Biology*, *16*, 595–601.

Lin, V. S., Lippert, A. R., & Chang, C. J. (2013). Cell-trappable fluorescent probes for endogenous hydrogen sulfide signaling and imaging H_2O_2-dependent H_2S production. *Proceedings of the National Academy of Sciences of the United States of America*, *110*, 7131–7135.

Lippert, A. R. (2014). Designing reaction-based fluorescent probes for selective hydrogen sulfide detection. *Journal of Inorganic Biochemistry*, *133*, 136–142.

Lippert, A. R., New, E. J., & Chang, C. J. (2011). Reaction-based fluorescent probes for the selective imaging of hydrogen sulfide in living cells. *Journal of the American Chemical Society*, *133*, 10078–10080.

Lippert, A. R., Van de Bittner, G. C., & Chang, C. J. (2011). Boronate oxidation as a bioorthogonal reaction approach for studying the chemistry of hydrogen peroxide in living systems. *Accounts of Chemical Research*, *44*, 793–804.

Liu, C., Chen, W., Shi, W., Peng, B., Zhao, Y., Ma, H., et al. (2014). Rational design and bioimaging applications of highly selective fluorescence probes for hydrogen polysulfides. *Journal of the American Chemical Society*, *136*, 7257–7260.

Liu, J., Sun, Y. Q., Zhang, J., Yang, T., Cao, J., Zhang, L., et al. (2013). A ratiometric fluorescent probe for biological signaling molecule H_2S: Fast response and high selectivity. *Chemistry: A European Journal*, *19*, 4717–4722.

Liu, T., Xu, Z., Spring, D. R., & Cui, J. (2013). A lysosome-targetable fluorescent probe for imaging hydrogen sulfide in living cells. *Organic Letters*, *15*, 2310–2313.

Miller, D. L., & Roth, M. B. (2007). Hydrogen sulfide increases thermotolerance and lifespan in *Caenorhabditis elegans*. *Proceedings of the National Academy of Sciences of the United States of America*, *104*, 20618–20622.

Minta, A., Kao, J. P., & Tsien, R. Y. (1989). Fluorescent indicators for cytosolic calcium based on rhodamine and fluorescein chromophores. *The Journal of Biological Chemistry*, *264*, 8171–8178.

Miranda, K. M., & Wink, D. A. (2014). Persulfides and the cellular thiol landscape. *Proceedings of the National Academy of Sciences of the United States of America*, *111*, 7505–7506.

Olson, K. R., DeLeon, E. R., & Liu, F. (2014). Controversies and conundrums in hydrogen sulfide biology. *Nitric Oxide*, *41*, 11–26. http://dx.doi.org/10.1016/j.niox.2014.05.012.

Papapetropoulos, A., Pyriochou, A., Altaany, Z., Yang, G., Marazioti, A., Zhou, Z., et al. (2009). Hydrogen sulfide is an endogenous stimulator of angiogenesis. *Proceedings of the National Academy of Sciences of the United States of America*, *106*, 21972–21977.

Papapetropoulos, A., Whiteman, M., & Giuseppe, C. (2014). Pharmacological tools for hydrogen sulfide research: A brief, introductory guide for beginners. *British Journal of Pharmacology*. http://dx.doi.org/10.1111/bph.12806, early view.

Parker, E. T., Cleaves, H. J., Dworkin, J. P., Glavin, D. P., Callahan, M., Aubrey, A., et al. (2011). Primordial synthesis of amines and amino acids in a 1958 Miller H_2S-rich spark discharge experiment. *Proceedings of the National Academy of Sciences of the United States of America*, *108*, 5526–5531.

Paul, B. D., Sbodio, J. I., Xu, R., Vandiver, M. S., Cha, J. Y., Snowman, A. M., et al. (2014). Cystathionine γ-lyase deficiency mediates neurodegeneration in Huntington's disease. *Nature*, *509*, 96–100.

Peng, H., Cheng, Y., Dai, C., King, A. L., Predmore, B. L., Lefer, D. J., et al. (2011). A fluorescent probe for fast and quantitative detection of hydrogen sulfide in blood. *Angewandte Chemie International Edition in English*, *50*, 9672–9675.

Peng, B., & Xian, M. (2014). Fluorescent probes for hydrogen sulfide detection. *Asian Journal of Organic Chemistry, 3*, 914–924.
Prescher, J. A., & Bertozzi, C. R. (2005). Chemistry in living systems. *Nature Chemical Biology, 1*, 13–21.
Shibuya, N., Tanaka, M., Yoshida, M., Ogasawara, Y., Togawa, T., Ishii, K., et al. (2009). 3-Mercaptopyruvate sulfurtransferase produces hydrogen sulfide and bound sulfane sulfur in the brain. *Antioxidants & Redox Signaling, 11*, 703–714.
Singh, S., Padovani, D., Leslie, R. A., Chiku, T., & Banerjee, R. (2009). Relative contributions of cystathionine β-synthase and γ-cystathionase to H_2S biogenesis via alternative transsulfuration reactions. *The Journal of Biological Chemistry, 284*, 22457–22466.
Stipnauk, M. H., & Beck, P. W. (1982). Characterization of the enzymic capacity for cysteine desulphhydration in liver and kidney of rat. *The Biochemical Journal, 206*, 267–277.
Szabó, C., Coletta, C., Chao, C., Módis, K., Szczesny, B., Papapetropoulos, A., et al. (2013). Tumor-derived hydrogen sulfide, produced by cystathionine-β-synthase, stimulates bioenergetics, cell proliferation, and angiogenesis in colon cancer. *Proceedings of the National Academy of Sciences of the United States of America, 110*, 12474–12479.
Vandiver, M., & Snyder, S. H. (2012). Hydrogen sulfide: A gasotransmitter of clinical relevance. *Journal of Molecular Medicine, 90*, 255–263.
Wan, Q., Song, Y., Li, Z., Gao, X., & Ma, H. (2013). *In vivo* monitoring of hydrogen sulfide using cresyl violet-based ratiometric fluorescence probe. *Chemical Communications, 49*, 502–504.
Wu, L., Yang, W., Jia, X., Yang, G., Duridanova, D., Cao, K., et al. (2009). Pancreatic islet overproduction of H_2S and suppressed insulin release in Zucker diabetic rats. *Laboratory Investigation, 89*, 59–67.
Yang, G., Wu, L., Jiang, B., Yang, W., Qi, J., Cao, K., et al. (2008). H_2S as a physiologic vasorelaxant: Hypertension in mice with deletion of cystathionine γ-lyase. *Science, 322*, 587–590.
Yu, F., Han, X., & Chen, L. (2014). Fluorescent probes for hydrogen sulfide detection and bioimaging. *Chemical Communications, 50*, 12234–12249.
Yu, F., Li, P., Song, P., Wang, B., Zhao, J., & Han, K. (2012). An ICT-based strategy to colorimetric and ratiometric fluorescence probe for hydrogen sulfide in living cells. *Chemical Communications, 48*, 2852–2854.
Zhao, W., & Wang, R. (2002). H_2S-induced vasorelaxation and underlying cellular and molecular mechanisms. *American Journal of Physiology. Heart and Circulatory Physiology, 283*, H474–H480.

CHAPTER FIVE

Chemiluminescent Detection of Enzymatically Produced H_2S

T. Spencer Bailey, Michael D. Pluth[1]
Department of Chemistry and Biochemistry, Institute of Molecular Biology, Materials Science Institute, University of Oregon, Eugene, Oregon, USA
[1]Corresponding author: e-mail address: pluth@uoregon.edu

Contents

Abstract

Hydrogen sulfide (H_2S) has emerged as an important biological signaling molecule. To better understand the multifaceted biological roles of H_2S, the development of selective and sensitive biocompatible assays for H_2S is becoming increasingly important. Motivated by these challenges, our laboratory is developing new methods to further detect and monitor biological H_2S. Here, we describe in detail our recent advances in the development and the use of chemiluminescence-based H_2S sensors to assist other investigators with use of these chemical tools. We highlight the use of these tools use by displaying their selectivity and high sensitivity toward H_2S and provide examples of assays we have developed to detect enzymatically produced H_2S.

Methods in Enzymology, Volume 554
ISSN 0076-6879
http://dx.doi.org/10.1016/bs.mie.2014.11.012

1. INTRODUCTION

Small-molecule reactive sulfur compounds are a recently discovered class of biological signaling molecules that play diverse roles in pathological and physiological processes. Chief among these species is hydrogen sulfide (H_2S), which has rapidly been linked to a myriad of diverse biological roles (Wallace, Dicay, McKnight, & Martin, 2007; Wang, 2009, 2012; Whiteman & Moore, 2009) since its initial report as a mediator of hippocampal long-term potentiation (Abe & Kimura, 1996). H_2S is produced primarily by three enzymes in mammals: cystathionine-γ-lyase (CSE), cystathionine-β-synthase (CBS), and 3-mercaptopyruvate sulfurtransferase (3-MST) (Wang, 2012). In addition to these primary pathways, H_2S genesis from D-Cys, catalyzed by D-amino acid oxidase and 3MST (Shibuya et al., 2013), as well as nonenzymatic processes are also possible. Based on the varied expression of these enzymes throughout different organs, it is clear that H_2S is involved in important processes throughout the cardiovascular, circulatory, respiratory, urinary, and nervous systems (Wang, 2012). This widespread importance is supported by studies linking misregulation of H_2S producing enzymes with disease phenotypes such as hypertension (Yang et al., 2008), diabetes (Wu et al., 2009), and various conditions of mental deficiency including Down's syndrome (Chen, Xin, & Zhu, 2007) and Alzheimer's disease (Qu, Lee, Bian, Low, & Wong, 2008). Such reports have motivated significant interest in developing H_2S-derived therapeutics (Chan & Wallace, 2013; Sparatore, Santus, Giustarini, Rossi, & Del Soldato, 2011; Szabó, 2007). Broadening the scope of this important small molecule, H_2S has also been shown to interact with cellular targets such as biological electrophiles (Nishida et al., 2012), heme proteins (Pietri, Roman-Morales, & Lopez-Garriga, 2011), cysteine residues in K_{ATP} channels (Jiang, Tang, Cao, Wu, & Wang, 2010), and reactive nitrogen species (Cortese-Krott et al., 2014; Filipovic et al., 2012).

To increase our understanding of the importance and differential roles of biological H_2S, an expanded toolbox of available methods to monitor H_2S is needed because it is unlikely that one specific method will be applicable or reliable under all circumstances. Specifically, chemical tools are needed to overcome current limitations that restrict effective biological H_2S detection. For example, efforts to determine the role of H_2S in inflammatory processes provide conflicting conclusions on whether H_2S generates proinflammatory or anti-inflammatory responses (Whiteman & Winyard, 2011). Traditional methods used for H_2S sensing including sulfide selective electrodes, gas chromatography, and the methylene blue test, all can be plagued by poor compatibility with live cells, often offer limited temporal resolution, and can require cumbersome sample preparation (DeLeon, Stoy, & Olson, 2012; Fogo & Popowsky, 1947; Lawrence, Davis, & Compton, 2000). In response to these challenges, new reaction-based methods for H_2S determination have emerged that offer higher cellular compatibility and spatiotemporal resolution by comparison to traditional methods. Such reaction-based systems make use of the unique chemical properties of H_2S to attack activated electrophiles (Chen, Zhu, et al., 2013; Liu et al., 2011; Montoya, Pearce, Hansen, Zakharov, & Pluth, 2013; Montoya & Pluth, 2014; Qian et al., 2011; Shen et al., 2011), precipitate metal salts (Sasakura et al., 2011), or to reduce azide or nitro groups on masked fluorophores (Lippert, New, & Chang, 2011; Montoya & Pluth, 2012; Peng et al., 2011; Thorson, Majtan, Kraus, & Barrios, 2013). More recent advances allow for the simultaneous measurement of both thiols and H_2S (Hammers & Pluth, 2014) and offer probes functionalized with targeting groups to report on H_2S in specific organelles (Bae et al., 2013; Liu, Xu, Spring, & Cui, 2013), proteins (Chen, Chen, Ren, & Ai, 2012), or cellular environments (Lin, Lippert, & Chang, 2013). The advent and elaboration of reaction-based H_2S detection methods has expanded the arsenal of tools available to further advance our understanding of biological H_2S.

Many recently developed methods for H_2S determination rely primarily upon the H_2S-mediated reduction azide-masked fluorophores. Such compounds typically mask a fluorogenic amine group of a fluorophore as an azide, thus generating a dark state of the fluorophore until unmasking of the fluorophore by H_2S. Under physiological conditions azide-quenched fluorophores are selectively reduced by H_2S, even in the presence of orders of magnitude larger concentrations of thiols such as GSH (Lin & Chang, 2012; Peng et al., 2011), to produce a fluorescent response. Although azide-masked fluorophores generally function as highly selective H_2S probes, the inherent photosensitivity of the azide functional group may limit their application. For example, although routine fluorescence measurements are unaffected, the high intensity light produced during confocal microscopy or HPLC measurements can result in the photodegradation of the azide functional group and decrease the accuracy of the method (Bailey & Pluth, 2013). As an example of this photosensitivity, we exposed

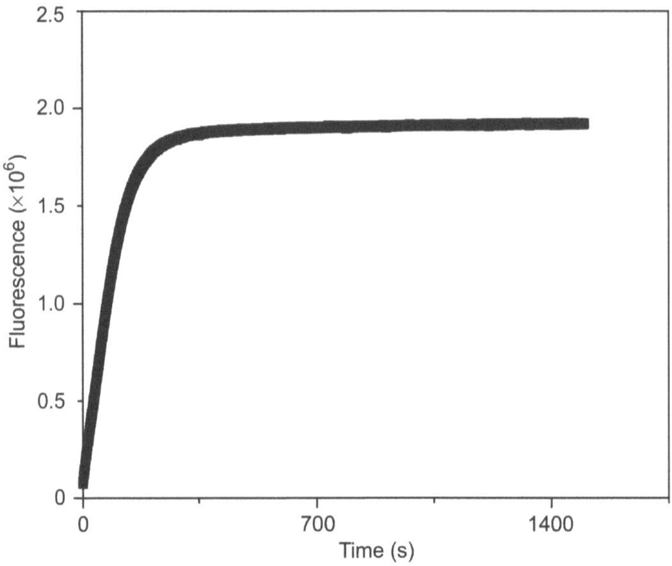

Figure 1 Photoactivation of C-7Az. Conditions: $\lambda_{ex} = 340$ nm, $\lambda_{em} = 445$ nm, 5 μ*M* probe in pH 7.4 PIPES buffer (50 mM PIPES, 100 m*M* KCl, pH 7.4). Slits: excitation = 5 nm, emission = 1.4 nm. *Adapted with permission from Bailey and Pluth (2013). Copyright 2013 American Chemical Society.*

C-7Az (Chen, Li, et al., 2013; Fig. 1) and other azide-based fluorescent H_2S sensors to continuous irradiation in the absence of H_2S and observed a strong signal solely deriving from photoactivation (Bailey & Pluth 2013).

To address this limitation of H_2S-mediated azide reduction fluorescent probes, we developed a chemiluminescent scaffold to avoid unwanted photoactivation. This method overcomes the photoinstability of azides in fluorescence-based methods by utilizing a chemiluminescent reporter that does not require an external excitation source to generate luminescence. Additionally, because most biological tissues do not spontaneously emit light, a chemiluminescent method should offer a much higher signal-to-noise ratio than that observed in fluorescence measurements. Chemiluminescence spectroscopy is a well-developed analytical method, which is widely employed to collect quantitative data in immunoassays (Williams & Campbell, 1986; Yuan et al., 2012) and chromatography (Garcia, Vinas, & Gil, 1993; Thurbide & Aue, 2002; Yan, 2006). Although chemiluminescent methods are being explored for various small molecules (McCutcheon, Paley, Steinhardt, & Prescher, 2012; Sun, Liu, Wang, Zhang, & Guo, 2012), the palette of chemiluminescent probes available for the detection of small molecules remains small by comparison to fluorescence analogues and is concentrated primarily on the detection of reactive oxygen species (Lee et al., 2008; Van de Bittner, Dubikovskaya, Bertozzi, & Chang, 2010; Yamaguchi et al., 2010). Based on the need to both increase the pool of available tools to study H_2S and to further expand the scope of chemiluminescent detection methods for small molecules, we have developed a chemiluminescent platform for H_2S, which we detail below.

2. CHEMILUMINESCENT PROBES FOR THE DETERMINATION OF SULFIDE

2.1. Probe design

The presented chemiluminescent H_2S probes are derivatives of the luminol scaffold, which has been used extensively in chemical-sensing methodologies (Marquette & Blum, 2006). Like other electron-withdrawn luminol derivatives, such as *N*-acyl (Drew & Garwood, 1939) or nitroluminol (Chen et al., 1998; Omote, Miyake, Ohmori, & Sugiyama, 1966), the azide-masked luminol compounds CLSS-1 and CLSS-2 (Fig. 2) show significantly reduced chemiluminescence by comparison to the parent compound. Unmasking of the azide to release the free amine liberates luminol (for CLSS-1) or isoluminol (for CLSS-2), both of which are chemiluminescent.

CLSS-1 and CLSS-2, which are not chemiluminescent prior to H_2S-mediated reduction of the azide, report by oxidation-induced chemiluminescence of the phthalhydrazide moiety. In a typical luminol assay, H_2O_2 is used as an oxidant in combination with horseradish peroxidase (HRP) as a catalyst to oxidize the phthalhydrazide functional group to produce a singlet carbonyl species (Fig. 3A). The transient excited phthalate intermediate then relaxes with concomitant liberation of N_2 to emit light with luminescence centered at 425 nm (White & Roswell, 1970). The chemiluminescent lifetime and

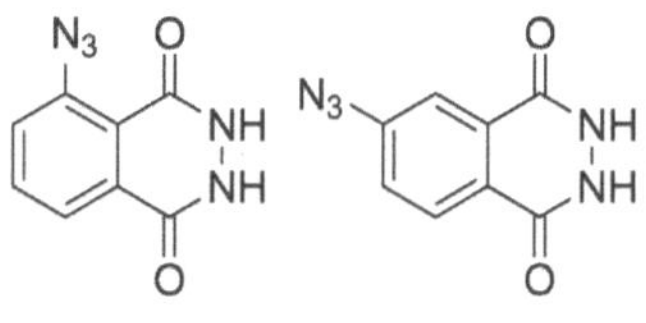

Figure 2 Molecular structures of ChemiLuminescent Sulfide Sensors CLSS-1 and CLSS-2, which are based on the luminol and isoluminol platforms, respectively.

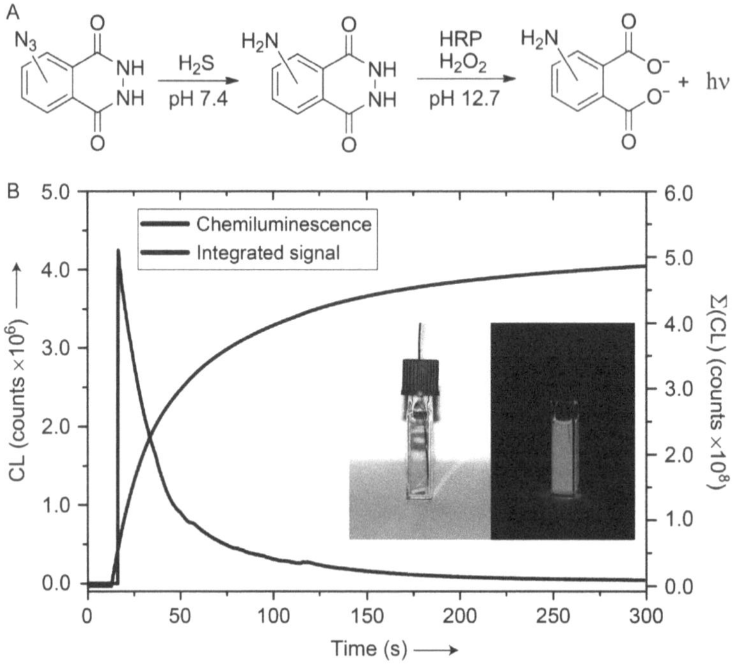

Figure 3 Chemiluminescence response profile for CLSS-1 and CLSS-2. (A) H_2S-mediated reduction of CLSS-1 and CLSS-2 followed by luminol visualization using HRP (horseradish peroxidase) and H_2O_2. (B) Chemiluminescence profile of 50 μ*M* CLSS-1 after incubation for 1 h with 33 equiv. of NaSH. Visual detection (inset) at 10 × concentration with 5 s camera exposure. Samples were incubated in pH 7.4 PIPES buffer at 37 °C prior to analysis. *Reprinted with permission from Bailey and Pluth (2013). Copyright 2013 American Chemical Society.* (See the color plate.)

brightness of this reaction can be enhanced in the presence of *p*-iodophenol or other small-molecule enhancers (Díaz, Sánchez, & González Garcia, 1998; Dodeigne, Thunus, & Lejeune, 2000). Because this reporting mechanism functions without the aid of optical excitation, the background signal prior to H_2S-mediated reduction is almost nonexistent. At an appropriate concentration, this method is sensitive enough to be visualized with the naked eye (Fig. 3B).

2.2. Reactivity

CLSS-1 and CLSS-2 are classified as chemodosimeters because they both react irreversibly with H_2S to generate the observed luminescent signal. The H_2S detection limit for CLSS-1 and CLSS-2 are 0.7 ± 0.3 and 4.6 ± 2.0 μ*M*, respectively (Fig. 4), with linear ranges extending into the high micromolar

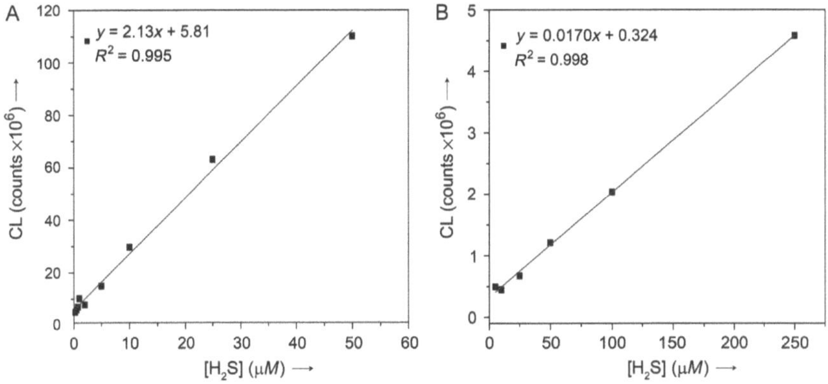

Figure 4 Concentration-dependent chemiluminescence response of (A) CLSS-1 and (B) CLSS-2. Values indicate the background corrected, integrated emission at $\lambda_{em} = 425$ nm and represent the average of at least three replicates. Samples were incubated for 60 min in pH 7.4 PIPES buffer at 37 °C prior to analysis. *Reprinted with permission from Bailey and Pluth (2013). Copyright 2013 American Chemical Society.*

concentration range. This detection limit range will be dependent on the instrument used for detection, with lower or higher ranges possible. The CLSS-1 platform provides a slightly lower H_2S detection limit than CLSS-2 due to the brighter luminosity of luminol by comparison to isoluminol. The dynamic range of CLSS-1 and CLSS-2 covers the micromolar concentrations relevant for H_2S therapeutics (Ali, Opere, & Singh, 2014) while also offering the sensitivity to possibly detect the much lower concentrations expected in cellular environments (Furne, Saeed, & Levitt, 2008).

CLSS-1 and CLSS-2 both produce robust luminescence when incubated with an excess of H_2S. CLSS-1 displays a 128-fold turn on, making it an excellent reporter for quantifying H_2S levels. In determining the response of CLSS-1 to H_2S and other biologically relevant reactive sulfur, oxygen, and nitrogen species (RSONS), CLSS-1 displays excellent selectivity for H_2S over nitrogen and oxygen species. Unexpectedly, CLSS-1 displays a low tolerance for amino acid containing thiols, especially GSH, making it ill suited for applications in which a large excess of thiols may be present. Although the 45-fold turn on from CLSS-2 is lower than that of CLSS-1, CLSS-2 displays excellent selectivity for H_2S over a wide variety of RSONS, including thiols. The improved selectivity of CLSS-2 for H_2S over amino acid containing thiols is achieved by moving the azide functionality further from the phthalhydrazide, which is discussed in detail in our initial report (Bailey & Pluth, 2013; Fig. 5).

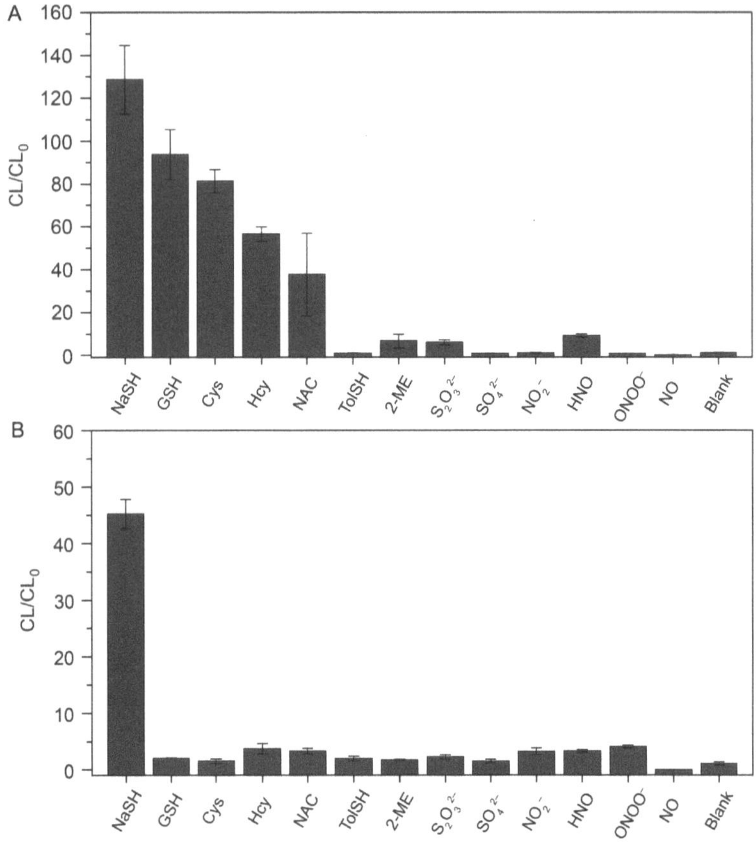

Figure 5 Selectivity of (A) CLSS-1 and (B) CLSS-2 with reactive oxygen, nitrogen, and sulfur species. Conditions: 50 μ*M* probe, 33 equiv. of RSONS, incubated 1 h at 37 °C. The reported intensities are background corrected, represent the integrated luminescence ($\lambda_{em}=425$ nm), and are the average of at least three replicates. Error bars represent ±SE. *Abbreviations*: Cys, cysteine; Hcy, homocysteine; NAC, *N*-acetylcysteine; GSH, reduced glutathione; $S_2O_3^{2-}$, thiosulfate; SO_4^{2-}, sulfate; NO, nitric oxide; HNO, nitroxyl; NO_2^-, nitrite. *Reprinted with permission from Bailey and Pluth (2013). Copyright 2013 American Chemical Society.*

Although the selectivity of the probes discussed here does not appear to be strongly dependent on pH between 7 and 8, the chemiluminescent signal generated by the HRP/luminol system employed is pH dependent. To ensure accurate measurements, the pH of analytical samples should be buffered. Because both the stability of the excited phthalate and the catalytic rate of HRP/H_2O_2 are influenced by the basicity of the solution, the pH needs to be adjusted to the same levels for comparable measurements.

2.3. Probe usage and storage

Although CLSS-1 and CLSS-2 are not sensitive to oxygen, H_2S, and other RSONS are known to react with oxygen, so measures to exclude air should be taken during sample preparation to obtain the most accurate results. Stock solutions of CLSS-1 and CLSS-2 should be prepared by dissolving a solvent-free aliquot of probe in an appropriate amount of air-free DMSO to produce a 10-m*M* stock solution. This DMSO stock solution should be stored in 50 μL aliquots and frozen at -20 °C. Care should be taken to avoid multiple free-thaw cycles of individual aliquots.

CLSS-1 and CLSS-2 are stable in the solid state at room temperature for over 6 months and do not display significant sensitivity to short-term light exposure. Thawed solutions of CLSS-1 and CLSS-2 do degrade slowly if exposed to light for extended periods of time and should remain frozen until prior to use. CLSS-1 and CLSS-2 are not sufficiently soluble in neutral pH buffers water to prepare aqueous stock solutions directly, but have moderate solubility in DMSO, DMF, methanol, ethanol, and alkaline H_2O. Stock solutions made in these solvents can be diluted with acetonitrile- or neutral-buffered H_2O without observed precipitation.

3. EXAMPLES OF ROUTINE PROBE USAGE

Imaging of any kind requires optimization of probe loading and instrument parameters. These parameters depend of the nature of the sample, the instruments being used, and the general nature of the experiment. CLSS-1 and CLSS-2 are well suited for a variety of experiment and instrument types and can be readily adapted for use with fluorimeters, fluorescence plate readers, and chemiluminometers. This section details the general approach used to determine the selectivity and sensitivity of both CLSS-1 and CLSS-2 described above.

3.1. Instrumentation and materials

3.1.1 Buffer

Piperazine-*N*,*N*′-bis(2-ethansulfonic acid) (PIPES, Aldrich) and potassium chloride (99.999%, Aldrich) were used to make buffered solutions (50 m*M* PIPES, 100 m*M* KCl, pH 7.4) with Millipore water. Buffered solutions were degassed by vigorous sparging with N_2 and stored in an inert atmosphere glove box.

3.1.2 Reactive species

Anhydrous sodium hydrogen sulfide (NaSH) was purchased from Strem Chemicals and handled under nitrogen. *S*-nitroso-*N*-acetyl-DL-penicillamine (SNAP), sodium peroxynitrite (NaO_2NO), and Angeli's salt (NaN_2O_3) were purchased from Cayman Chemical and stored either at −30 or −80 °C prior to use. L-cysteine, *N*-acetyl-L-cysteine, and DL-homocysteine were purchased from TCI. Reduced glutathione was purchased from Aldrich. Stock solutions of the reactive species were prepared in either buffer or DMSO under nitrogen immediately prior to use. Type VI-A peroxidase from horseradish 2000 U/mg was purchased from Aldrich, dissolved in pH 7.4 PIPES buffer, and stored in aliquots in the glove box freezer (1000 U/mL).

3.1.3 Instrumentation

Chemiluminescence measurements were measured on a Photon Technology International Quanta Master 40 spectrofluorimeter equipped with an USHIO UXL-75XF Xenon short arc lamp, a LPS-100 lamp power supply, a 914 photomultiplier detection system, and a Quantum Northwest TLC-50 temperature controller set to 37.0 ± 0.05 °C. To ensure the minimum possible baseline, the excitation shutter was closed during all measurements. Emissions slits were opened to 25 nm to ensure all possible light was collected. All chemiluminescent measurements were made under an inert atmosphere in 1.0 cm path length septum-sealed cuvettes obtained from Starna Scientific and were repeated at least in triplicate. Samples were incubated using a Fischer Isotemp cuvette block stored in an insulated box.

3.2. Preparation of CLSS-1 and CLSS-2

CLSS-1 and CLSS-2 are prepared as described below and as depicted in Scheme 1.

3.2.1 CLSS-1

Dry reagent grade DMSO over 4 Å molecular sieves for a period of at least 24 h prior to use. No further precautions are taken to exclude water or air. Luminol (100 mg, 0.56 mmol) is dissolved in 5 mL DMSO in a 20 mL glass scintillation vial to produce a yellow solution. After wrapping the scintillation vial in aluminum foil, *tert*-butylnitrite (*tert*-BuONO, 100 μL, 0.85 mmol) is added dropwise to the luminol solution, followed immediately by azidotrimethylsilane (TMS-N_3, 95 μL, 0.68 mmol). The solution is stirred at room temperature for 1 h, after which the volatile reagents and reaction byproducts are removed under vacuum (1 torr) to produce

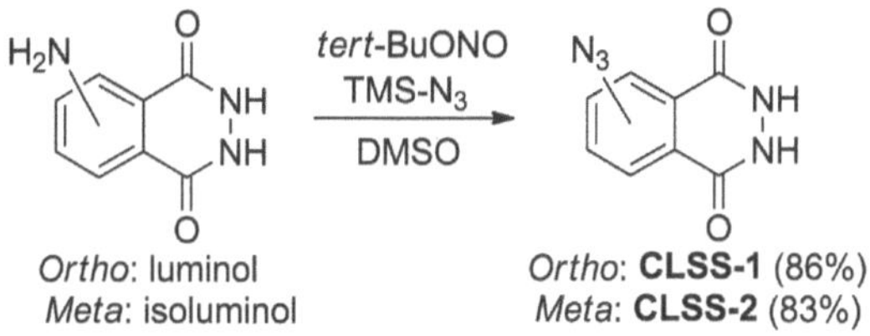

Scheme 1 Synthesis of CLSS-1 and CLSS-2. *Reprinted with permission from Bailey and Pluth (2013). Copyright 2014 American Chemical Society.*

a brown solution. Addition of 15 mL of 5% dichloromethane in hexanes precipitates a gray solid, which is collected by vacuum filtration and rinsed with cold dichloromethane to yield CLSS-1 (95 mg, 83%). In the solid-phase CLSS-1 slowly forms a hydrate complex with water if left under ambient atmosphere for extended periods of time.

3.2.2 CLSS-2

Dry reagent grade DMSO over 4 Å molecular sieves for a period of at least 24 h prior to use. No further precautions are taken to exclude water or air. Isoluminol (100 mg, 0.56 mmol) is dissolved in 5 mL DMSO in a 20 mL glass scintillation vial to produce a yellow solution. After wrapping the scintillation vial in aluminum foil, *tert*-butylnitrite (*tert*-BuONO, 100 μL, 0.85 mmol) is added dropwise to the luminol solution, followed immediately by azidotrimethylsilane (TMS-N_3, 95 μL, 0.68 mmol). The solution is stirred at room temperature for 1 h, after which the volatile reagents and reaction byproducts are removed under vacuum (1 torr) to produce a brown–orange solution. Addition of 15 mL of 5% dichloromethane in hexanes precipitates an orange solid, which is collected by vacuum filtration and rinsed with cold dichloromethane to yield CLSS-2 (100 mg, 87%). In the solid-phase CLSS-2 slowly forms a hydrate complex with water if left under ambient atmosphere for extended periods of time, resulting in a purple solid.

3.3. Sensing method

In an oxygen-free glove box, 3.00 mL of PIPES buffer is transferred into each of three quartz cuvettes with stir bars. Although we used a glove box to ensure anaerobic conditions, any apparatus that excludes atmospheric oxygen, such as a glove bag, should be sufficient to generate reproducible results. A 50 μL aliquot of a 10 mM DMSO stock solution of the desired probe is thawed immediately prior to use, and 15 μL is transferred into each of the three cuvettes. The cuvettes are sealed with septa and removed from the glove box. Each cuvette is charged with 50 μL of a 100 mM anaerobic stock solution of the RSON in PIPES buffer (SNAP, Angeli's salt, and NaOONO excluded), and incubated for 60 min in the absence of light. For H_2S sensitivity measurements, the appropriate volume of stock solution required to generate the desired concentration of sulfide replaces the 50 μL addition.

After the 60-min incubation with the reactive species, the cuvettes are basified with 40 μL of 6 *M* NaOH and charged with 10 μL of a solution containing 1000 U/mL HRP and 0.2 μ*M* *p*-iodophenol. Data collection is initiated (monitoring at 425 nm with 4 scans/s and a 0.05 s integration time) and the first 60 s of data points are used as a background measurement. After 60 s, 50 μL of H_2O_2 (35%, Aldrich) is injected into the cuvette and the luminescence is monitored for an additional 300 s. In addition to providing an oxidant source for the luminol reaction, the large excess of H_2O_2 also ensures that any remaining reductants in the cuvette are quenched.

3.4. Data processing and analysis

Data are collected in FelixGX v.4.0.4 (the standard software for the PTI fluorimeter) and exported into MS Excel for analysis. To correct for the detector background, the average luminescence value of the first 60 s of the data acquisition prior to H_2O_2 addition is subtracted from each data point measured during data collection. The combined luminescence curve is then integrated and normalized to the blank for quantification purposes.

4. DETECTION OF ENZYMATICALLY PRODUCED H_2S

In this section, we highlight the effectiveness of CLSS-2 to detect and measure the concentration of H_2S is complex biological mixtures by measuring enzymatically produced H_2S. CSE is one of three H_2S generating enzymes present in mammalian systems and is especially prevalent in the kidneys and liver. CSE converts cysteine or homocysteine into free H_2S using PLP as a cofactor, and can be inhibited by β-cyanoalanine (Pfeffer & Ressler, 1967). This section outlines an assay for the determination of CSE activity in C6 cells, which express CSE and produce H_2S endogenously.

4.1. Instrumentation and materials

4.1.1 Instrumentation

Chemiluminescence measurements were obtained on a Photon Technology International Quanta Master 40 spectrofluorimeter equipped with an USHIO UXL-75XF Xenon short arc lamp, a LPS-100 lamp power supply, a 914 photomultiplier detection system, and a Quantum Northwest TLC-50 temperature controller at 37.0 ± 0.05 °C. To ensure the minimum possible baseline measurements, the excitation beam shutter was closed during measurements. Emission slits were opened to 25 nm to ensure that all possible light was collected. All chemiluminescent measurements were made under an inert atmosphere in 1.0 cm path length septum-sealed cuvettes obtained from Starna Scientific and were repeated at least in triplicate. Samples were incubated using a Fischer Isotemp cuvette block stored in an insulated box. Cells were counted Bio RAD TC20 automated cell counter.

4.1.2 Media

Dulbecco's modified Eagle's medium (DMEM, Cellgro, MediaTek, Inc.) supplemented with 10% fetal bovine serum (FBS, HyClone), and 1% penicillin/streptomycin.

4.1.3 Probe

A 10 mM CLSS-2 stock solution in DMSO.

4.1.4 Materials

Isolated CSE and β-cyanoalanine were purchased from Cayman Chemicals.

4.1.5 Reactive species

Ten micrograms of CSE, 20 m*M* Hcy, 20 m*M* BCA, 25 μ*M* pyruvate, 25 μ*M* NH_3, luminescence measurements on cell lysates were made using 100 μL of lysate solution (2×10^6 cells per experiment) under ambient atmosphere.

4.2. Cell culture and lysing

C6 cells were obtained from ATCC and cultured in DMEM. Cells were passed and plated into T-75 flasks containing 10 mL of DMEM, and incubated at 37 °C with 5% CO_2. For luminescence studies, the cells were washed with 1 × phosphate-buffered saline (PBS), trypsinized with 5 mL of trypsin, and then centrifuged to form a cell pellet. The cell pellet was resuspended in 5 mL of 1 × PBS and the cells were counted. Cells were centrifuged at 1000 RPM for 5 min at room temperature, placed on ice and lysed using 100 μL of RIPA buffer (pH 7.5, 10 m*M* Tris–HCl, 150 m*M* NaCl, 1.0% Nonidet P-40, 0.1% SDS, 0.1% sodium deoxycholate) containing protease inhibitor (PhosSTOP, Roche) for every 2×10^6 cells in the pellet.

4.3. Assay for enzymatically produced H_2S

Isolated CSE enzyme or C6 cell lysates were added to a cuvette containing 3.00 mL pH 7.4 PIPES buffer under an anaerobic atmosphere and incubated for 48 h. This incubation period can be changed depending on the amount of enzyme added as well as the enzyme activity. After the initial incubation period, introduce 15 μL of 10 m*M* CLSS-2 and incubate for an additional 60 min. Upon completion of the second incubation, 40 μL of 6 *M* NaOH is added to increase the pH to 12.7, an optimal level for luminol chemiluminescence. After pH adjustment, 10 μL of 1000 U/mL (HRP) containing 0.2 μ*M* *p*-iodophenol is added. A background reading is acquired for 60 s, after which 50 μL of H_2O_2 (35%) is added. The sample luminosity at 425 nm is then integrated for a time suitable to capture the sample luminescence. The data reported are the average of at least three independent experiments.

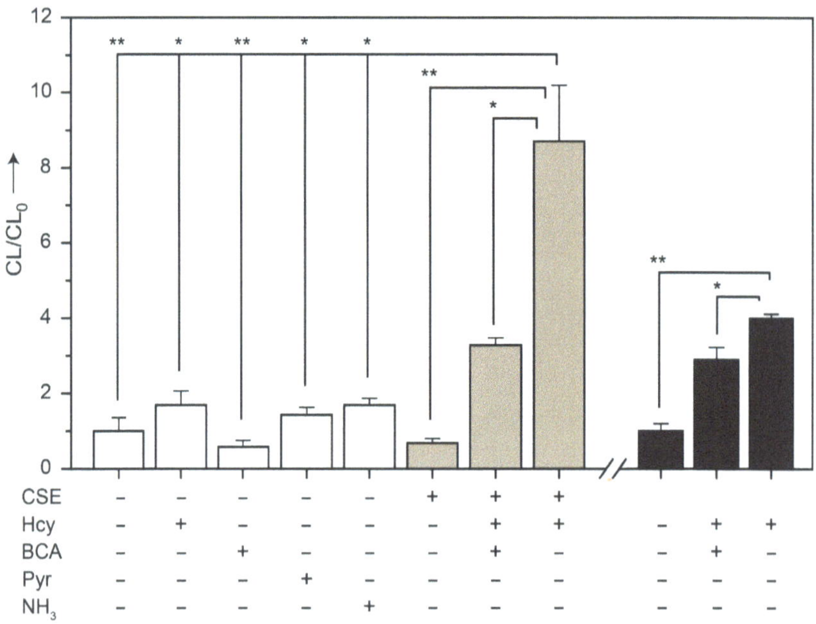

Figure 6 Detection of CSE-produced H_2S using CLSS-2. Conditions: absence of enzyme (white), 10 μg CSE (light gray), 20 m*M* Hcy, 20 m*M* BCA, 25 μ*M* pyruvate, 25 μ*M* NH_3; incubated at in 3.0 mL buffer at 37 °C for 48 h prior to detection with 50 μ*M* CLSS-2. Comparison with C6 cell lysates containing 2×10^6 cells (dark gray). Each data point represents the mean ± SE derived from at least three independent experiments; $*p < 0.005$, $**p < 0.001$. *Reprinted with permission from Bailey and Pluth (2013). Copyright 2013 American Chemical Society.*

4.4. Results and controls

CLSS-2 is nonreactive ($p<0.05$) with the CSE substrate Hcy, the CSE inhibitor BCA, and the CSE reaction byproducts pyruvate (Pyr) and NH_3 (Fig. 6, white bars). Similarly, CLSS-2 does not produce a significant signal in the presence of isolated CSE alone. Upon introduction of Hcy as a CSE substrate in the presence of both CLSS-1 and CSE; however, a strong signal is observed when compared with CSE alone ($p<0.001$) or when BCA is used to inhibit the H_2S-forming reaction ($p<0.005$) (Fig. 6, light gray bars). The known kinetic parameters for the CSE-mediated conversion of Hcy to H_2S (Faccenda, Wang, & Mutus, 2012) are in agreement with H_2S quantification values using the chemiluminescent response curve for CLSS-2 in Fig. 4B. These results show that CLSS-2 can be used to quantify H_2S enzymatically generated by CSE and to differentiate between the inhibited and uninhibited enzyme states.

C6 cells, which express CSE and produce H_2S endogenously in the presence of CSE substrates (Kandil, Brennan, & McBean, 2010), were also used to demonstrate this assay. Introduction of CLSS-2 to C6 cell lysates in the absence of Cys or Hcy results in minimal luminescent (Fig. 6, dark gray), consistent with limited cellular H_2S production in the absence of CSE substrates. When CLSS-2 is added to C6 cell lysates preincubated with Hcy as a CSE substrate, the luminescence increases significantly ($p<0.001$) by comparison to lysates lacking substrate, signifying that CSE present in the cell lysates produces sufficient H_2S to be detected by CLSS-2. Furthermore, addition of both Hcy and BCA to C6 cell lysates abrogated the luminescent response ($p<0.005$), which is consistent with CSE inhibition. These results build upon the isolated CSE experiments and demonstrate that CLSS-2 can detect endogenously produced H_2S even in the presence of other biological species. Most importantly, this result confirms that other biological species in the cellular milieu do not activate CLSS-2.

5. CONCLUSIONS

Utilizing a chemiluminescent platform for detecting H_2S allows for the determination of H_2S at physiologically relevant levels. Not only are these platforms the first example of reaction-based chemiluminescent probes for H_2S, but they also offer insight into new strategies to separate the reactivity of H_2S from other biological sulfhydryl-containing species. In addition to detecting exogenous H_2S, CLSS-2 can detect enzymatically produced H_2S from both isolated CSE enzymes and also from C6 cell lysates and can also differentiate inhibited and native states of the enzyme.

ACKNOWLEDGMENTS

This work was supported by the NIGMS (R00 GM092970) and funding from the University of Oregon (UO). We thank Leticia Montoya for assistance with cell culture work.

REFERENCES

Abe, K., & Kimura, H. (1996). The possible role of hydrogen sulfide as an endogenous neuromodulator. *Journal of Neuroscience*, *16*, 1066–1071.

Ali, H., Opere, C., & Singh, S. (2014). In vitro-controlled release delivery system for hydrogen sulfide donor. *AAPS PharmSciTech*, *15*, 910–919.

Bae, S. K., Heo, C. H., Choi, D. J., Sen, D., Joe, E.-H., Cho, B. R., et al. (2013). A ratiometric two-photon fluorescent probe reveals reduction in mitochondrial H_2S production in Parkinson's disease gene knockout astrocytes. *Journal of the American Chemical Society*, *135*, 9915–9923.

Bailey, T. S., & Pluth, M. D. (2013). Chemiluminescent detection of enzymatically produced hydrogen sulfide: Substrate hydrogen bonding influences selectivity for H_2S over biological thiols. *Journal of the American Chemical Society*, *135*, 16697–16704.

Chan, M. V., & Wallace, J. L. (2013). Hydrogen sulfide-based therapeutics and gastrointestinal diseases: Translating physiology to treatments. *American Journal of Physiology. Gastrointestinal and Liver Physiology*, *305*, G467–G473.

Chen, S., Chen, Z.-J., Ren, W., & Ai, H.-W. (2012). Reaction-based genetically encoded fluorescent hydrogen sulfide sensors. *Journal of the American Chemical Society*, *134*, 9589–9592.

Chen, B., Li, W., Lv, C., Zhao, M., Jin, H., Jin, H., et al. (2013). Fluorescent probe for highly selective and sensitive detection of hydrogen sulfide in living cells and cardiac tissues. *The Analyst*, *138*, 946–951.

Chen, G. N., Lin, R. E., Zhuang, H. S., Zhao, Z. F., Xu, X. Q., & Zhang, F. (1998). Study of electrogenerated chemiluminescence of *N*-(β-carboxyl-propionyl)luminol. *Analytica Chimica Acta*, *375*, 269–275.

Chen, C.-Q., Xin, H., & Zhu, Y.-Z. (2007). Hydrogen sulfide: Third gaseous transmitter, but with great pharmacological potential. *Acta Pharmacologica Sinica*, *28*, 1709–1716.

Chen, Y., Zhu, C., Yang, Z., Chen, J., He, Y., Jiao, Y., et al. (2013). A ratiometric fluorescent probe for rapid detection of hydrogen sulfide in mitochondria. *Angewandte Chemie International Edition*, *52*, 1688–1691.

Cortese-Krott, M. M., Fernandez, B. O., Santos, J. L. T., Mergia, E., Grman, M., Nagy, P., et al. (2014). Nitrosopersulfide ($SSNO^-$) accounts for sustained NO bioactivity of S-nitrosothiols following reaction with sulfide. *Redox Biology*, *2*, 234–244.

DeLeon, E. R., Stoy, G. F., & Olson, K. R. (2012). Passive loss of hydrogen sulfide in biological experiments. *Analytical Biochemistry*, *421*, 203–207.

Díaz, A. N., Sánchez, F. G., & González Garcia, J. A. (1998). Phenol derivatives as enhancers and inhibitors of luminol–H_2O_2–horseradish peroxidase chemiluminescence. *Journal of Bioluminescence and Chemiluminescence*, *13*, 75–84.

Dodeigne, C., Thunus, L., & Lejeune, R. (2000). Chemiluminescence as diagnostic tool. A review. *Talanta*, *51*, 415–439.

Drew, H. D. K., & Garwood, R. F. (1939). 177. Chemiluminescent organic compounds. Part VII. Substituted phthalaz-1: 4-Diones. Effect of substituents on the luminescent power. *Journal of the Chemical Society*, 836–837.

Faccenda, A., Wang, J., & Mutus, B. (2012). Polydimethylsiloxane permeability-based method for the continuous and specific detection of hydrogen sulfide. *Analytical Chemistry*, *84*, 5243–5249.

Filipovic, M. R., Miljkovic, J. L., Nauser, T., Royzen, M., Klos, K., Shubina, T., et al. (2012). Chemical characterization of the smallest S-nitrosothiol, HSNO; cellular cross-talk of H_2S and S-nitrosothiols. *Journal of the American Chemical Society*, *134*, 12016–12027.

Fogo, J. K., & Popowsky, M. (1947). Spectrophotometric determination of hydrogen sulfide. *Analytical Chemistry*, *21*, 732–734.

Furne, J., Saeed, A., & Levitt, M. D. (2008). Whole tissue hydrogen sulfide concentrations are orders of magnitude lower than presently accepted values. *American Journal of Physiology. Regulatory, Integrative and Comparative Physiology*, *295*, R1479–R1485.

Garcia, I. L., Vinas, P., & Gil, J. A. M. (1993). FIA titrations of sulfide, cysteine and thiol-containing drugs with chemiluminescent detection. *Fresenius Journal of Analytical Chemistry*, *345*, 723–726.

Hammers, M. D., & Pluth, M. D. (2014). Ratiometric measurement of hydrogen sulfide and cysteine/homocysteine ratios using a dual-fluorophore fragmentation strategy. *Analytical Chemistry*, *86*, 7135–7140.

Jiang, B., Tang, G. H., Cao, K., Wu, L. Y., & Wang, R. (2010). Molecular mechanism for H_2S-induced activation of K-ATP channels. *Antioxidants and Redox Signaling*, *12*, 1167–1178.

Kandil, S., Brennan, L., & McBean, G. J. (2010). Glutathione depletion causes a JNK and p38MAPK-mediated increase in expression of cystathionine-γ-lyase and upregulation of the transsulfuration pathway in C6 glioma cells. *Neurochemistry International*, *56*, 611–619.

Lawrence, N. S., Davis, J., & Compton, R. G. (2000). Analytical strategies for the detection of sulfide: A review. *Talanta*, *52*, 771–784.

Lee, D. W., Erigala, V. R., Dasari, M., Yu, J. H., Dickson, R. M., & Murthy, N. (2008). Detection of hydrogen peroxide with chemiluminescent micelles. *International Journal of Nanomedicine*, *3*, 471–476.

Lin, V. S., & Chang, C. J. (2012). Fluorescent probes for sensing and imaging biological hydrogen sulfide. *Current Opinion in Chemical Biology*, *16*, 595–601.

Lin, V. S., Lippert, A. R., & Chang, C. J. (2013). Cell-trappable fluorescent probes for endogenous hydrogen sulfide signaling and imaging H_2O_2-dependent H_2S production. *Proceedings of the National Academy of Sciences*, *110*, 7131–7135.

Lippert, A. R., New, E. J., & Chang, C. J. (2011). Reaction-based fluorescent probes for selective imaging of hydrogen sulfide in living cells. *Journal of the American Chemical Society*, *133*, 10078–10080.

Liu, C. R., Pan, J., Li, S., Zhao, Y., Wu, L. Y., Berkman, C. E., et al. (2011). Capture and visualization of hydrogen sulfide by a fluorescent probe. *Angewandte Chemie International Edition*, *50*, 10327–10329.

Liu, T., Xu, Z., Spring, D. R., & Cui, J. (2013). A lysosome-targetable fluorescent probe for imaging hydrogen sulfide in living cells. *Organic Letters*, *15*, 2310–2313.

Marquette, C., & Blum, L. (2006). Applications of the luminol chemiluminescent reaction in analytical chemistry. *Analytical and Bioanalytical Chemistry*, *385*, 546–554.

McCutcheon, D. C., Paley, M. A., Steinhardt, R. C., & Prescher, J. A. (2012). Expedient synthesis of electronically modified luciferins for bioluminescence imaging. *Journal of the American Chemical Society*, *134*, 7604–7607.

Montoya, L. A., Pearce, T. F., Hansen, R. J., Zakharov, L. N., & Pluth, M. D. (2013). Development of selective colorimetric probes for hydrogen sulfide based on nucleophilic aromatic substitution. *Journal of Organic Chemistry*, *78*, 6550–6557.

Montoya, L. A., & Pluth, M. D. (2012). Selective turn-on fluorescent probes for imaging hydrogen sulfide in living cells. *Chemical Communications*, *48*, 4767–4769.

Montoya, L. A., & Pluth, M. D. (2014). Hydrogen sulfide deactivates common nitrobenzofurazan-based fluorescent thiol labeling reagents. *Analytical Chemistry*, *86*, 6032–6039.

Nishida, M., Sawa, T., Kitajima, N., Ono, K., Inoue, H., Ihara, H., et al. (2012). Hydrogen sulfide anion regulates redox signaling via electrophile sulfhydration. *Nature Chemical Biology*, *8*, 714–724.

Omote, Y., Miyake, T., Ohmori, S., & Sugiyama, N. (1966). The chemiluminescence of acyl luminols. *Bulletin of the Chemical Society of Japan*, *39*, 932–935.

Peng, H., Cheng, Y., Dai, C., King, A. L., Predmore, B. L., Lefer, D. J., et al. (2011). A fluorescent probe for fast and quantitative detection of hydrogen sulfide in blood. *Angewandte Chemie International Edition*, *50*, 9672–9675.

Pfeffer, M., & Ressler, C. (1967). Beta-cyanoalanine an inhibitor of rat liver cystathionase. *Biochemical Pharmacology*, *16*, 2299–2308.

Pietri, R., Roman-Morales, E., & Lopez-Garriga, J. (2011). Hydrogen sulfide and hemeproteins: Knowledge and mysteries. *Antioxidants and Redox Signaling*, *15*, 393–404.

Qian, Y., Karpus, J., Kabil, O., Zhang, S. Y., Zhu, H. L., Banerjee, R., et al. (2011). Selective fluorescent probes for live-cell monitoring of sulphide. *Nature Communications*, *2*, 495.

Qu, K., Lee, S. W., Bian, J. S., Low, C. M., & Wong, P. T. H. (2008). Hydrogen sulfide: Neurochemistry and neurobiology. *Neurochemistry International*, *52*, 155–165.

Sasakura, K., Hanaoka, K., Shibuya, N., Mikami, Y., Kimura, Y., Komatsu, T., et al. (2011). Development of a highly selective fluorescence probe for hydrogen sulfide. *Journal of the American Chemical Society*, *133*, 18003–18005.

Shen, X., Pattillo, C. B., Pardue, S., Bir, S. C., Wang, R., & Kevil, C. G. (2011). Measurement of plasma hydrogen sulfide in vivo and in vitro. *Free Radical Biology and Medicine*, *50*, 1021–1031.

Shibuya, N., Koike, S., Tanaka, M., Ishigami-Yuasa, M., Kimura, Y., Ogasawara, Y., et al. (2013). A novel pathway for the production of hydrogen sulfide from D-cysteine in mammalian cells. *Nature Communications*, *4*, 1366.

Sparatore, A., Santus, G., Giustarini, D., Rossi, R., & Del Soldato, P. (2011). Therapeutic potential of new hydrogen sulfide-releasing hybrids. *Expert Review of Clinical Pharmacology*, *4*, 109–121.

Sun, Y. Q., Liu, J., Wang, P., Zhang, J., & Guo, W. (2012). D-luciferin analogues: A multicolor toolbox for bioluminescence imaging. *Angewandte Chemie International Edition*, *51*, 8428–8430.

Szabó, C. (2007). Hydrogen sulphide and its therapeutic potential. *Nature Reviews Drug Discovery*, *6*, 917–935.

Thorson, M. K., Majtan, T., Kraus, J. P., & Barrios, A. M. (2013). Identification of cystathionine β-synthase inhibitors using a hydrogen sulfide selective probe. *Angewandte Chemie International Edition*, *52*, 4641–4644.

Thurbide, K. B., & Aue, W. A. (2002). Chemiluminescent emission spectra of lead, chromium, ruthenium, iron, manganese, rhenium, osmium and tungsten in the reactive flow detector. *Spectrochimica Acta Part B: Atomic Spectroscopy*, *57*, 843–852.

Van de Bittner, G. C., Dubikovskaya, E. A., Bertozzi, C. R., & Chang, C. J. (2010). In vivo imaging of hydrogen peroxide production in a murine tumor model with a chemoselective bioluminescent reporter. *Proceedings of the National Academy of Sciences*, *107*, 21316–21321.

Wallace, J. L., Dicay, M., McKnight, W., & Martin, G. R. (2007). Hydrogen sulfide enhances ulcer healing in rats. *FASEB Journal: Official Publication of the Federation of American Societies for Experimental Biology*, *21*, 4070–4076.

Wang, R. (2009). Hydrogen sulfide: A new EDRF. *Kidney International*, *76*, 700–704.

Wang, R. (2012). Physiological implications of hydrogen sulfide: A whiff exploration that blossomed. *Physiological Reviews*, *92*, 791–896.

White, E. H., & Roswell, D. F. (1970). Chemiluminescence of organic hydrazides. *Accounts of Chemical Research*, *3*, 54–62.

Whiteman, M., & Moore, P. K. (2009). Hydrogen sulfide and the vasculature: A novel vasculoprotective entity and regulator of nitric oxide bioavailability? *Journal of Cellular and Molecular Medicine*, *13*, 488–507.

Whiteman, M., & Winyard, P. G. (2011). Hydrogen sulfide and inflammation: The good, the bad, the ugly and the promising. *Expert Review of Clinical Pharmacology*, *4*, 13–32.

Williams, E. J., & Campbell, A. K. (1986). A homogeneous assay for biotin based on chemiluminescence energy transfer. *Analytical Biochemistry*, *155*, 249–255.

Wu, L. Y., Yang, W., Jia, X. M., Yang, G. D., Duridanova, D., Cao, K., et al. (2009). Pancreatic islet overproduction of H_2S and suppressed insulin release in Zucker diabetic rats. *Laboratory Investigation*, *89*, 59–67.

Yamaguchi, S., Kishikawa, N., Ohyama, K., Ohba, Y., Kohno, M., Masuda, T., et al. (2010). Evaluation of chemiluminescence reagents for selective detection of reactive oxygen species. *Analytica Chimica Acta*, *665*, 74–78.

Yan, X. (2006). Unique selective detectors for gas chromatography: Nitrogen and sulfur chemiluminescence detectors. *Journal of Separation Science*, *29*, 1931–1945.

Yang, G. D., Wu, L. Y., Jiang, B., Yang, W., Qi, J. S., Cao, K., et al. (2008). H_2S as a physiologic vasorelaxant: Hypertension in mice with deletion of cystathionine gamma-lyase. *Science*, *322*, 587–590.

Yuan, H., Chong, H., Wang, B., Zhu, C., Liu, L., Yang, Q., et al. (2012). Chemical molecule-induced light-activated system for anticancer and antifungal activities. *Journal of the American Chemical Society*, *134*, 13184–13187.

CHAPTER SIX

Quantification of Hydrogen Sulfide Concentration Using Methylene Blue and 5,5′-Dithiobis(2-Nitrobenzoic Acid) Methods in Plants

Zhong-Guang Li[1]

School of Life Sciences; Engineering Research Center of Sustainable Development and Utilization of Biomass Energy, Ministry of Education; Key Laboratory of Biomass Energy and Environmental Biotechnology, Yunnan Normal University, Kunming, Yunnan Province, PR China

[1]Corresponding author: e-mail address: zhongguang_li@163.com

Contents

Methods in Enzymology, Volume 554
ISSN 0076-6879
http://dx.doi.org/10.1016/bs.mie.2014.11.031

Abstract

Hydrogen sulfide (H_2S) has long been considered as a phytotoxin but, nowadays, as a novel signaling molecule at low concentration in plants, which is involved in plant growth, development, and the acquisition of tolerance to abiotic and biotic stresses. H_2S is commonly found only in very small amounts in most biological systems. To further understand H_2S functions, accurate measurement of H_2S concentration in plants is very important. In this chapter, quantificational methods of H_2S in plant tissues, namely, methylene blue (MB) and 5,5′-dithiobis(2-nitrobenzoic acid) (DTNB) methods, are introduced; the former has higher sensitivity than that of the latter, and the determination limits of MB and DTNB methods are 1.5×10^{-7} *M* (0.005 μg/ml) and 3.7×10^{-6} *M* (0.13 μg/ml), respectively, hoping to stimulate more research into H_2S signal molecule in plants.

1. THEORY

Hydrogen sulfide (H_2S) is considered as the third gaseous signal molecule after nitric oxide (NO) and carbon monoxide (CO) in animal system. In plant system, H_2S has long been considered as a phytotoxin but, nowadays, as a novel signaling molecule in plants at low concentrations, which plays very important roles in growth, development, and the acquisition of stress tolerance in plants. H_2S dissociates in aqueous solutions to HS^- and S^{2-} with HS^- as the major form. Because hydrosulfide HS^- exhibits both high nucleophilicity and reducing ability, the development of reaction-based detection methods for H_2S mainly relies on these properties.

H_2S is very volatile and prone to oxidation. This has brought great challenge in the accurate measurement of this molecule. As a result, sample preparation using Zn^{2+} ($ZnCl_2$ or $Zn(OAc)_2$) is usually needed for stabilization of H_2S (Peng, Chen, & Wang, 2012). Although new methods are emerging and traditional methods such as methylene blue (MB) and 5,5′-dithiobis(2-nitrobenzoic acid) (DTNB) are being displaced, they are still commonly found in publications most likely because of the commercial availability of the reagents. In MB method, a strongly acidic condition is used to liberate H_2S from zinc sulfide complex. H_2S then reacts with *N,N*-dimethyl-*p*-phenylenediamine dihydrochloride (DMPD) in the presence of $FeCl_3$ as

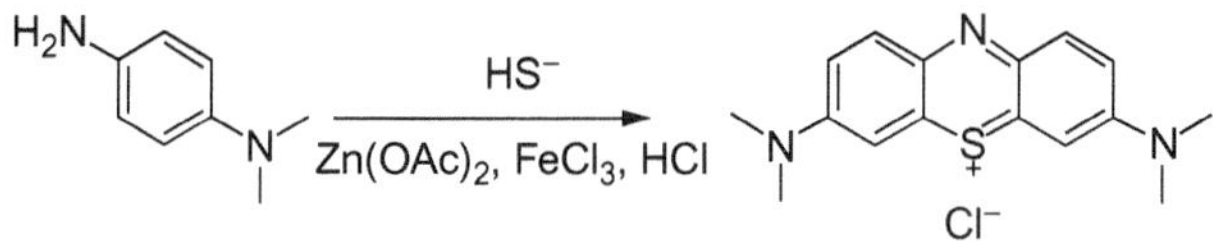

Figure 1 Reaction in the methylene blue method for sulfide detection.

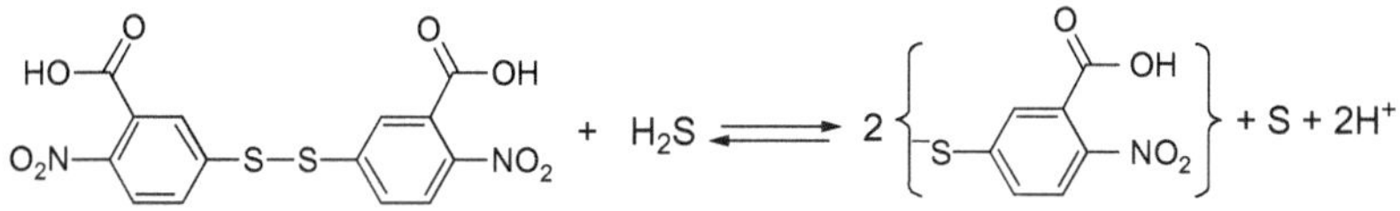

Figure 2 Reaction in the 5,5′-dithiobis(2-nitrobenzoic acid) method for sulfide detection.

an oxidizing agent, yielding MB that absorbs strongly at 670 nm (Peng et al., 2012; Fig. 1).

In addition, in DTNB method, H_2S reacts with 1 mol of DTNB and produces 2 mol of the thiol anion, 5-thio-2-nitrobenzoate (Fig. 2), which absorbs strongly at 412 nm. Concentrations of H_2S down to 3.7×10^{-6} M (0.13 μg/ml) can be easily determined with DTNB. In the presence of proteins containing different amounts of sulfhydryls reacting with DTNB, 78–94% of the added H_2S can be recovered (Christou, Manganaris, Papadopoulos, & Fotopoulos, 2013; Nashet, Osuga, & Feeney, 1977).

2. EQUIPMENT

Electronic balance
Centrifuge
5 ml centrifuge tubes
Spectrophotometer
Disposable polystyrene cuvettes or glass cuvettes (1 cm path length)
Micropipettors
Micropipettor tips
Pipettes
Voter mixer
Test tube
Mortar and pestle
Volumetric flasks
Beakers

3. MATERIALS

Tris hydroxy methyl aminomethan (Tris)
concentrated hydrochloric acid (HCl)
Ethylenediaminetetraacetic acid (EDTA)
5,5′-Dithiobis(2-nitrobenzoic acid) (DTNB)
Ferric chloride ($FeCl_3$)
N,*N*-dimethyl-*p*-phenylenediamine dihydrochloride
Concentrated hydrochloride acid
Sodium hydrosulfide (NaHS) or sodium sulfide (Na_2S)
Zinc acetate ($Zn(OAc)_2$)
Methylene blue
Phosphopyridoxal
Liquid nitrogen

3.1. Solution and buffer

1. Extraction buffer: 20 m*M* Tris-HCl buffer, pH 8.0

Component	Final concentration (m*M*)	Stock (m*M*)	Amount/l
Tris	20	200	10 ml
HCl	10.72	200	5.36 ml
EDTA	10	0.5	20 ml
$Zn(OAc)_2$	20	0.5	40 ml

Add water to 100 ml.

2. 20 m*M* DTNB

Component	Final concentration (m*M*)	Stock (m*M*)	Amount/100 ml
DTNB	20	100	20 ml

Add water to 100 ml.

3. 30 m*M* $FeCl_3$

Component	Final concentration	Stock	Amount/100 ml
$FeCl_3$	30 m*M*	100 m*M*	30 ml
HCl	1.2 M	12 *M* (concentrated HCl)	10 ml

Add water to 100 ml.

4. 20 m*M* DMPD

Component	Final concentration	Stock	Amount/100 ml
DMPD	20 m*M*	100 m*M*	20 ml
HCl	7.2 *M*	12 *M* (concentrated HCl)	60 ml

Add water to 100 ml.

4. PROTOCOL 1

4.1. Duration

Preparation	1 h
Protocol	About 2 h

4.2. Preparation

Prepare solution and buffer according to Section 3.1. Sampling and extract H_2S from plant tissues in accordance with the flowchart of the complete Protocol 1 (see Fig. 3).

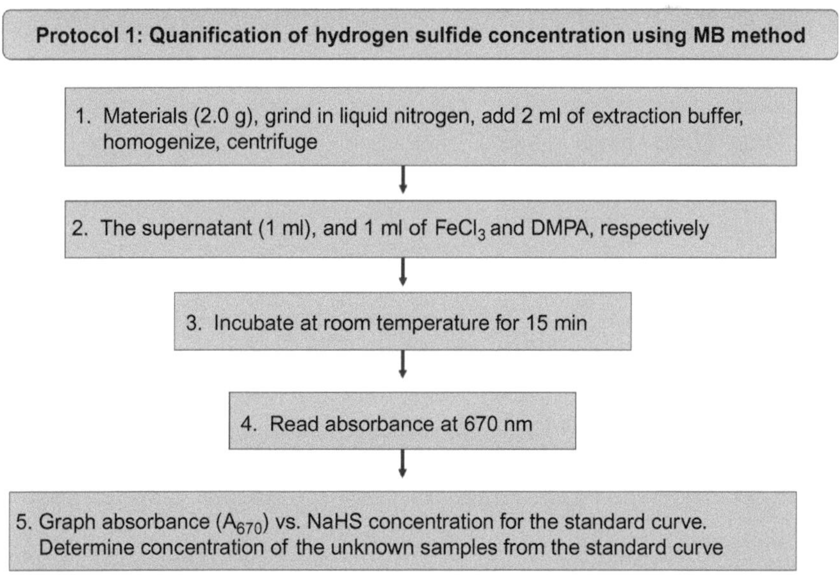

Figure 3 Flowchart for Protocol 1.

5. STEP 1: QUANTIFICATION OF H_2S CONCENTRATION USING MB METHOD

5.1. Overview

H_2S is determined by measuring absorbance at 670 nm using MB method after being extracted from plant tissues. Estimate the H_2S concentration of the unknown sample solution from a standard curve of a suitable H_2S donor (e.g., NaHS or Na_2S) or extinction coefficient of MB ($15 \times 10^6\ M^{-1}\ cm^{-1}$) at 670 nm, and H_2S content is expressed as nmol g^{-1} FW (fresh weight).

5.2. Duration

2 h

1. Two grams of plant material is ground into fine powder with a mortar and pestle under liquid nitrogen and then homogenized in 2 ml of extraction buffer: 20 m*M* Tris-HCl buffer (pH 8.0) containing 10 m*M* EDTA and 20 m*M* $Zn(OAc)_2$.
2. The homogenate is centrifuged at $15{,}000 \times g$ for 15 min at 4 °C.
3. One milliliter of the supernatant is added to 1 ml of 30 m*M* $FeCl_3$ dissolved in 1.2 *M* HCl and 1 ml of 20 m*M* DMPD dissolved in 7.2 *M* HCl mix.
4. The assay mixture is incubated at room temperature for 15 min, and the absorbance is read at 670 nm.
5. Graph the standard curve as absorbance (A_{670}) versus NaHS or Na_2S concentration (μg/ml). Determine the H_2S concentrations of the unknown samples from the graph, and the content of H_2S in plant tissues is expressed as nmol g^{-1} FW.

5.3. Tip

The spectrophotometer should be warmed up 15–20 min before use and zeroed using assay mixture without the supernatant, which is replaced by the equal volume of extraction buffer. After reading the samples, rezero to determine any instrument drift. Ideally, the absorbance should fall between 0.1 and 1 units, depending on the performance of the spectrophotometer.

5.4. Tip

Graph the standard curve as absorbance (A_{670}) versus concentration (μg/ml). Determine the concentrations of the unknown samples from the graph. Alternatively, the concentration can be determined using MB molar extinction coefficient of $15 \times 10^6\ M^{-1}\ \text{cm}^{-1}$ and the Beer–Lambert equation:

$$A = \varepsilon cl$$

where ε is the molar extinction coefficient, c is the concentration of MB, and l is the path length in cm (Papenbrock & Schmidt, 2000).

5.5. Tip

In this protocol, $FeCl_3$ and DMPD are dissolved in 1.2 and 7.2 *M* HCl, respectively. In addition, both can be dissolved in 3.5 and 100 m*M* H_2SO_4 and form 5 m*M* DMPD and 50 mM $FeCl_3$, respectively (Sekiya, Schmidt, Wilson, & Filner, 1982).

5.6. Tip

To accurately determine the amount of emission of H_2S, the homogenate is mixed with a solution containing 100 m*M* phosphate-buffered saline (pH 7.0), 10 m*M* L-cysteine, and 2 m*M* phosphopyridoxal in a test tube at room temperature, and the released H_2S is absorbed in a zinc acetate trap, a small glass tube containing zinc acetate (20 m*M*), and fixed in the bottom of the test tube, and then, the equal volume of $FeCl_3$ and DMPD solution, respectively, is added to measure the content of H_2S as the abovementioned methods (Riemenschneider, Nikiforova, Hoefgen, DeKok, & Papenbrock, 2005).

5.7. Tip

To accurately determine the content of H_2S, the volume of supernatant should be measured or brought up to 2 ml with extraction buffer.

6. PROTOCOL 2

6.1. Duration

Preparation	**1 h**
Protocol	About 2 h

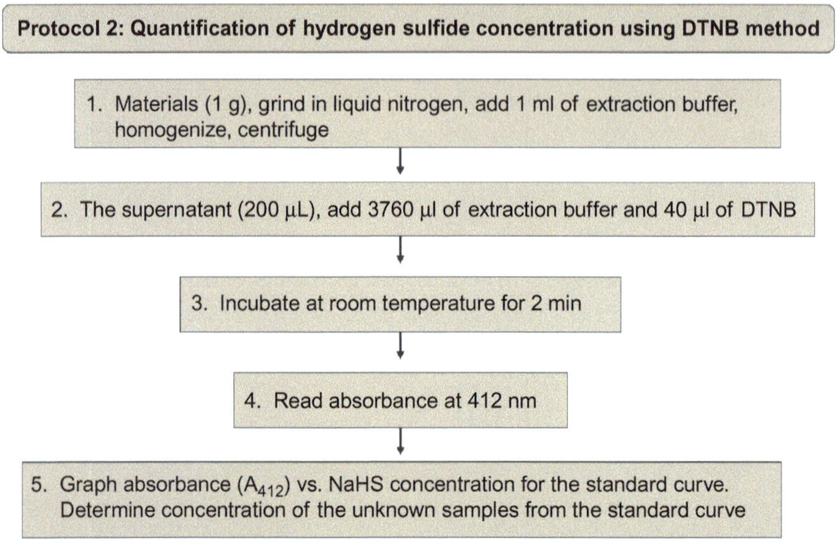

Figure 4 Flowchart for Protocol 2.

6.2. Preparation

Prepare solution and buffer according to Section 3.1. Sample and extract H_2S from plant tissues in accordance with the flowchart of the complete Protocol 2 (see Fig. 4).

7. STEP 1: QUANTIFICATION OF H_2S CONCENTRATION USING 5,5′-DITHIOBIS (2-NITROBENZOICACID) METHOD

7.1. Overview

H_2S is determined by measuring absorbance at 412 nm using DTNB method after extracted from plant tissues. Calculate the H_2S concentration of the unknown solution from a standard curve of a suitable H_2S donor (e.g., NaHS or Na_2S) or extinction coefficient of 5-thio-2-nitrobenzoate ($13.6 \times 10^3\ M^{-1}\ cm^{-1}$) at 412 nm, and H_2S content is expressed as nmol g^{-1} FW.

7.2. Duration

2 h

1. One gram of plant material is ground into fine powder with a mortar and pestle under liquid nitrogen and then homogenized in 1 ml of extraction

buffer: 20 m*M* Tris-HCl buffer (pH 8.0) containing 10 m*M* EDTA and 20 m*M* $Zn(OAc)_2$.

2. The homogenate is centrifuged at 15,000 × *g* for 15 min at 4 °C.
3. Two hundred microliter of the supernatant is used for the quantification of H_2S, in an assay mixture containing also 3760 μl extraction buffer and 40 μl of 20 m*M* 5,5′-dithiobis(2-nitrobenzoic acid), in a total volume of 4 ml.
4. The assay mixture is incubated at room temperature for 2 min and the absorbance is read at 412 nm.
5. Graph the standard curve as absorbance (A_{412}) versus NaHS or Na_2S concentration (μg/ml). Determine the H_2S concentrations of the unknown samples from the graph, and the content of H_2S is expressed as nmol g^{-1} FW.

7.3. Tip

Spectrophotometer should be warmed up 15–20 min before use; see Section 5.3.

7.4. Tip

Graph the standard curve as absorbance (A_{412}) versus concentration (μg/ml). Determine the concentrations of the unknown samples from the graph. Alternatively, the concentration can be determined using molar extinction coefficient of 5-thio-2-nitrobenzoate of $13.6 \times 10^3\ M^{-1}\ \mathrm{cm}^{-1}$ and the Beer–Lambert equation:

$$A = \varepsilon c l$$

where ε is the molar extinction coefficient, c is the concentration of 5-thio-2-nitrobenzoate, and l is the path length in cm (Nashet et al., 1977).

7.5. Tip

There are essentially no differences observed when the determinations are done at pH 8.0 or 7.0; pH 7.0 is chosen so that H_2S will be liberated from its alkaline medium. Alternatively, because EDTA can stabilize the color value of the anion 5-thio-2-nitrobenzoate, it might best be added as recommended for general use, but the use of double-deionized water is unnecessary and can shorten the periods of color development before reading (Nashet et al., 1977).

7.6. Tip

The same as Section 5.5

7.7. Tip

The same as Section 5.7

ACKNOWLEDGMENT

This work is supported by the National Natural Science Foundation of China (31360057).

REFERENCES

Christou, A., Manganaris, G. A., Papadopoulos, I., & Fotopoulos, V. (2013). Hydrogen sulfide induces systemic tolerance to salinity and non-ionic osmotic stress in strawberry plants through modification of reactive species biosynthesis and transcriptional regulation of multiple defence pathways. *Journal of Experimental Botany*, *64*, 1953–1966.

Nashet, A. S., Osuga, D. T., & Feeney, R. E. (1977). Determination of hydrogen sulfide with 5,5'-dithiobis-(2-nitrobenzoic acid), N-ethylmaleimide, and para-chloromercuribenzoate. *Analytical Biochemistry*, *79*, 394–405.

Papenbrock, J., & Schmidt, A. (2000). Characterization of a sulfurtransferase from Arabidopsis thaliana. *European Journal of Biochemistry*, *267*, 145–154.

Peng, H. J., Chen, W. X., & Wang, B. H. (2012). Methods for the detection of gasotransmitters. In A. Hermann, G. F. Sitdikova, & T. M. Weiger (Eds.), *Gasotransmitters: Physiology and pathophysiology* (pp. 99–137). Heidelberg: Springer.

Riemenschneider, A., Nikiforova, V., Hoefgen, R., DeKok, L. J., & Papenbrock, J. (2005). Impact of elevated H_2S on metabolite levels, activity of enzymes and expression of genes involved in cysteine metabolism. *Plant Physiology and Biochemistry*, *43*, 473–483.

Sekiya, J., Schmidt, A., Wilson, L. G., & Filner, P. (1982). Emission of hydrogen sulfide by leaf tissue in response to l-cysteine. *Plant Physiology*, *70*, 430–436.

CHAPTER SEVEN

H_2S Analysis in Biological Samples Using Gas Chromatography with Sulfur Chemiluminescence Detection

Victor Vitvitsky, Ruma Banerjee[1]
Department of Biological Chemistry, University of Michigan Medical School, Ann Arbor, Michigan, USA
[1]Corresponding author: e-mail address: rbanerje@umich.edu

Contents

Abstract

Hydrogen sulfide (H_2S) is a metabolite and signaling molecule in biological tissues that regulates many physiological processes. Reliable and sensitive methods for H_2S analysis are necessary for a better understanding of H_2S biology and for the pharmacological modulation of H_2S levels *in vivo*. In this chapter, we describe the use of gas chromatography coupled to sulfur chemiluminescence detection to measure the rates of H_2S production and degradation by tissue homogenates at physiologically relevant concentrations of substrates. This method allows separation of H_2S from other sulfur compounds and provides sensitivity of detection to ~15 pg (or 0.5 pmol) of H_2S per injected sample.

Methods in Enzymology, Volume 554
ISSN 0076-6879
http://dx.doi.org/10.1016/bs.mie.2014.11.013

1. INTRODUCTION

Reliable detection of hydrogen sulfide (H_2S) in biological samples is challenging due to its volatility, redox lability, and most importantly, its low steady-state concentrations (Kabil & Banerjee, 2010). These difficulties combined with the use of methods that are not specific for measurement of free sulfide at physiological pH, have resulted in reports in the literature of H_2S concentrations spanning five orders of magnitude ranging from the tens of nanomolar to hundreds of micromolar (Olson, 2009). This ambiguity in the physiologically relevant concentration of H_2S has, in turn, resulted in the use of widely varying concentrations of sulfide donors to elicit physiological effects. H_2S is a product of sulfur metabolism (Kabil & Banerjee, 2010; Kabil, Vitvitsky, & Banerjee, 2014) and is synthesized and degraded by mammalian tissues at relatively high rates (Jurkowska et al., 2014; Vitvitsky, Kabil, & Banerjee, 2012). The methods used for H_2S detection have been discussed in Olson (2012) and include colorimetric analysis monitoring methylene blue formation, use of a sulfide ion-selective or a polarographic electrode, gas chromatography (GC) with flame photometric or sulfur chemiluminescence detection, ion chromatography, HPLC analysis of the monobromobimane derivative of sulfide with fluorescence detection, and the use of sulfide-sensitive fluorescent dyes. Using the GC-coupled sulfur chemiluminescence detection method for H_2S, recent studies have converged on low nanomolar concentrations of H_2S in biological samples (Furne, Saeed, & Levitt, 2008; Kabil, Vitvitsky, Xie, & Banerjee, 2011; Levitt, Abdel-Rehim, & Furne, 2011; Vitvitsky et al., 2012). In this chapter, we describe the application of the GC-based sulfur chemiluminescence method for H_2S detection in biological samples.

2. PRINCIPLE OF THE GC-COUPLED SULFUR CHEMILUMINESCENCE METHOD

Analysis of H_2S levels in biological samples using GC with sulfur chemiluminescence detection relies on the equilibration of H_2S between the liquid sample and gas phases in a hermetically sealed sample chamber. H_2S present in the gas phase is measured following GC separation from other sulfur compounds and detected using a 355 sulfur chemiluminescence detector (Agilent Technologies, Santa Clara, CA). In the detector, the sample undergoes combustion in the furnace generating sulfur monoxide, which reacts with ozone in the reaction cell to form sulfur dioxide emitting light (Fig. 1). The light intensity is measured using a photomultiplier tube and the signal is linearly proportional to the quantity of sulfur in the sample. The GC method allows separation of H_2S from different natural sulfur compounds (Fig. 2) (Levitt, Furne, Springfield, Suarez, & DeMaster, 1999). The sensitivity of the instrument allows detection of 0.5 pg of sulfur and the linear range spans approximately four orders of magnitude based on the vendor's information.

2.1. Limitations of the GC method

The sensitivity of the method is influenced by parameters such as column type, injection volume, injection mode, and gas flow rate, which affect the amount of sample that reaches the detector. Hence, these parameters must be adjusted to optimize sensitivity prior to sample analysis. A second issue that influences the sensitivity of this method is that low H_2S concentrations necessitate injection of relatively large samples volumes. Modern

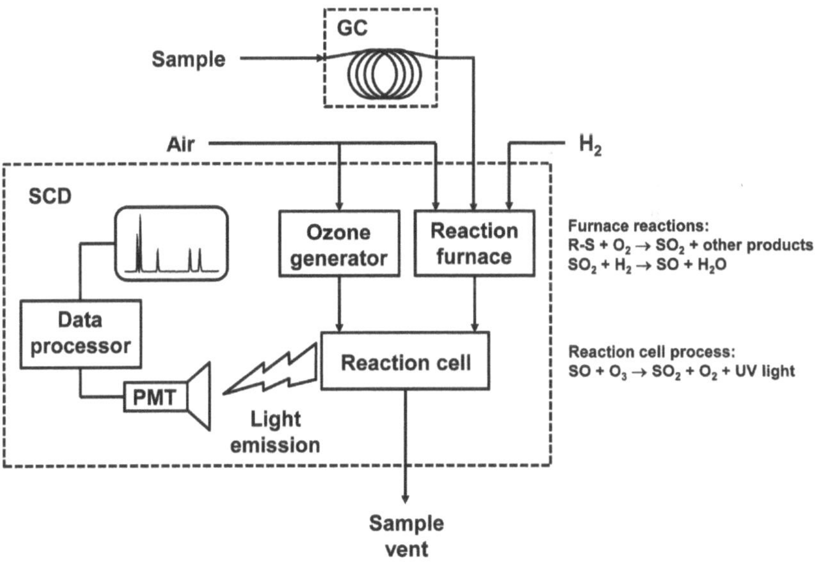

Figure 1 Scheme showing analysis of sulfur compounds using GC coupled to a sulfur chemiluminescence detector (SCD). Sulfur compounds in the sample are separated on a GC column prior to entering the SCD. In the reaction furnace, the samples undergo combustion in an air and hydrogen mixture producing sulfur monoxide. In the reaction cell, sulfur monoxide reacts with ozone to produce sulfur dioxide and light emission. The latter is detected using a photomultiplier tube (PMT).

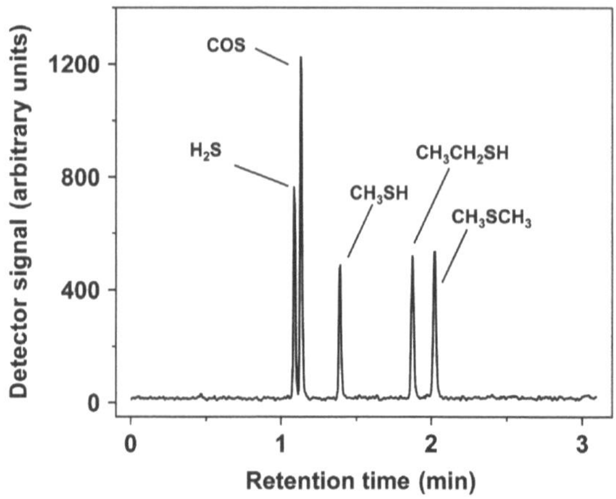

Figure 2 Analysis of a mixture of sulfur gases using GC–SCD. The mixture (200 μl total injection volume) contained 2.7 ppm H_2S, 5 ppm carbonyl sulfide (COS), 5 ppm 3-methanethiol (CH_3SH), 5 ppm ethanethiol (CH_3CH_2SH), and 5 ppm 5-dimethylsulfide (CH_3SCH_3) in N_2.

GC capillary columns with a maximal available inner diameter of 0.53 mm are not suitable for injection of more than 1 ml of a gas sample. Packed columns allow injection of up to 10 ml of gas samples and significantly enhance sensitivity of H_2S analysis. Unfortunately, the newest 355 sulfur chemiluminescence detector does not interface with packed columns.

3. PROTOCOL FOR GC-COUPLED SULFUR CHEMILUMINESCENCE DETECTION OF H_2S

3.1. Materials

- NORM-JECT polypropylene syringes (Henke Sass Wolf, Tuttlingen, Germany)
- Plastic three-way stopcocks (Smiths Medical ASD, Dublin, OH)
- Sleeve stopper septa for 5 mm NMR tubes (white, 1.5 × 3.9 i.d. × o.d., Sigma Aldrich, St. Louis, MO)
- Tedlar gas sample bag (Jensen Inert Products, Coral Spring, FL)
- Gas-tight glass syringe with PTFE plunger and stainless steel needle (Agilent Technologies, Santa Clara, CA)
- H_2S calibration mixture, 40 ppm H_2S in N_2 (Cryogenic Gases, Detroit, MI)
- Mixture of natural sulfur gases in N_2 (Cryogenic Gases, Detroit, MI)

- Ultrahigh purity grade helium, H_2, and N_2 cylinders (Cryogenic Gases, Detroit, MI)
- Zero grade air cylinder (Cryogenic Gases, Detroit, MI)
- DB-1 capillary column (30 m × 0.53 mm × 1.0 μm, Agilent Technologies, Santa Clara, CA)

3.2. Calibration standards

For instrument calibration, gas cylinders containing either 40 ppm H_2S in N_2 or a mixture of 10 ppm each of carbonyl sulfide (COS), methanethiol (CH_3SH), ethanethiol (CH_3CH_2SH), and dimethylsulfide (CH_3SCH_3) in N_2 are obtained from Cryogenic Gases (Detroit, MI, USA). For routine instrument calibration, ~1 l of 40 ppm H_2S in N_2 was kept in a Tedlar gas sample bag in which the valve was sealed with sleeve stopper septum. The H_2S concentration in the bag was stable over the course of the day and the bag was refilled daily with fresh calibration standards.

3.3. Sample manipulation

H_2S disappears rapidly if kept in glass containers or if exposed to rubber (e.g., stoppers). Hence, disposable polypropylene syringes were utilized for sample handling. The sample is placed in the barrel of a syringe and the syringe is sealed with a plunger. The syringe tip is fitted with a three-way stopcock, which is used to make the sample anaerobic by flushing the syringe with N_2. During sample incubation, the top of the three-way stopcock is sealed with a small-sleeve stopper septum. Gas aliquots for measurement of H_2S levels are collected from the syringe through the septa and injected into the GC using a gas-tight glass syringe with a PTFE plunger and stainless steel needle. Since H_2S levels are stable in glass syringes for at least 5 min, they are suitable for sample injection. Aliquots of standard H_2S mixture (40 ppm in N_2) are added when needed, through the septa into the syringes filled with N_2 to obtain different dilutions for instrument calibration.

3.4. Chromatography conditions

Gas samples were injected into an Agilent 6890 Series Gas Chromatograph equipped with a DB-1 capillary column. The following chromatographic conditions were found to provide optimal sensitivity and resolution: manual splitless injection, inlet temperature 105 °C, total gas flow rate 29.9 ml min^{-1}, column gas flow rate 2 ml min^{-1}, detector outlet temperature 150 °C. Helium is used as a carrier gas. The oven temperature is maintained at 30 °C during sample injection and analysis, increased to 110 °C at the end of the run at a rate of 20 °C min^{-1}, and maintained at 110 °C for 2 min before returning to the initial conditions. The oven temperature is increased to 110 °C in order to clean the column from possible contamination with organic substances and water present in the injected samples.

H_2S analysis in a single sample takes ~2 min. The total run cycle including the temperature gradient and restoration to initial conditions takes ~20 min. In our hands, the linear range of this method extends over two orders of magnitude (Fig. 3), and the sensitivity of detection is ~15 pg (i.e., 0.5 pmol) of H_2S per injection. We estimate that sensitivity could be enhanced ~5-fold by direct sample injection onto the column and even further, using a packed column.

4. ANALYSIS OF BIOLOGICAL SAMPLES

4.1. Monitoring H_2S production in tissue homogenate

Substrates for H_2S production (cysteine, homocysteine, or both for cystathionine β-synthase and γ-cystathionase and mercaptopyruvate and a reductant such as dithiothreitol or dihydrolipoic acid for mercaptopyruvate

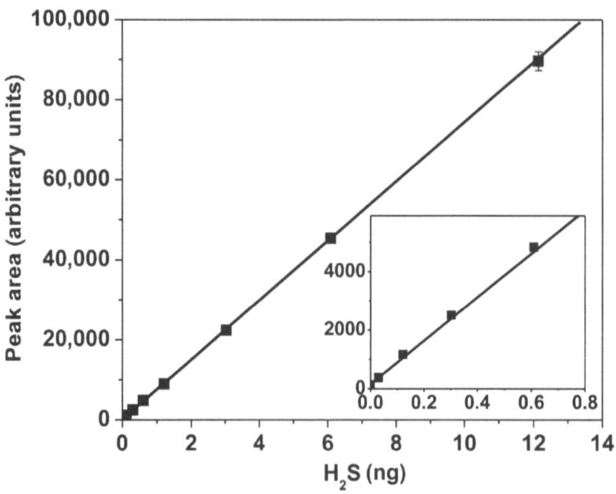

Figure 3 Linear dependence of the H_2S peak area on H_2S amount. The symbols and line represent the experimental data and linear fit, respectively. Each experimental point represents the mean ± SD of 2–4 independent measurements. In most cases, the standard deviation is equal to or less than the symbol size. Samples (200 μl) containing different amounts of H_2S were prepared by dilution of the stock solution (40 ppm H_2S) with N_2.

sulfurtransferase) are added to tissue homogenates prepared as described (Kabil et al., 2011; Vitvitsky et al., 2012). The sample with a total volume of 0.5 ml is placed in the barrel of a 20-ml polypropylene syringe attached to a three-way stopcock. The syringe is sealed with a plunger and made anaerobic by flushing the headspace with N_2 five times using the three-way stopcock (Fig. 4) and filled with N_2 to a total volume (liquid + gas) of 20 ml. Then, the stopcock is disconnected from the source of N_2 and sealed with a sleeve stopper septum, and the syringe is incubated for 20 min at 37 °C with gentle shaking (75 rpm). Control samples in which the tissue homogenate is replaced by buffer are prepared and incubated in parallel. Aliquots (200 μl) from the gas phase of the reaction syringes are collected using a gas-tight syringe through the sleeve stopper septum attached to the stopcock, and injected into the GC. For samples containing a high concentration of H_2S, a 20-fold dilution with N_2 is used to prevent column overloading. For this, 0.5 ml of the sample is injected into a syringe filled with 9.5 ml of N_2. Then, 200 μl of the diluted sample is injected into the GC.

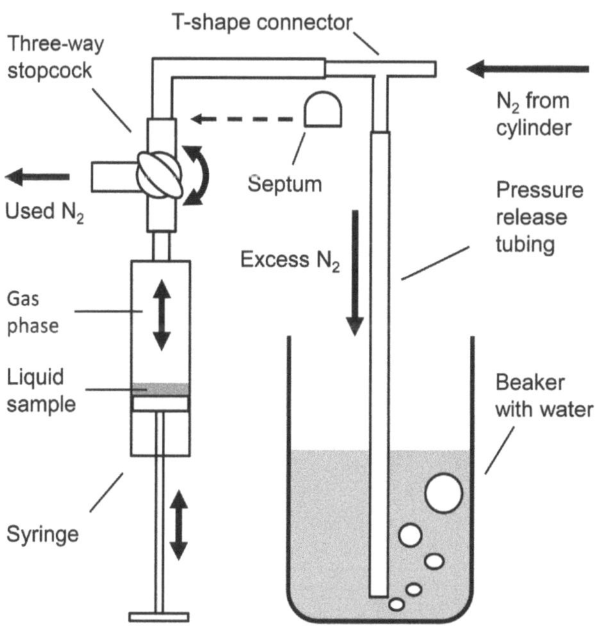

Figure 4 Scheme showing setup for sample preparation. The solid arrows depict the direction of N_2 flow, the movement of the syringe plunger and movement of the three-way stopcock valve during flushing of the syringe with N_2. The dashed arrow indicates the position where the sleeve stopper septum is attached during sample incubation.

Different dilutions of the H_2S gas calibration mixture (40 ppm in N_2) are prepared using N_2. Aliquots (200 μl) of the varying H_2S dilutions in N_2 are injected into the GC and the corresponding peak areas are used to generate a standard curve. The concentration of H_2S in the injected sample is calculated from the peak area using the calibration coefficient obtained from the standard curve. The total amount of H_2S in syringes needed to estimate the H_2S production rates in tissues is calculated as described below in Section 4.3. It is important for H_2S concentration estimations and for instrument response linearity that the injection volume of the samples and calibration gas mixture are the same.

Typically, since peaks other than H_2S are not observed in these samples, multiple injections can be made during a single run cycle (Fig. 5). For this, the oven temperature in the GC is maintained at 30 °C for 20 min during which several (up to 10) injections of sample and calibration gas aliquots can be made. Following this, the temperature is increased to complete the run cycle. Usually, the calibration mixture containing 40 ppm H_2S in N_2 is injected with each analytical run to control for variability.

4.2. Monitoring H_2S degradation in tissue homogenates

Tissue homogenate (0.5 ml) is placed in the barrel of a 20-ml polypropylene syringe attached to a three-way stopcock and the syringe is sealed with a

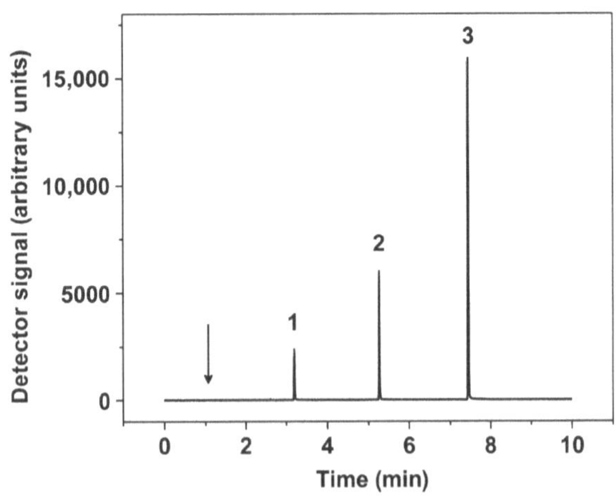

Figure 5 A representative chromatogram of H_2S production by murine liver homogenate at different cysteine concentrations. Peaks 1–3 were obtained after anaerobic incubation for 20 min of the homogenate (pH 7.4, 37 °C) with 0.1, 0.2, and 0.5 m*M* cysteine, respectively. The arrow indicates the control sample lacking cysteine.

plunger. For anaerobic assays, the syringe headspace is flushed five times with N_2 using a three-way stopcock (Fig. 4) and then filled with N_2 to a total volume (liquid + gas) of 10 ml. Then, 10 ml of the H_2S gas mixture (40 ppm in N_2) corresponding to 17.9 nmoles H_2S is added to the syringe using the three-way stopcock to give a final volume of 20 ml. For aerobic assays, the syringe headspace is filled with air to a total volume (liquid + gas) of 10 ml and 10 ml of the H_2S gas mixture (40 ppm in N_2) is added using a three-way stopcock, to give a final volume of 20 ml. The syringes are incubated at room temperature (25 °C) with stirring and 200 μl aliquots are collected from the gas phase over 20 min and as described above, multiple samples and the calibration standard are injected per GC run cycle (Fig. 6). The concentration of H_2S in the samples is determined from the peak area using a calibration coefficient as described in the next section.

4.3. Estimation of H_2S production and degradation rates

To determine the rates of H_2S production and degradation, the total amount of H_2S in the reaction syringe needs to be calculated taking into account its equilibration between the gas and liquid phases and its dissociation at the pH of the reaction mixture. In the samples described above, the amount of H_2S

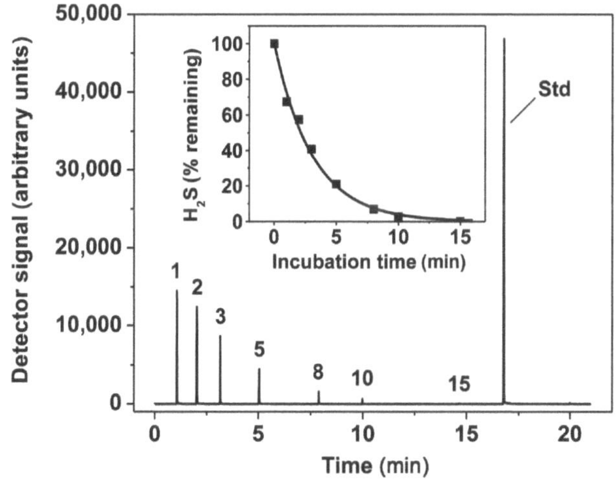

Figure 6 A representative chromatogram showing the kinetics of aerobic H_2S degradation by murine liver homogenate. The numbers correspond to samples (200 μl) removed at 1, 2, 3, 5, 8, 10, and 15 min following incubation of liver homogenate with H_2S at 25 °C. The calibration peak (12.2 ng H_2S) is denoted by Std. The inset shows the kinetics of H_2S disappearance. The amount of H_2S at $t = 0$, shows the amount of H_2S added to the reaction mixture.

in the gas phase is equal to the H_2S concentration in the sample aliquots multiplied by the total gas phase volume (i.e., 19.5 ml). The amount of dissolved H_2S present in the liquid phase is calculated by multiplying the concentration of H_2S in the gas phase by 1.6, which is the equilibrium ratio between the concentration of H_2S in the gas and liquid phases at 37 °C (Furne et al., 2008) and by the volume of the liquid (0.5 ml). The amount of dissolved sulfide anion (HS^-) in the liquid phase is calculated using a pKa of 6.8 for ionization of H_2S in water at 37 °C (Hershey, Plese, & Millero, 1988). The total amount of H_2S in the sample is then the sum of the amounts of H_2S in the gas phase, dissolved H_2S, and HS^-. The pKa for H_2S is influenced by the ionic strength (Almgren, Dyrssen, Elgquist, & Johansson, 1976; Hershey et al., 1988; Millero, 1986) and is expected to decrease from 6.8 in water to ~6.6 at physiologically relevant salt concentrations. Under our experimental conditions with a gas:liquid ratio of 39:1, the majority (~80%) of the H_2S in the reaction mixture is present in the gas phase at pH 7.4. Since adjusting the pKa value from 6.8 to 6.6 changes the total H_2S value by <10%, a pKa value of 6.8 was employed in our calculations. The rate of H_2S production is then determined by subtracting the sum of H_2S amounts in the gas and liquid phases in the control from the sample and dividing that value by the incubation time and the wet weight of the tissue. Since the H_2S degradation kinetics are measured at 25 °C a value of 2.0 for the equilibrium ratio for H_2S in the gas versus liquid phase (Furne et al., 2008) and a pKa value of 7.0 (Hershey et al., 1988) are used in the calculations.

The total amount of sulfide in the reaction mixture (S_T) needed to calculate the rate of H_2S production and degradation is the sum of the sulfide amounts in the gas (S_{Gas}) and liquid (S_{Liq}) phases of the mixture (Eq. 1).

$$S_T = S_{Gas} + S_{Liq} \tag{1}$$

From the peak area of the sample injected into the GC, one can calculate the concentration ($S_{[Gas]}$) and the amount (S_{Gas}) of H_2S in the gas phase of the reaction mixture using Eqs. (2) and (3).

$$S_{[Gas]} = A/C \tag{2}$$

$$S_{Gas} = S_{[Gas]} * V_{Gas} = A * V_{Gas}/C \tag{3}$$

Here A is the peak area of H_2S in the sample, C is the calibration coefficient (i.e., the peak area corresponding to 1 M H_2S) obtained from the standard curve and V_{Gas} is the volume of the gas phase in the reaction mixture. The total sulfide amount in the liquid phase of the reaction mixture (S_{Liq}) is the sum of the amount of dissolved undissociated (S_{H_2S}) and ionized (S_{HS^-}) forms (Eq. 4).

$$S_{Liq} = S_{H_2S} + S_{HS^-} \quad (4)$$

The amount of dissolved sulfide depends on the concentration in the liquid phase of the undissociated ($S_{[H_2S]}$) and dissociated ($S_{[HS^-]}$) forms and on the liquid volume (V_{Liq}) (Eqs. 5 and 6).

$$S_{H_2S} = S_{[H_2S]} * V_{Liq} \quad (5)$$

$$S_{HS^-} = S_{[HS^-]} * V_{Liq} \quad (6)$$

The concentration of undissociated H_2S in the liquid phase is determined using the equilibrium ratio (R) between the gas and liquid phases (Eq. 7) and depends on the temperature.

$$S_{[H_2S]} = R * S_{[Gas]} \quad (7)$$

Similarly, the concentration of the dissolved ionized sulfide ($S_{[HS^-]}$) is determined using the pKa value for the dissociation of H_2S to sulfide anion (Eq. 8) taking into account the pH of the liquid phase.

$$S_{[HS^-]} = S_{[H_2S]} * 10^{(pH \quad pKa)} \quad (8)$$

The total amount of sulfide in the reaction mixture is obtained from the sample peak area (A), the calibration coefficient (C), the volumes of the gas (V_{Gas}) and liquid (V_{Liq}) phases, the pH of the liquid phase, the equilibrium ratio for H_2S between the gas and liquid phase, and the pKa value of H_2S (Eq. 9), which simplifies to Eq. (10).

$$\begin{aligned} S_T &= S_{Gas} + S_{Liq} = S_{Gas} + S_{H_2S} + S_{HS^-} \\ &= S_{[Gas]} * V_{Gas} + \left(S_{[H_2S]} + S_{[HS^-]}\right) * V_{Liq} \\ &= S_{[Gas]} * V_{Gas} + \left(R * S_{[Gas]} + R * S_{[Gas]} * 10^{(pH - pKa)}\right) * V_{Liq} \\ &= S_{[Gas]} * V_{Gas} + R * S_{[Gas]}\left(1 + 10^{(pH - pKa)}\right) V_{Liq} \\ &= A * V_{Gas}/C + (R * A/C)\left(1 + 10^{(pH - pKa)}\right) * V_{Liq} \\ &= (A/C)\left(V_{Gas} + R * V_{Liq}\left(1 + 10^{(pH - pKa)}\right)\right) \end{aligned} \quad (9)$$

$$S_T = (A/C)\left(V_{Gas} + R * V_{Liq}\left(1 + 10^{(pH - pKa)}\right)\right) \quad (10)$$

5. ADDITIONAL TECHNICAL DETAILS

5.1. Column conditioning

The protocol recommended for conditioning a new capillary column includes heating it to 250 °C. However, this procedure is very detrimental for the catalytic element in the sulfur chemiluminescence detector and leads to a significant drop in detector sensitivity. Hence, the column must be disconnected from the detector and the inlet to the detector must be plugged with an external nut and no-hole ferrule before the column is heated to 250 °C. In our experience, heating the capillary column to >200 °C causes a significant loss in the detector sensitivity and requires replacement of the catalytic element in the reaction furnace, i.e., the small ceramic tube, if the sulfur chemiluminesce detector is not disconnected from the column.

5.2. Additional gas purification

The sensitivity of the detector declines rapidly if the helium and hydrogen carrier gases are not purified prior to their use due to the presence of trace contaminants, particularly hydrocarbons. To increase the lifetime of the small ceramic tube, inline cartridges can be used to remove contaminants from the gases. Use of the Big Universal Trap RMSH-2 cartridge for helium and the Big Universal Trap RMSHY-2 cartridge (Agilent Technologies) for hydrogen helps maintain detector sensitivity with regular usage over a period of a year. Zero grade air does not require additional purification.

ACKNOWLEDGMENT

This work was supported in part by the National Institutes of Health (HL58984 and GM112455).

REFERENCES

Almgren, T., Dyrssen, D., Elgquist, B., & Johansson, O. (1976). Dissociation of hydrogen sulphide in seawater and comparison of pH scales. *Marine Chemistry*, *4*, 289–297.

Furne, J., Saeed, A., & Levitt, M. D. (2008). Whole tissue hydrogen sulfide concentrations are orders of magnitude lower than presently accepted values. *American Journal of Physiology. Regulatory, Integrative and Comparative Physiology*, *295*(5), R1479–R1485.

Hershey, J. P., Plese, T., & Millero, F. J. (1988). The pK1 for the dissociation of H_2S in various ionic media. *Geochimica et Cosmochimica Acta*, *52*, 2047–2051.

Jurkowska, H., Roman, H. B., Hirschberger, L. L., Sasakura, K., Nagano, T., Hanaoka, K., et al. (2014). Primary hepatocytes from mice lacking cysteine dioxygenase show increased cysteine concentrations and higher rates of metabolism of cysteine to hydrogen sulfide and thiosulfate. *Amino Acids*, *46*(5), 1353–1365.

Kabil, O., & Banerjee, R. (2010). The redox biochemistry of hydrogen sulfide. *The Journal of Biological Chemistry*, *285*, 21903–21907.
Kabil, O., Vitvitsky, V., & Banerjee, R. (2014). Sulfur as a signaling nutrient through hydrogen sulfide. *Annual Review of Nutrition*, *34*, 171–205.
Kabil, O., Vitvitsky, V., Xie, P., & Banerjee, R. (2011). The quantitative significance of the transsulfuration enzymes for h_2s production in murine tissues. *Antioxidants & Redox Signaling*, *15*, 363–372.
Levitt, M. D., Abdel-Rehim, M. S., & Furne, J. (2011). Free and acid-labile hydrogen sulfide concentrations in mouse tissues: Anomalously high free hydrogen sulfide in aortic tissue. *Antioxidants & Redox Signaling*, *15*, 373–378.
Levitt, M. D., Furne, J., Springfield, J., Suarez, F., & DeMaster, E. (1999). Detoxification of hydrogen sulfide and methanethiol in the cecal mucosa. *The Journal of Clinical Investigation*, *104*(8), 1107–1114.
Millero, F. J. (1986). The thermodynamics and kinetics of the hydrogen sulfide system in natural waters. *Marine Chemistry*, *18*, 121–147.
Olson, K. R. (2009). Is hydrogen sulfide a circulating "gasotransmitter" in vertebrate blood? *Biochimica et Biophysica Acta*, *1787*(7), 856–863.
Olson, K. R. (2012). A practical look at the chemistry and biology of hydrogen sulfide. *Antioxidants & Redox Signaling*, *17*(1), 32–44.
Vitvitsky, V., Kabil, O., & Banerjee, R. (2012). High turnover rates for hydrogen sulfide allow for rapid regulation of its tissue concentrations. *Antioxidants & Redox Signaling*, *17*(1), 22–31.

SECTION II

Hydrogen Sulfide Donors and Their Pharmacological Activity

CHAPTER EIGHT

Use of Phosphorodithioate-Based Compounds as Hydrogen Sulfide Donors

Chung-Min Park, Ming Xian[1]
Department of Chemistry, Washington State University, Pullman, Washington, USA
[1]Corresponding author: e-mail address: mxian@wsu.edu

Contents

Abstract

Synthetic hydrogen sulfide (H_2S) donors are useful research tools as well as potential therapeutic agents. In this chapter, we report the detailed protocols for the synthesis and evaluation of a series of phosphorodithioate-based H_2S donors. Fluorescence assays were used to determine H_2S release from the donors in both aqueous buffers and in cultured cells. These donors were found to be slow-release donors, much like the well-known GYY4137. These donors also showed some protective effects against hydrogen peroxide (H_2O_2)-induced oxidative damage in myocytes.

Methods in Enzymology, Volume 554
ISSN 0076-6879
http://dx.doi.org/10.1016/bs.mie.2014.11.032

1. INTRODUCTION

Hydrogen sulfide (H_2S) has been known as a colorless, flammable, extremely hazardous gas with a rotten-egg smell for many years. However, this gaseous molecule is recently classified as an important cell signaling molecule with numerous physiological functions (Fukuto et al., 2012; Lavu, Bhushan, & Lefer, 2011; Li, Rose, & Moore, 2011; Wang, 2012). The production of H_2S in mammalian systems has been involved in transsulfuration pathways by at least three endogenous enzymes (Kimura, 2011; Kimura, Shibuya, & Kimura, 2012; Olson, 2012; Predmore, Lefer, & Gojon, 2012): cystathionine β-synthase, cystathionine γ-lyase, and 3-mercaptopyruvate sulfur-transferase. L-cysteine (Cys) or L-cysteine derivatives are nature substrates of these enzymes. In addition, nonenzymatic H_2S production pathways are also known which utilize labile sulfane sulfurs (compounds containing sulfur atoms bound only to other sulfur atoms) to generate H_2S under reducing conditions or simple degradation (Jacob, Anwar, & Burkholz, 2008; Nagy et al., 2014). Although the reported functions of H_2S are numerous, the chemical mechanisms behind its activities are not well established. Nevertheless, several biochemical reactions of H_2S are suggested to be important. These include aerobic redox reactions, reactions with endogenously generated oxidants like peroxynitrite, superoxide, and hydrogen peroxide (Chang et al., 2008; Geng et al., 2004; Whiteman et al., 2004), S-sulfhydration on protein cysteine residues (Krishnan, Fu, Pappin, & Tonks, 2011; Mustafa et al., 2009; Sen et al., 2012), etc. Increasing literature evidences have demonstrated that H_2S, both endogenously produced and exogenously administered, can have protective effects in many pathologies such as vasodilation and anti-inflammation (Fukuto et al., 2012; Lavu et al., 2011; Li et al., 2011; Wang, 2012; Whiteman et al., 2004). These results strongly suggest that modulation of H_2S levels could have potential therapeutic values.

In H_2S research, researchers often use sulfide salts (NaHS or Na_2S) as H_2S sources. Sulfide salts are short-lasting H_2S donors as they release H_2S quickly (within seconds). Noncontrollable and high-dose exposures may cause insufficient cardiac output, irregular heartbeat, and conduction abnormalities. In addition, sulfide concentrations in aqueous solution can rapidly decrease due to volatilization (DeLeon, Stoy, & Olson, 2012), thus significantly limiting the utility of these chemical precursors.

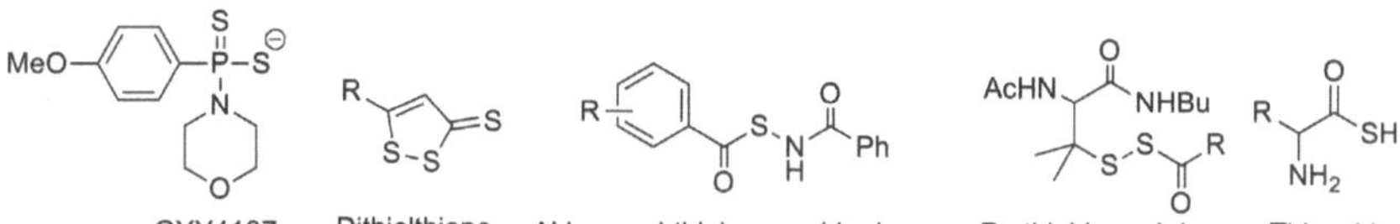

Scheme 1 Structures of several representative synthetic H_2S donors.

Due to these considerations, researchers have started to use synthetic H_2S-releasing agents (i.e., H_2S donors) to explore the biological functions of H_2S (Caliendo, Cirino, Santagada, & Wallace, 2010; Hughes, Centelles, & Moore, 2009; Kashfi & Olson, 2013; Zhao, Tyler, & Xian, 2014). Several representative donors are shown in Scheme 1, including (1) GYY4137 (Li et al., 2008), (2) the dithiolthione derivatives (Wallace, Caliendo, Santagada, Cirino, & Fiorucci, 2007), (3) *N*-(benzoyl)-thiobenzamide derivatives (Zhao, Wang, & Xian, 2011), (4) *S*-acylated perthiol derivatives (Zhao et al., 2013), and (5) amino acid-based thioacids (von Wantoch Rekowski et al., 2010).

GYY4137 is perhaps the most often used donor among these synthetic donors. The core structure of GYY4137 is the phosphorodithioate template, which is believed to release H_2S upon hydrolysis in aqueous buffers. Most importantly H_2S release from GYY4137 is very slow and only less than 10% H_2S can be generated even after 7 days (Lee et al., 2011). GYY4137 is thus considered as a slow-release donor. Some studies have revealed that GYY4137 can exhibit H_2S-relevant biological activities such as vasodilation and antihypertension. (Lee et al., 2011; Li et al., 2008; Yong et al., 2011). To further expand the scope of GYY4137-like H_2S donors, we have recently developed a series of *O*-substituted phosphorodithioate-based donors (**PDDs**) (Park et al., 2013). In this chapter, we provide a detailed description of the preparation and evaluation of these donors.

2. SYNTHESIS OF PHOSPHORODITHIOATE-BASED DONORS

2.1. Materials

Reagents and solvents employed were of the highest grade available. Reagent grade solvents were used for either chromatography or extraction without further purification before use. Carbon disulfide (CS_2) and 1,8-diazo[5.4.0]-bicycloundec-7-ene (DBU) were purchased from Aldrich and used as received. Elemental Sulfur (S_8) was purchased from Alfar Asar and used directly. All alcohols, amines, and 1,2-ethanedithiol were purchased from Aldrich and used without further purification. Phosphorous trichloride (PCl_3) was purchased from Aldrich and freshly distilled before use. Redistilled diisopropyl ethylamine was purchased from Aldrich and used without further purification. Biograde dimethyl sulfoxide (DMSO) and hydrogen peroxide (H_2O_2, 3% solution) was purchased from biomedical LLC and J.T. Baker, respectively, and used without further purification.

2.2. Synthetic route

PDDs were prepared from PCl_3 in three steps (Scheme 2). Briefly freshly distilled PCl_3 was treated with 1,2-dithioethane to provide 2-chloro-1,3,2-dithiaphospholane **1**. Compound **1** was then treated with aniline ($PhNH_2$) and followed by sulfurization to give the key intermediate **2**. Finally different phenol derivatives were reacted with **2** to furnish the donors (**PDD-1** to **PDD-5**).

2.3. Protocols and data

2-Chloro-1,3,2-dithiaphospholane ***1***: Commercially available 1,2-ethanedithiol (3.56 mL, 42.5 mmol) was added to freshly distilled PCl_3 (1.75 g, 127 mmol) at room temperature under argon atmosphere. The mixture was stirred for 3 h (HCl bubbles were generated and observed in this reaction). Then the excess of PCl_3 was removed by a simple distillation to yield the product, 2-chloro-1,3,2-dithiaphospholane, as pale yellow oil (6.4 g, 95%). This oil was stored under argon atmosphere and used for the next step without further purification. ^{1}H NMR (300 MHz, $CDCl_3$) δ 3.77–3.67 (m, 2H),

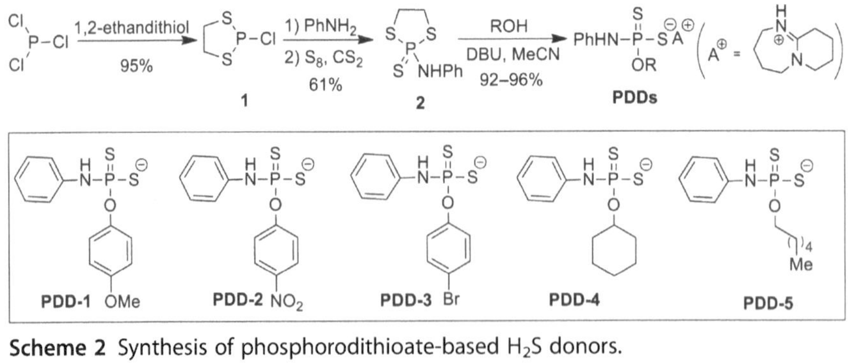

Scheme 2 Synthesis of phosphorodithioate-based H_2S donors.

3.62–3.52 (m, 2H); ^{31}P NMR (122 MHz, $CDCl_3$) δ 168.39; ^{13}C NMR (75 MHz, $CDCl_3$) δ 42.99; MS (ESI) *m/z* calcd for $C_2H_4ClPS_2$ $[M]^+$ 158.0, found 158.0.

2-(Phenylamino)-1,3,2-dithiaphospholane 2-sulfide **2**: 2-Chloro-1,3,2-dithiophospholane (1.1 g, 6.94 mmol) in deoxygenized MeCN (10 mL) was added dropwise to a stirred solution of aniline (0.63 mL, 6.94 mmol) and diisopropylethylamine (1.33 mL, 7.63 mmol) in dry, degassed MeCN (4 mL) at −40 °C (dry-ice/MeCN bath). After stirring for 1 h under the same temperature, the reaction mixture was warmed to room temperature and stirred for additional 1 h (monitored by TLC). A solution of elemental sulfur S_8 (667 mg, 7.63 mmol) dissolved in CS_2 (10 mL) was then added into the reaction mixture. The resultant light yellow heterogeneous mixture was stirred vigorously overnight. The reaction mixture was filtered and rinsed with EtOAc. Precipitate formed in the filtrate was filtered again to remove excess sulfur. Finally, the filtrate was concentrated under reduced pressure and the crude product was purified by flash chromatography on preneutralized (1% TEA in hexanes) silica gel to give the desired product as a white solid (1.04 g, 61%). ^{1}H NMR (300 MHz, $CDCl_3$) δ 7.47–7.01 (m, 5H), 5.52 (d, *J*=9.5 Hz, 1H), 3.75–3.28 (m, 4H); ^{31}P NMR (122 MHz, $CDCl_3$) δ 96.91; ^{13}C NMR (75 MHz, $CDCl_3$) δ 139.11, 129.44, 124.91, 124.89, 123.22, 123.14, 42.09, 42.07; IR (thin film) 3247.0, 2963.6, 1596.5, 1494.6, 1374.5, 1273.0, 1219.8, 1029.6, 912.2, 747.0 cm^{-1}; MS (ESI) *m/z* calcd for $C_8H_{10}NNaPS_3$ $[M+Na]^+$ 270.0, found 270.1.

General procedure of ***PDD*** *donors*: To a solution of intermediate **2** (1 equiv.) and the corresponding phenol (1 equiv.) in MeCN (total *c*=0.1–0.2 M) at room temperature was added DBU (1 equiv.). The mixture was stirred for 1 h at room temperature. Upon completion, the solvent was removed under reduced pressure. The crude product was frozen in diethyl ether using a dry-ice acetone bath then the solid was crashed with a specular. Solid product was formed under this condition. Diethyl ether was decanted and the remaining solid was washed with cold diethyl ether and dried to afford the desire product.

Compound ***PDD-1*** *was obtained as white solid (96% yield)*: mp 110 °C; ^{1}H NMR (300 MHz, $CDCl_3$) δ 9.60 (s, 1H), 7.34–7.04 (m, 6H), 6.83 (t, *J*=6.9 Hz, 1H), 6.73 (d, *J*=9.0 Hz, 2H), 5.75 (d, *J*=10.0 Hz, 1H), 3.70 (s, 3H), 3.34 (m, 6H), 2.78 (, *J*=4.9 Hz, 2H), 1.87 (q, *J*=5.9 Hz, 2H), 1.72–1.50 (m, 6H); ^{31}P NMR (122 MHz, $CDCl_3$) δ 94.95; ^{13}C NMR (75 MHz, $CDCl_3$) δ 166.24, 155.99, 146.67, 142.45, 128.92,

123.46, 123.40, 120.19, 118.20, 118.11, 113.99, 113.96, 55.75, 54.67, 48.88, 38.61, 33.03, 29.12, 26.88, 24.11, 19.64.; IR (thin film) 3224.7, 3100.8, 2924.8, 1637.5, 1601.6, 1575.5, 1503.8, 1438.6, 1389.8, 1282.2, 1190.9, 1102.9, 1034.5, 904.1 871.5, 832.4, 777.0, 679.3 cm^{-1}; MS (ESI, negative mode) *m/z* calcd for $C_{13}H_{13}NO_2PS_2$ $[M-DBU]^-$ 310.0, found 310.2.

Compound ***PDD-2*** *was obtained in 92% yield*: ^{1}H NMR (300 MHz, $CDCl_3$) δ 9.88 (br-s, 1H), 8.12 (d, *J*=9.1 Hz, 1H), 8.04 (d, *J*=9.1 Hz, 1H), 7.55–7.41 (m, 1H), 7.32–7.05 (m, 4H), 6.99–6.78 (m, 2H), 5.85 (d, *J*=10.3 Hz, 1H), 3.41 (m, 6H), 2.90–2.70 (m, 2H), 1.94 (q, 2H), 1.80–1.49 (m, 6H); ^{31}P NMR (122 MHz, $CDCl_3$) δ 94.70; ^{13}C NMR (75 MHz, $CDCl_3$) δ 166.42, 166.16, 129.09, 126.48, 125.05, 122.82, 122.75, 120.86, 118.39, 118.30, 116.54, 54.77, 48.92, 38.62, 33.10, 29.14, 26.90, 24.12, 19.65; IR (thin film) 3228.1, 3105.1, 3036.6, 2932.0, 2855.2, 1643.5, 1586.1, 1513.2, 1488.2, 1383.2, 1335.9, 1285.6, 1220.4, 1158.3, 1106.3, 906.3, 872.2, 753.0, 680.1 cm^{-1}; MS (ESI, negative mode) *m/z* calcd for $C_{12}H_{10}N_2O_3PS_2$ $[M-DBU]^-$ 325.0, found 325.1.

Compound ***PDD-3*** *was obtained as white solid* (*95% yield*): mp 108°C; ^{1}H NMR (300 MHz, $CDCl_3$) δ 8.92 (br-s, 1H), 7.45–7.03 (m, 8H), 6.86 (t, *J*=6.9 Hz, 1H), 5.77 (d, *J*=10.1 Hz, 1H), 3.37 (m, 6H), 2.81 (m, 2H), 1.92 (m, 2H), 1.81–1.44 (m, 6H); ^{31}P NMR (122 MHz, $CDCl_3$) δ 94.85, ^{13}C NMR (75 MHz, $CDCl_3$) δ 166.35, 142.12, 131.87, 128.99, 124.71, 124.65, 120.48, 118.29, 118.20, 54.71, 48.92, 38.61, 33.03, 29.17, 26.94, 24.12, 19.67; IR (thin film) 3269.9, 3222.6, 3106.5, 2939.2, 2857.7, 1640.4, 1597.7, 1578.0, 1479.0, 1388.2, 1319.8, 1288.9, 1200.7, 1158.7, 1066.3, 1009.2, 907.7, 866.3, 838.2, 767.7, 723.5, 699.8, 690.9 cm^{-1}; MS (ESI, negative mode) *m/z* calcd for $C_{12}H_{10}BrNOPS_2$ $[M-DBU]^-$ 357.9, found 358.0.

Compound ***PDD-4*** *was obtained as sticky white solid* (*92% yield*): ^{1}H NMR (300 MHz, $CDCl_3$) δ 10.41 (br-s, 1H), 7.33–7.00 (m, 4H), 6.74 (m, 1H), 3.61–3.19 (m, 8H), 3.19–2.48 (m, 5H), 2.44–2.08 (m, 1H), 1.83 (m, 3H), 1.71–1.41 (m, 9H), 1.48–0.97 (m, 2H); ^{31}P NMR (122 MHz, $CDCl_3$) δ 91.02; ^{13}C NMR (75 MHz, $CDCl_3$) δ 166.14, 142.50, 128.79, 128.52, 119.99, 118.85, 118.07, 70.38, 54.63, 48.87, 42.20, 41.13, 38.49, 35.75, 32.84, 31.83, 29.15, 26.93, 25.68, 24.23, 19.68; IR (thin film) 3219.4, 3091.3, 3035.4, 2928.8, 2856.4, 1642.0, 1598.2, 1495.7, 1444.4, 1382.5, 1321.2, 1285.7, 1204.8, 1105.1, 895.4, 751.5, 692.2 cm^{-1}; MS (ESI, negative mode) *m/z* calcd for $C_{12}H_{17}NOPS_2$ $[M-DBU]^-$ 286.4, found 286.3.

Compound ***PDD-5*** *was obtained as sticky white solid (93% yield)*: ^{1}H NMR (300 MHz, $CDCl_3$) δ 7.25 (m, 1H), 7.21–6.97 (m, 3H), 6.73 (m, 1H), 5.54 (d, J=10.2 Hz, 1H), 3.86 (m, 1H), 3.35 (m, 6H), 2.82 (m, 3H), 2.80–2.66 (m, 2H), 1.85 (m, 2H), 1.58 (m, 6H), 1.36–0.98 (m, 6H), 0.98–0.62 (m, 4H); ^{31}P NMR (122 MHz, $CDCl_3$) δ 95.10; ^{13}C NMR (75 MHz, $CDCl_3$) δ 166.14, 142.60, 128.76, 119.65, 117.67, 117.58, 65.63, 54.63, 48.88, 38.50, 32.84, 31.77, 30.59, 29.14, 26.93, 25.89, 24.21, 22.82, 19.68, 14.29; IR (thin film) 3224.6, 3095.9, 3034.0, 2928.2, 2856.7, 2799.2, 1642.1, 1598.4, 1495.9, 1381.8, 1321.1, 1285.4, 1228.2, 1204.8, 1105.3, 894.8, 750.7, 692.2 cm^{-1}; MS (ESI, negative mode) m/z calcd for $C_{12}H_{19}NOPS_2$ $[M-DBU]^-$ 288.1, found 288.2.

3. MEASUREMENTS OF H_2S GENERATION FROM THE DONORS USING FLUORESCENCE METHODS

In previous studies of phosphorodithioate-type donors like GYY4137, the standard methylene blue assay (Fogo & Popowsky, 1949; Lawrence, Davis, & Compton, 2000) was often used to determine H_2S production. In this assay, H_2S released from donors is first rapidly trapped with $Zn(OAc)_2$ to form ZnS. The trapped H_2S is then released under highly acidic conditions and reacts with *N*,*N*-dimethyl-*p*-phenylenediamine in the presence of $FeCl_3$ to form the methylene blue dye which can be easily detected by UV/vis at 670 nm. Although this method is convenient it is not appropriate for the evaluation of phosphorodithioate-based donors. It has been reported that the hydrolysis of phosphorodithioates is pH-dependent. Acidic conditions can significantly enhance hydrolysis (Ora, Järvi, Oivanen, & Lönnberg, 2000) so methylene blue assay is likely to give false positive results for phosphorodithioate-based donors. Because of this problem, we envisioned that fluorescence methods would be useful. Recently, a number of H_2S-specific fluorescent probes have been reported (Lin & Chang, 2012; Lippert, 2014; Peng et al., 2012; Peng & Xian, 2014; Pluth, Baisley, Hammer, & Montoya, 2013) and these detections usually occur under neutral pH. In this study, we used two H_2S fluorescent probes (**DNS-Az** and **WSP-1**, Scheme 3) to measure H_2S generation from the donors.

3.1. Materials and instrument

DNS-Az was synthesized using the known method (Peng et al., 2011). WSP-1 was synthesized as previously described (Liu et al., 2011). WSP-1

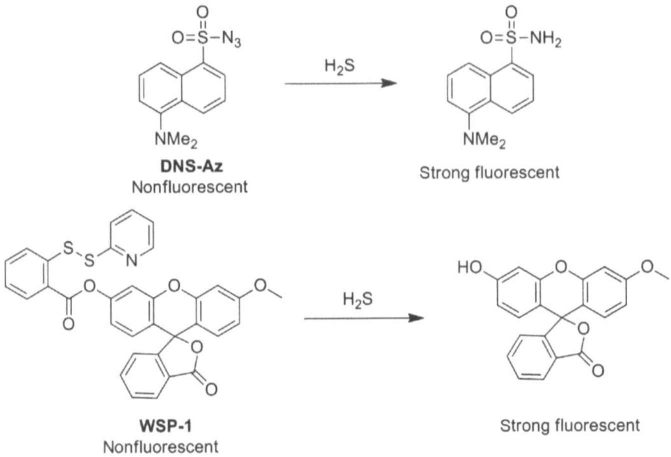

Scheme 3 Fluorescent probes used in this study.

can also be purchased from Cayman Chemicals Co. Sodium sulfide nonahydrate ($Na_2S{\cdot}9H_2O$) (>98.0%) was purchased from Acros and used without further purification. Sodium phosphate buffer was freshly prepared and used. The stock solution of DNS-Az in absolute ethanol (20 m*M*) was freshly prepared and stored in a refrigerator. The stock solution of Na_2S (2 m*M*) in sodium phosphate buffer (20 m*M*, pH 7.4) was freshly prepared, sealed, and kept in ice-bath. The stock solution of each donor (20 m*M*) in biograde DMSO was freshly prepared, sealed, and kept in a refrigerator. Cary Eclipse Fluorescence spectrophotometer (Agilent Technologies) was used to measure fluorescence changes.

3.2. Calibration curve for fluorescence measurements

A 4-mL cuvette was charged with 3.94 mL mixture solution of sodium phosphate buffer (20 m*M*, pH 7.4) and CH_3CN (v/v, 1:1). To this solution was then added 40 μL stock solution of DNS-Az (20 m*M*) and a requisite volume of Na_2S stock solution (4, 8, 12, 16, and 20 μL, respectively). The final volume of the reaction solution was adjusted to 4 mL with sodium phosphate buffer (20 m*M*, pH 7.4). After mixing thoroughly and then standing for 2 min at room temperature, the fluorescence intensity of reaction solution was measured with $\lambda_{ex} = 340$ nm. In the meantime, a blank solution containing no testing species sample was prepared and measured under the same conditions for comparison (Fig. 1).

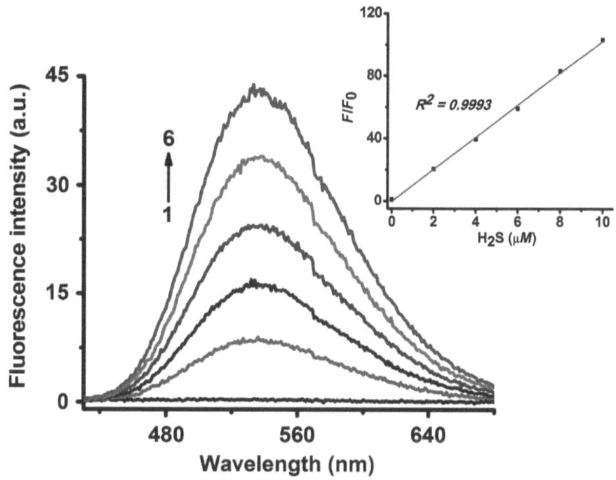

Figure 1 Fluorescence emission spectra of DNS-Az (200 μ*M*) with varied concentrations of Na_2S_2 (0, 2, 4, 6, 8, 10 μ*M* for curves 1–6, respectively). The reactions were carried out for 2 min at room temperature in sodium phosphate buffer (20 m*M*, pH 7.4) with 50% CH_3CN. Data were acquired with excitation at 340 nm. The inset depicts the plot of relative fluorescence intensity (F/F_0) of the reaction system at $\lambda_{ex/em}$ = 340/535 nm against the corresponding reagent blank (without Na_2S), which is the calibration curve for this study.

3.3. Determination of H_2S release from donors

A 4-mL cuvette with a PTFE stopper was charged with 3.94 mL mixture of sodium phosphate buffer (20 m*M*, pH 7.4) and CH_3CN (v/v, 1:1) followed by the addition of 40 μL the stock solution of DNS-Az. The solution was gently shaken followed by the addition of 20 μL of each donor's stock solution. After thoroughly mixing and standing for 2 min at room temperature, the changes in fluorescence emission spectra at 535 nm (λ_{ex} = 340 nm) were then monitored at 15 min intervals for 3 h at room temperature. Finally, the fluorescence signals were converted to H_2S concentrations based on the reference curve obtained with a series of Na_2S standard solutions (Fig. 2).

4. H_2S RELEASE FROM THE DONORS IN CULTURED CELLS

These donors' H_2S release capabilities were evaluated in H9c2 cells. In this study, we first examined cytotoxicity of three donors **PDD-1**, **PDD-2**, and **PDD-3**. For comparison, NaHS and GYY4137 were also tested. Based on cytotoxicity results, **PDD-1** was used in cell imagine to prove its H_2S release in cells.

A

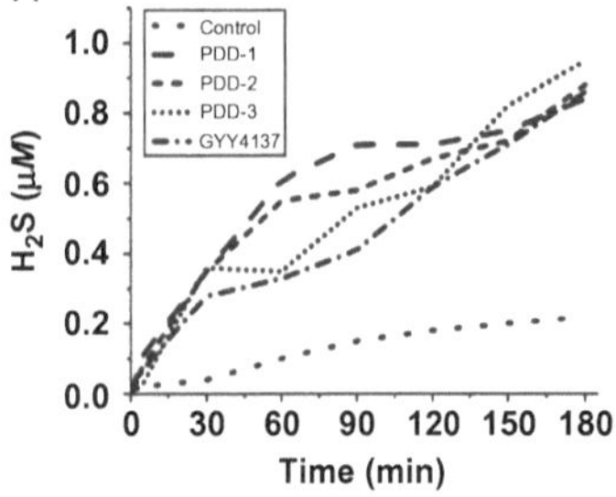

B

Time (min)	H_2S (μM)			
	PDD-1	PDD-2	PDD-3	GYY4137
0	0	0	0	0
5	0.09	0.09	0.01	0.05
30	0.31	0.31	0.32	0.24
60	0.51	0.51	0.25	0.23
90	0.56	0.56	0.38	0.26
120	0.53	0.53	0.41	0.41
150	0.55	0.55	0.62	0.51
180	0.62	0.62	0.73	0.64

Figure 2 (A) Time-dependent H_2S releasing profiles of donors **PDD-1**, **PDD-2**, **PDD-3**, and GYY4137. (B) Data of H_2S concentration changes of each donor.

4.1. Materials

Cells: Rat embryonic cardiomyoblasts cell line H9c2 was purchased from the American Tissue Culture Collection and cultured in Dulbecco's modified Eagle's Medium (DMEM) supplemented with 10% fetal bovine serum, 4 m*M* glutamine, 100 IU/mL penicillin, and 100 μg/mL streptomycin at 37 °C and with 5% CO_2.

Cell viability Kit: Cell Counting Kit-8 (CCK-8) was purchased from Dojindo Molecular Technologies, Inc. 10% CCK-8 solution in FBS-free DMEM with 20 U/mL penicillin and 20 μg/mL streptomycin was directly added to the cells. The cell viability was measured by a microplate reader (Infinite M1000, TECAN Group Ltd.).

4.2. Cell viability test protocol and results

Cell viability was measured using CCK-8. H9c2 cells were cultured in 96-well plates, with four duplicate wells in each group. When 80–90% confluence was reached, cells were treated with each donor (20 m*M* stock solutions were prepared in DMSO) at different concentrations in 1% DMSO in medium and incubated for 24 h. Then CCK-8 solution (10 μL) at a 1:10 dilution with FBS-free DMEM/F12 (100 μL) was added to each well followed by a further 3 h incubation under 5% CO_2 at 37 °C. Absorbance was automatically measured at 450 nm with a microplate reader (Infinite M1000, TECAN Group Ltd.). The mean optical density (OD, absorbance) of four wells in the indicated groups was used to calculate the percentage of cell viability as follows: percentage of cell viability $= (A_{\text{treatment}} - A_{\text{blank}})/(A_{\text{control}} - A_{\text{blank}}) \times 100\%$ (where, A = absorbance). Values were plotted by averaging duplicate wells. Figure 3 shows the results of three donors

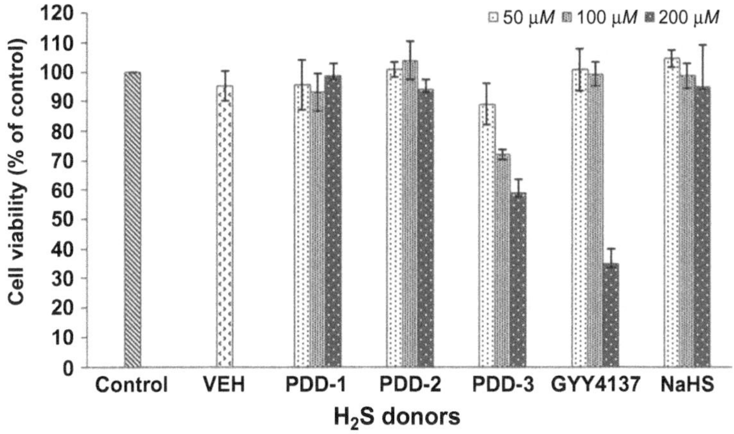

Figure 3 Effects of H_2S donors on cell viability. H9c2 cells were treated with different concentrations of **PDD-1**, **PDD-2**, **PDD-3**, GYY4137, and NaHS for 24 h. CCK-8 assay was performed to measure cell viability. *Reprinted with permission from Park et al. (2013). Copyright 2013 Molecular BioSystems.*

(**PDD-1**, **PDD-2**, and **PDD-3**). For comparison, GYY4137 and NaHS were also tested and their results are included in Fig. 3.

4.3. Images of H_2S release in cells

PDD-1 was selected as the example due to its low cytotoxicity. H9c2 cells were incubated with **PDD-1** at different concentrations (0, 100, 200 and 400 μ*M*) for 24 h. Then a H_2S-specific fluorescent probe, WSP-1 (Liu et al., 2011), was introduced into the cells to monitor the production of H_2S. As expected, donor-treated cells (Fig. 4B–D) showed much enhanced fluorescent signals compared to vehicle treated cells (Fig. 4A).

Method and instrument: H9c2 cells were incubated in 24-well plates in medium for 3 days (> 90% confluence). After removing the fetal-serum containing medium, the cells in each well were treated with different concentrations (0, 100, 200, and 400 μ*M*) of **PDD-1** (20 m*M* stock solution in DMSO) in FBS-free DMEM (0.5 mL) medium and then incubated under 5% CO_2 at 37 °C for 24 h. The old medium was removed and then the cells were incubated for 1 h with 0.5 mL of a fresh medium containing a fluorescence probe WSP-1 (final c = 100 μ*M*) and cetyltrimethylammonium bromide (final c = 1 m*M*). After removing the medium, the cells were carefully rinsed with PBS buffer (1 mL × 1). To each well was added 0.5 mL of PBS and cell images were taken on EVOS fl microscope (Fig. 4). This experiment was conducted in triplicate.

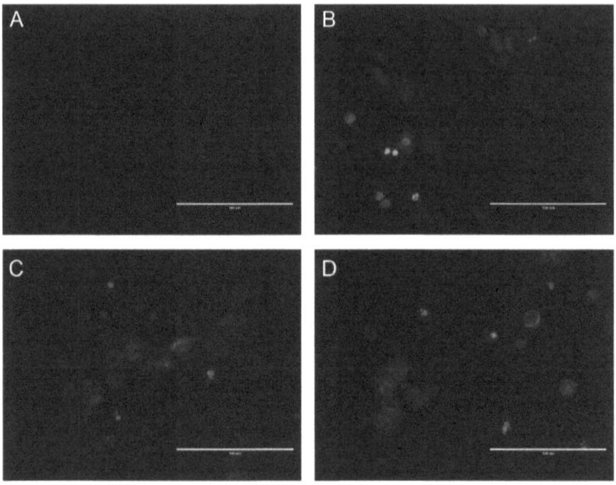

Figure 4 H_2S production of **PDD-1** in H9c2 cells. Cells were incubated with vehicle (A), 100 μ*M* (B), 200 μ*M* (C), and 400 μ*M* (D) of donor for 24 h. After removal of excess donor, 100 μ*M* of WSP-1 was added. Images were taken afterwards. *Reprinted with permission from Park et al. (2013). Copyright 2013 Molecular BioSystems.* (See the color plate.)

5. DONOR'S ACTIVITY AGAINST H_2O_2-INDUCED CELL DAMAGE

This experiment was designed to investigate the protective effects of **PDD-1**, **PDD-2**, GYY4137, and NaSH against H_2O_2-induced cell damage in H9c2 cardiomyoblasts.

5.1. Materials

30% H_2O_2 was purchased from J.T. Baker and used without further purification. Biograde DMSO was purchased from biomedical LLC and directly used. CCK-8 was purchased from Dojindo Molecular Technologies, Inc. 10% CCK-8 solution in FBS-free DMEM with 20 U/mL penicillin and 20 μg/mL streptomycin was prepared and directly added to the cell. The cell viability was measured by a microplate reader (Infinite M1000, TECAN Group Ltd.).

5.2. The optimal concentration of H_2O_2 for cell damage experiments

To examine H_2O_2 induced injury, cells were exposed to different concentrations of H_2O_2 (50, 100, 150, 200, 300, 400, 500, 600, 700, 1000 μ*M*) in

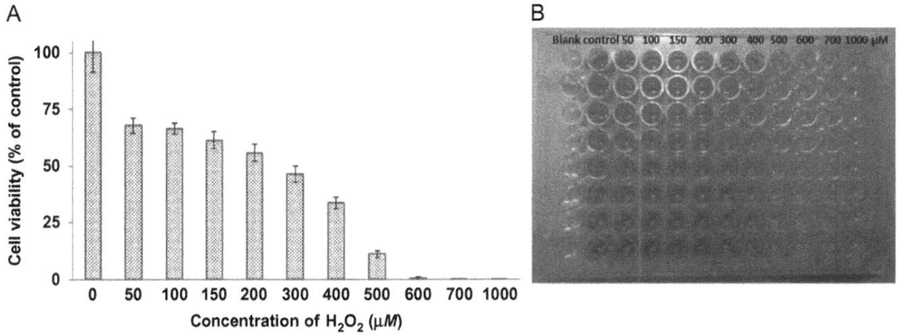

Figure 5 Effects of H_2O_2 on H9c2 cell viability. The cells were treated with increasing concentrations of H_2O_2 (0–1000 μ*M*) for 5 h. (A) H9c2 cell viability measured by CCK-8 assay. (B) Picture of H9c2 cell viability in a 96-well plate.

FBS-free DMEM medium and incubated with 5% CO_2 at 37 °C for 5 h. Cells exposed to highly concentrated H_2O_2 (>400 μ*M*, see graph) were not healthy, detached, and dead. This observation was similar to a literature (Sun, Sun, Wang, Xiao, & Sun, 2011). Following the same procedure for the CCK-8 kit as stated above, the mean OD (absorbance) was obtained and used to calculate the cell viability (Fig. 5). The optimal concentration of H_2O_2 was decided as 150 μ*M*.

5.3. Evaluation of donor's protective effects against H_2O_2 damage

When cell confluence (cultured in 96-well plates, four duplicates) reached 80–90%, old media were removed and cells were washed with PBS buffer (×1). The cells were treated with 100 μL of freshly prepared donor solutions in medium which were prepared as follows: the stock solutions of **PDD-1**, **PDD-2**, GYY4137 (20 m*M* in DMSO) were diluted with FBS-free DMEM media to prepare solutions in three different concentrations (50, 100, and 200 μ*M*). After incubation for 24 h under 5% CO_2 at 37 °C, 3.75 μL of H_2O_2 (4 m*M* stock solution in PBS buffer) was directly added to each well to afford 150 μ*M* H_2O_2 final concentration. After incubation for additional 5 h, CCK-8 solution (10 μL) at a 1:10 dilution with FBS-free DMEM/F12 (100 μL) was added to each well and incubated for additional 3 h under 5% CO_2 at 37 °C. Without further treatment, the absorbance was recorded at 450 nm using a microplate reader. The percentage of cell viability was calculated according to a following equation ($p = (A_{treatment} - A_{blank})/(A_{control} - A_{blank}) \times 100\%$ (where, A = absorbance)). Values were plotted by averaging wells duplicated four times. As shown

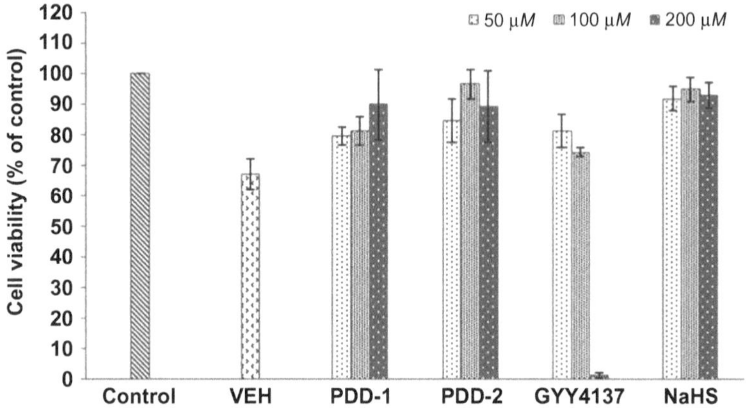

Figure 6 H9c2 cell viability after oxidative injury by H_2O_2 in the presence of donors **PDD-1**, **PDD-2**, GYY4137, and NaHS. *Reprinted with permission from Park et al. (2013). Copyright 2013 Molecular BioSystems.*

in Fig. 6, cell viability was reduced by 35% when treated with H_2O_2 only. In the presence of donors, however, cell viability increased significantly. These results suggest that H_2S donors indeed have some protective effects against H_2O_2-induced injury.

6. SUMMARY

In this chapter, we provide detailed experimental protocols for the synthesis and evaluation of a series of phosphorodithioate-based H_2S donors. A fluorescence method was used to determine their H_2S release capability. These donors were found to release H_2S slowly in aqueous buffers, very similar to the well-known GYY4137. In addition, these donors exhibited effective H_2S production in cultured cells (as monitored by a fluorescent probe) and protective effects against H_2O_2 induced oxidative damage in cells.

ACKNOWLEDGMENTS

This work was supported by NIH (R01HL116571) and the ACS-Teva USA Scholar Award.

REFERENCES

Caliendo, G., Cirino, G., Santagada, V., & Wallace, J. L. (2010). Synthesis and biological effects of hydrogen sulfide (H_2S): Development of H_2S-releasing drugs as pharmaceuticals. *Journal of Medicinal Chemistry*, *53*(17), 6275–6286.

Chang, L., Geng, B., Yu, F., Zhao, J., Jiang, H., Du, J., et al. (2008). Hydrogen sulfide inhibits myocardial injury induced by homocysteine in rats. *Amino Acids*, *34*(4), 573–585.

DeLeon, E. R., Stoy, G. F., & Olson, K. R. (2012). Passive loss of hydrogen sulfide in biological experiments. *Analytical Biochemistry, 421*(1), 203–207.
Fogo, J. K., & Popowsky, M. (1949). Spectrophotometric determination of hydrogen sulfide. *Analytical Chemistry, 21*(6), 732–734.
Fukuto, J. M., Carrington, S. J., Tantillo, D. J., Harrison, J. G., Ignarro, L. J., Freeman, B. A., et al. (2012). Small molecule signaling agents: The integrated chemistry and biochemistry of nitrogen oxides, oxides of carbon, dioxygen, hydrogen sulfide, and their derived species. *Chemical Research in Toxicology, 25*(4), 769–793.
Geng, B., Yang, J., Qi, Y., Zhao, J., Pang, Y., Du, J., et al. (2004). H_2S generated by heart in rat and its effects on cardiac function. *Biochemical and Biophysical Research Communications, 313*(2), 362–368.
Hughes, M. N., Centelles, M. N., & Moore, K. P. (2009). Making and working with hydrogen sulfide: The chemistry and generation of hydrogen sulfide in vitro and its measurement in vivo: A review. *Free Radical Biology & Medicine, 47*(10), 1346–1353.
Jacob, C., Anwar, A., & Burkholz, T. (2008). Perspective on recent developments on sulfur-containing agents and hydrogen sulfide signaling. *Planta Medica, 74*(13), 1580–1592.
Kashfi, K., & Olson, K. R. (2013). Biology and therapeutic potential of hydrogen sulfide and hydrogen sulfide-releasing chimeras. *Biochemical Pharmacology, 85*(5), 689–703.
Kimura, H. (2011). Hydrogen sulfide: Its production, release and functions. *Amino Acids, 41*(1), 113–121.
Kimura, H., Shibuya, N., & Kimura, Y. (2012). Hydrogen sulfide is a signaling molecule and a cytoprotectant. *Antioxidants & Redox Signaling, 17*(1), 45–57.
Krishnan, N., Fu, C., Pappin, D. J., & Tonks, N. K. (2011). H_2S-induced sulfhydration of the phosphatase PTP1B and its role in the endoplasmic reticulum stress response. *Science Signaling, 4*(203), ra86.
Lavu, M., Bhushan, S., & Lefer, D. J. (2011). Hydrogen sulfide-mediated cardioprotection: Mechanisms and therapeutic potential. *Clinical Science, 120*(6), 219–229.
Lawrence, N. S., Davis, J., & Compton, R. G. (2000). Analytical strategies for the detection of sulfide: A review. *Talanta, 52*(5), 771–784.
Lee, Z. W., Zhou, J., Chen, C. S., Zhao, Y., Tan, C. H., Li, L., et al. (2011). The slow-releasing hydrogen sulfide donor, GYY4137, exhibits novel anti-cancer effects in vitro and in vivo. *PLoS One, 6*(6), e21077.
Li, L., Rose, P., & Moore, P. K. (2011). Hydrogen sulfide and cell signaling. *Annual Review of Pharmacology and Toxicology, 51*, 169–187.
Li, L., Whiteman, M., Guan, Y. Y., Neo, K. L., Cheng, Y., Lee, S. W., et al. (2008). Characterization of a novel, water-soluble hydrogen sulfide-releasing molecule (GYY4137): New insights into the biology of hydrogen sulfide. *Circulation, 117*(18), 2351–2360.
Lin, V. S., & Chang, C. J. (2012). Fluorescent probes for sensing and imaging biological hydrogen sulfide. *Current Opinion in Chemical Biology, 16*(5–6), 595–601.
Lippert, A. R. (2014). Designing reaction-based fluorescent probes for selective hydrogen sulfide detection. *Journal of Inorganic Biochemistry, 133*, 136–142.
Liu, C., Pan, J., Li, S., Zhao, Y., Wu, L. Y., Berkman, C. E., et al. (2011). Capture and visualization of hydrogen sulfide by a fluorescent probe. *Angewandte Chemie International Edition, 50*(44), 10327–10329.
Mustafa, A. K., van Rossum, D. B., Patterson, R. L., Maag, D., Ehmsen, J. T., Gazi, S. K., et al. (2009). Glutamatergic regulation of serine racemase via reversal of PIP2 inhibition. *Proceedings of the National Academy of Sciences of the United States of America, 106*(8), 2921–2926.
Nagy, P., Palinkas, Z., Nagy, A., Budai, B., Toth, I., & Vasas, A. (2014). Chemical aspects of hydrogen sulfide measurements in physiological samples. *Biochimica et Biophysica Acta, 1840*(2), 876–891.

Olson, K. R. (2012). A practical look at the chemistry and biology of hydrogen sulfide. *Antioxidants & Redox Signaling, 17*(1), 32–44.
Ora, M., Järvi, J., Oivanen, M., & Lönnberg, H. (2000). Hydrolytic reactions of the phosphorodithioate analogue of uridylyl(3',5')uridine: Kinetics and mechanisms for the cleavage, desulfurization, and isomerization of the internucleosidic linkage. *The Journal of Organic Chemistry, 65*(9), 2651–2657.
Park, C. M., Zhao, Y., Zhu, Z., Pacheco, A., Peng, B., Devarie-Baez, N. O., et al. (2013). Synthesis and evaluation of phosphorodithioate-based hydrogen sulfide donors. *Molecular BioSystems, 9*(10), 2430–2434.
Peng, H., Chen, W., Cheng, Y., Hakuna, L., Strongin, R., & Wang, B. (2012). Thiol reactive probes and chemosensors. *Sensors, 12*(11), 15907–15946.
Peng, H., Cheng, Y., Dai, C., King, A. L., Predmore, B. L., Lefer, D. J., et al. (2011). A fluorescent probe for fast and quantitative detection of hydrogen sulfide in blood. *Angewandte Chemie International Edition, 50*(41), 9672–9675.
Peng, B., & Xian, M. (2014). Fluorescent probes for hydrogen sulfide detection. *Asian Journal of Organic Chemistry, 3*(9), 914–924.
Pluth, M. D., Baisley, T. S., Hammer, M. D., & Montoya, L. A. (2013). Chemical tools for studying biological hydrogen sulfide. In Craig A. Bayse & Julia L. Brumaghim (Eds.), Vol. 1152. *Biochalcogen chemistry: The biological chemistry of sulfur, selenium, and tellurium* (pp. 15–32). USA: American Chemical Society Symposium Series eBooks.
Predmore, B. L., Lefer, D. J., & Gojon, G. (2012). Hydrogen sulfide in biochemistry and medicine. *Antioxidants & Redox Signaling, 17*(1), 119–140.
Sen, N., Paul, B. D., Gadalla, M. M., Mustafa, A. K., Sen, T., Xu, R., et al. (2012). Hydrogen sulfide-linked sulfhydration of NF-kappaB mediates its antiapoptotic actions. *Molecular Cell, 45*(1), 13–24.
Sun, X., Sun, G.-b, Wang, M., Xiao, J., & Sun, X.-b. (2011). Protective effects of cynaroside against H_2O_2-induced apoptosis in H9c2 cardiomyoblasts. *Journal of Cellular Biochemistry, 112*(8), 2019–2029.
von Wantoch Rekowski, M., Pyriochou, A., Papapetropoulos, N., Stossel, A., Papapetropoulos, A., & Giannis, A. (2010). Synthesis and biological evaluation of oxadiazole derivatives as inhibitors of soluble guanylyl cyclase. *Bioorganic & Medicinal Chemistry, 18*(3), 1288–1296.
Wallace, J. L., Caliendo, G., Santagada, V., Cirino, G., & Fiorucci, S. (2007). Gastrointestinal safety and anti-inflammatory effects of a hydrogen sulfide-releasing diclofenac derivative in the rat. *Gastroenterology, 132*(1), 261–271.
Wang, R. (2012). Physiological implications of hydrogen sulfide: A whiff exploration that blossomed. *Physiological Reviews, 92*(2), 791–896.
Whiteman, M., Armstrong, J. S., Chu, S. H., Jia-Ling, S., Wong, B. S., Cheung, N. S., et al. (2004). The novel neuromodulator hydrogen sulfide: An endogenous peroxynitrite 'scavenger'? *Journal of Neurochemistry, 90*(3), 765–768.
Yong, Q. C., Cheong, J. L., Hua, F., Deng, L. W., Khoo, Y. M., Lee, H. S., et al. (2011). Regulation of heart function by endogenous gaseous mediators-crosstalk between nitric oxide and hydrogen sulfide. *Antioxidants & Redox Signaling, 14*(11), 2081–2091.
Zhao, Y., Bhushan, S., Yang, C., Otsuka, H., Stein, J. D., Pacheco, A., et al. (2013). Controllable hydrogen sulfide donors and the activity against myocardial ischemia-reperfusion injury. *ACS Chemical Biology, 8*(6), 1283–1290.
Zhao, Y., Tyler, D. B., & Xian, M. (2014). Hydrogen sulfide (H_2S) releasing agents: Chemistry and biological applications. *Chemical Communications, 50*(80), 11788–11805.
Zhao, Y., Wang, H., & Xian, M. (2011). Cysteine-activated hydrogen sulfide (H_2S) donors. *Journal of the American Chemical Society, 133*(1), 15–17.

CHAPTER NINE

GYY4137, a Novel Water-Soluble, H_2S-Releasing Molecule

Peter Rose*, Brian W. Dymock†, Philip K. Moore‡,1

*University of Lincoln, Lincoln, Lincolnshire, United Kingdom
†Department of Pharmacy, National University of Singapore, Singapore
‡Neurobiology Program, Life Science Institute and Department of Pharmacology, National University of Singapore, Singapore
[1]Corresponding author: e-mail address: dprmpk@nus.edu.sg

Contents

Abstract

Hydrogen sulfide (H_2S) is now recognized as the so called "third gasotransmitter" taking its place alongside nitric oxide and carbon monoxide. In recent years, H_2S has been reported to exhibit a diverse range of pharmacological effects in biological systems. Much of this evidence is derived from a combination of conventional pharmacological and genetic approaches coupled with the use of chemical compounds such as sodium hydrosulfide, a rapid H_2S releasing donor. Developments in the design of new drug entities which attempt to take into account physicochemical properties, targeting to specific cellular organelles, triggering of H_2S release upon specific chemical reactions in the cell, and controlling the release of H_2S over extended periods of time have been described. For most of these molecules, little or no work has been conducted to determine their biological activity or possible therapeutic effects. It is therefore not clear whether such molecules have therapeutic potential which highlights the need for further *in vivo* studies. One exception to the general rule is GYY4137 (morpholin-4-ium

Methods in Enzymology, Volume 554
ISSN 0076-6879
http://dx.doi.org/10.1016/bs.mie.2014.11.014

4-methoxyphenyl(morpholino) phosphinodithioate), a slow releasing H_2S donor, which has been evaluated for activity in a range of pharmacological models both *in vitro* and *in vivo*. GYY4137 was first reported to release H_2S and exhibit vasodilator activity over 5 years ago and, to date, GYY4137 is becoming increasingly employed as a pharmacological "tool" to explore the biological functions of H_2S.

1. INTRODUCTION

In recent years much attention has been focussed on elucidating the functional roles of the gaseous mediator hydrogen sulfide (H_2S) in biological systems (for reviews see, Guo, Cheng, & Zhu, 2013; Li, Rose, & Moore, 2011; Olas, 2014; Wang, 2012). It is now clear that biologically active quantities of H_2S are synthesized naturally by a number of cell types in a range of different tissues in mammals, nonmammalian species, and even plants. Moreover, evidence is growing that its principal role is to act as a signaling molecule bringing about a distinct set of physiological and biochemical effects.

Considerable emphasis has been placed on determining the role of this gas in normal physiological processes and, at the same time, whether changes in the homeostatic regulation of the H_2S system can trigger, contribute to, or exacerbate disease. A better understanding of the physiology and pathophysiology of H_2S will be key to the development of novel therapeutics based on the general principles of either, (i) reducing endogenous H_2S biosynthesis or, as will become apparent in this review, more likely, (ii) replenishing deficient production of this evanescent gas.

Much progress has been made in understanding the biology of H_2S in the last decade. Thus, without going into specifics which will be dealt with elsewhere in this volume, H_2S has been reported to play a part in a range of biologically fundamental processes which affect most cells/tissues including mitochondrial function, cytoprotection, inflammation, vascular perfusion, tissue preservation and repair, apoptosis, cell cycle regulation, and aging. Interestingly, many of these effects of H_2S in biological systems are not restricted to mammalian species. For example, biosynthesis of this gas has been detected in the tissues from organisms found across several taxonomic kingdoms (Calderwood & Kopriva, 2014; Clarke, 1953; Julian, Statile, Wohlgemuth, & Arp, 2002; Olson et al., 2008; Papenbrock, Riemenschneider, Kamp, Schulz-Vogt, & Schmidt, 2007). In the plant kingdom, exogenous application of H_2S has been reported to increase crop yields, germination rates, influence root development, and alleviate heavy metal ion induced stress responses (Calderwood & Kopriva, 2014; Lisjak, Teklic, Wilson, Whiteman, & Hancock, 2013). Certainly, a considerable body of work now exists describing the functional roles of H_2S in bacterial, animal, and plant systems.

Much attention has, not surprisingly, been focussed on the mechanism(s) by which cells generate H_2S. By and large, the synthesis and production of this molecule is largely mediated *via* enzymatic routes. Recent evidence supports the liberation of H_2S from endogenous persulfides and polysulfide species (Greiner et al., 2013; Ida et al., 2014; Miranda & Wink, 2014). Intracellular levels of glutathione persulfide were quantified at over 100 μM in mouse brain leading to the hypothesis that exogenous H_2S is taken up primarily into the "persulfide pool." The biological equilibrium is hence heavily weighted in favor of persulfides which could be major sulfur transfer signaling molecules. The major biosynthetic origins for H_2S in mammals are *via* the catabolic activity of cystathione β synthetase (CBS, EC 4.2.1.22), cystathionine γ lyase (CSE, EC 4.4.1.1) on L-cysteine and 3-mercaptopyruvate sulfurtransferase (3-MST, EC 2.8.1.2) on 3-mercaptopyruvate. Worth noting is that additional systems are also likely involved in H_2S homeostasis yet their relative contribution in maintaining tissue H_2S levels remains largely unknown and, undoubtedly, other pathways responsible for H_2S biosynthesis and degradation will likely be discovered.

Perturbations in endogenous H_2S biosynthesis *via* one or other of these enzymatic pathways, may cause or play a part in disease states. For the most part, these disease states have been linked to loss of H_2S production due to reduced activity or expression of CSE, CBS, 3-MST, or a combination thereof. Although somewhat simplistic, a variety of conditions including hypertension, inflammation, atherosclerosis, sexual dysfunction and reperfusion injury, and perhaps also cancer, are associated with reduced H_2S generation or availability. Clearly, multiple pathologies and mechanisms underpin these diseases but, intriguingly, a lack of H_2S production seems to be at least one common thread. Much effort has therefore been expended to understand the processes whereby both the activity and the expression of CSE, CBS, and 3-MST are controlled. This work is ongoing and will help to unravel the complex biological significance of H_2S.

2. WHY SLOW RELEASING H_2S DONORS?

As noted above it is becoming increasingly clear that a number of disease states have been linked to what might be described as a state of "relative H_2S deficiency" in different cells, tissues and/or organs. As such, the expeditious use of H_2S donor molecules may, logically, provide a possible common therapeutic approach to a wide range of disease states by replenishing H_2S. A number of H_2S donors have been reported in recent years. In addition to H_2S gas itself, by far the most commonly used are the simple sulfide salts such as NaHS and Na_2S. The use of these salts has been widespread in the literature over the last decade and has undoubtedly added greatly to our understanding of the complex biological effects of H_2S. It could even be argued that the widespread availability of, for example, NaHS, which is both inexpensive to purchase and readily soluble in water, has been a crucial factor in creating the interest and giving impetus to researchers to study H_2S, especially so in the early days. However, a major concern with NaHS is that it does not allow for the controlled release of H_2S at rates that would be expected to mimic physiological conditions. Indeed, once solubilized, NaHS releases large amounts of H_2S spontaneously, i.e., a bolus effect. It is very unlikely that endogenous H_2S, made enzymatically in cells and tissues as detailed above, could ever be generated at the very high concentrations and over the very short time period, which is the case when cells or tissue are exposed to NaHS. One caveat in this argument is that the concentration of endogenous H_2S generated enzymatically within individual tissues or indeed within individual subcellular components inside the target cell is not known and will require the development of much more sensitive and selective methods to detect this gas than are currently available.

With this in mind, a consequence of the explosive H_2S release profile of NaHS is that the biological effects elicited by this, or like, compounds when introduced into a biological system (i.e., a conscious or anesthetized animal, cells maintained in culture or tissues in organ baths) reflects the pharmacological (perhaps toxicological) activity of H_2S and not its physiological role within that tissue/cell. As such, the use of compounds like NaHS, whilst valuable, may in some cases, obscure rather than illuminate, the biology of this gas. Clearly, using NaHS to probe for physiological roles of H_2S is problematic. One approach to overcome this shortcoming has been the development of organic molecules which slowly release H_2S at rates which are more akin to the rate expected from cells *in vivo* (Song et al., 2014).

A number of H_2S donors of different types are now available. Many of these have been designed with the specific intent of delivering controlled levels of H_2S gas to tissues. Of particular interest are a number of structurally divergent molecule classes encompassing H_2S releasing nonsteroidal anti-inflammatory drugs (Isenberg et al., 2007; Sparatore et al., 2009), L-cysteine activated arylthioamide H_2S donor molecules (Martelli et al., 2013), caged gem-dithiol derivatives that release H_2S upon light stimulation (Devarie-Baez et al., 2013) and molecules that release H_2S in aqueous systems. Many of these compounds, that are now available commercially, have been reviewed elsewhere (Kashfi & Olson, 2013; Song et al., 2014). More recently, additional H_2S donors have been introduced including poly(ethylene glycol)-ADT (PEG-ADT) (Hasegawa & van der Vlies, 2014), and *S*-aroylthiooximes (Foster, Powell, Radzinski, & Matson, 2014). In some cases, molecules which target delivery of H_2S to specific organelles within cells, notably mitochondria, have been reported (Le Trionnaire, Perry, Whatmore, Wood, & Whiteman, 2013; Le Trionnaire et al., 2014). The advantage of all of these new drugs (c.f. NaHS) is their potential at least to provide a means to manipulate endogenous H_2S levels within biological systems over a longer time course and within the range of concentrations likely to be achieved naturally. This approach will provide researchers with a wider range of "tools" to probe the biological significance of H_2S and, perhaps in due course, will provide clinicians with alternative therapeutics.

3. THE DEVELOPMENT AND CHARACTERIZATION OF GYY4137

GYY4137 (morpholin-4-ium 4-methoxyphenyl (morpholino) phosphinodithioate) was described as a slow releasing H_2S donor with vasorelaxant activity both *in vitro* and *in vivo* in 2008 (Li et al., 2008). Interestingly, GYY4137 was initially developed as a chemical entity for use as an accelerator in the vulcanization of natural rubber back in the late 1950s but it was not until several decades later that its use as a potential drug candidate was envisaged. GYY4137 was the first organic small molecule, to be reported and characterized as a slow releasing H_2S donor (Fig. 1). GYY4137 is highly water soluble (>1 mg/mL at pH7.4), releases H_2S at a slow rate (unlike sulfide salts) and is relatively easy to synthesize chemically or can be obtained commercially. To date, the biological effects of

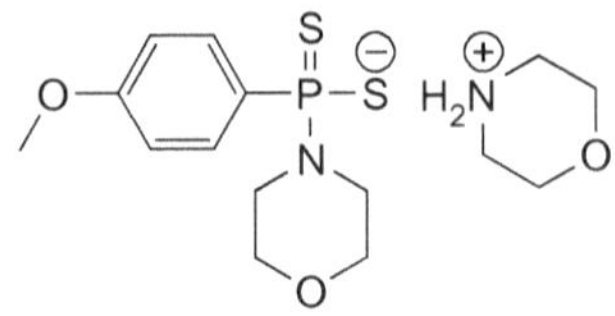

Figure 1 The chemical structure of GYY4137.

GYY4137 have been examined both *in vitro* and *in vivo* and a number of these studies will be described later in this article.

Initial work on the solution chemistry of GYY4137 indicated that it released low quantities of H_2S over a sustained period (hours to days) in aqueous solution (pH 7.4, 37 °C), (Li et al., 2008). Moreover, the release of H_2S from GYY4137 was pH and temperature dependent, with less H_2S released at 4 °C and greater release at pH 3.0. Confirmation of the H_2S releasing properties of GYY4137 within biological settings has also been demonstrated. Release of H_2S from GYY4137 has now being independently verified using a range of assay procedures including use of the fluorescent probe 2,6-dansyl azide and 5,5′-dithiobis(2-nitrobenzoic acid) (Qabazard et al., 2014) alongside more refined procedures such as the H_2S selective polarographic electrode (Kolluru, Shen, Bir, & Kevil, 2013) and detection of H_2S using fluorescent HPLC analysis of monobromobimane derivatives (Shen, Peter, Bir, Wang, & Kevil, 2012).

4. FACILE SYNTHESIS AND CHEMICAL CHARACTERIZATION OF GYY4137

GYY4137 can be conveniently prepared by ring opening of Lawesson's Reagent (0.8 g, 2 mmol) with morpholine (0.87 g, 1 mmol) in dichloromethane (20 ml) solvent at room temperature. Upon stirring for about 2 h, the product slowly precipitates from the reaction mixture forming an ionic salt. This salt is filtered, washed with dichloromethane (10 ml, twice) to good purity. The solid is dried under vacuum at room temperature avoiding heat. GYY4137 is highly water soluble and care must be taken to avoid any aqueous work-up or washing with polar solvents. Once the solid is isolated, it is characterized *via* spectroscopic methods. Nuclear Magnetic Resonance (NMR) in D_2O solvent confirms the identity and structure of GYY4137, (^1H NMR (400 MHz, D_2O) δ 8.02–7.84 (m, 2H), 7.05–6.94 (m, 2H), 3.93–3.84 (m, 4H), 3.81 (s, 3H), 3.71–3.54 (m, 4H), 3.23–3.21

(m, 4H), 2.86 (dd, $J=9.6$, 5.5 Hz, 4H). ^{13}C NMR (101 MHz, D_2O) δ 161.03 (d, J (C–P) = 3.0 Hz), 132.48 (d, J (C–P) = 12.8 Hz), 131.43 (d, J (C–P) = 109.7 Hz), 113.56 (d, J (C–P) = 14.6 Hz), 66.66 (d, J (C–P) = 11.7 Hz), 63.64, 55.49, 44.72 (d, J (C–P) = 1.8 Hz), 43.19. ^{31}P NMR (162 MHz, D_2O) δ 89.00 (s)). In the infrared spectrum a stretch at 650 cm^{-1} can be seen for the P=S double bond.

GYY4137 should not be heated or exposed to moisture over time. It is recommended to store the solid material under nitrogen at −20 °C. For biological testing, a fresh solid sample should be weighed after allowing to warm to room temperature. For each assay, a solution in water or DMSO as required can be prepared and used freshly for the assay.

H_2S release from GYY4137 can conveniently be quantified using a dansyl azide probe *in vitro* (Wang, Peng, Ni, Dai, & Wang, 2014). The recommended assay uses 2, 6-dansyl azide with a specific procedure in aqueous acetonitrile to ensure a good signal:noise ratio. Upon release of H_2S from GYY4137 the sulfonyl azide group of the probe molecule rapidly reduces to a fluorescent primary sulfonamide product. Fluorescence accumulates over the time of the assay and can be measured using a plate reader at frequent time intervals up to days or even weeks. The limit of detection for this assay is around 1 μ*M* of released H_2S. Concentrations of GYY4137 above 100 μ*M*, preferably incubated at 37 °C, are required to obtain a clear response.

5. BIOLOGICAL EFFECTS OF GYY4137: AN OVERVIEW AND POTENTIAL ROLE IN DISEASE?

A plethora of publications are now available describing one or other of the biological effects of NaHS in cells, tissues, and organs in health as well as in a range of experimentally induced disease states in animals. Any description of this body of work is beyond the scope of the current review but will be found elsewhere in this volume. In general, there is much less information available about the biological activity of any of the new classes of slow releasing H_2S donors noted above and, indeed, for some of these compounds biological characterization is lacking altogether. At present, GYY4137 is without doubt the best studied of these H_2S donors in biological terms and it is the findings from these studies that will be described here with an overview of the major biological effects of GYY4137 provided in Table 1.

Table 1 Chronological summary of the major experimental evidence documenting the pharmacological effects of the novel H_2S releasing molecule, GYY4137

Experimental model	Reported effect	Outcome	Reference
Rat aortic rings and perfused rat kidney *in vitro* and in the anesthetized rat *in vivo*	Vasorelaxant	Slow relaxation of rat aortic rings and dilated the perfused rat renal vasculature by opening vascular smooth muscle K_{ATP} channels	Li et al. (2008)
Lipopolysaccharide (LPS; 4 mg/kg, i.v.) administered anesthetized rats	Anti-inflammatory	Inhibited LPS induced TNF-α production in rat blood. Decreased proinflammatory cytokines levels, TNF-α, IL-1β, IL-6, nitrite/nitrate, C-reactive protein, and L-selectin *in vivo*	Li, Salto-Tellez, Tan, Whiteman, and Moore (2009)
NO and H_2S cross talk within the heart of mice	Contractition	Heart contractility induced by GYY4137, in the presence of the NO donor molecule sodium nitroprusside	Yong et al. (2011)
Anticancer activity in a murine xenograph model, subcutaneous transplantation of either HL-60 or MV4–11 cells	Inhibition of tumor growth	Administered daily—reduced tumor volume in HL-60 and MV4–11 injected animals	Lee et al. (2011)
A mouse model of acute joint inflammation	Anti-inflammatory	GYY4137 reduced synovial fluid myeloperoxidase (MPO) and *N*-acetyl-β-D-glucosaminidase (NAG) activity and decreased TNF-α, IL-1β, IL-6, and IL-8 concentration with joints	Li et al. (2013) and Li, Yang, et al. (2013)
Platelet activation and microvascular thrombus formation	Antithrombotic activities	GYY4137 TRAP-induced adhesion molecule expression. Prolongation of venular thrombus formation and tail-vein bleeding time	Grambow et al. (2014)

High fat fed apolipoprotein E(-/-) mice	Antiatherosclerotic activity	Decreased aortic atherosclerotic plaque formation and partially restored aortic endothelium-dependent relaxation. Decreased ICAM-1, TNF-α, and IL-6 mRNA expression, and superoxide ($O_2^{\cdot-}$) generation in aorta. Increased aortic eNOS, phosphorylation expression of PI3K, enhanced Akt Ser(473) phosphorylation, and downregulated the expression of LOX-1.	Liu et al. (2013)
High fat fed C57BL/6J mice	Prevention of insulin resistance	Decreased phosphorylation of perilipin 1 and hormone sensitive lipase	Geng et al. (2013)
In vitro cell culture models across a range of immortalized and nonimmortalized cell lines	Anticancer properties	Exposure to NaHS or GYY4137 decreased cell survival more effectively in cancer cells. GYY4137, significantly increased glycolysis leading to overproduction of lactate, decreased anion exchanger, and sodium/proton exchanger activity. Combined effects leading to increased metabolic acid production, defective pH regulation, and cell death.	Lee et al. (2014)

5.1. Cardiovascular system: Vascular smooth muscle and platelet function

One of the defining moments which spawned interest in the field of H_2S biology was the finding that this gaseous molecule dilates blood vessels *in vitro* and *in vivo* (Hosoki, Matsuki, & Kimura, 1997). Since this discovery a number of the newly introduced H_2S donor molecules, including GYY4137, have been found to exert similar vascular effects. The first report using GYY4137 identified a persistent and slowly developing relaxation of precontracted rat aortic rings which was antagonized by glibenclamide indicating that, like NaHS, this effect of GYY4137 is mediated *via* opening of vascular smooth muscle K_{ATP} channels (Li et al., 2008). In a separate study, GYY4137 also relaxed precontracted mouse aortic rings (Bucci et al., 2012). In addition, GYY4137 dilated the perfused preconstricted rat renal vasculature indicating that it acts not only on large capacitance vessels (e.g., aorta) but also on small resistance blood vessels which govern organ perfusion and blood pressure (Li et al., 2008). Finally, this report demonstrated that chronic treatment of rats with GYY4137 resulted in a lowering of blood pressure both in normotensive and, to a greater extent, in spontaneously hypertensive rats (Li et al., 2008). The effect of GYY4137 on cardiac motility has yet to be determined but one report did indicate that this H_2S donor (like NaHS) could reduce the negative inotropic effect of NO (Yong et al., 2011). It should be noted whilst chronic treatment with GYY4137, using a dose regimen which reduced blood pressure for a period of several days, caused no apparent deleterious side effects in this study, a full screen for possible toxic side effects of GYY4137 has yet to be reported.

Apart from its effect on vascular smooth muscle, GYY4137 also has additional actions within the cardiovascular system. These include antithrombotic activity *via* inhibition of thrombin receptor agonist peptide (TRAP)-induced adhesion molecule expression and associated platelet activation (Grambow et al., 2014) as well as antiatherosclerotic activity in high fat fed, apolipoprotein $E^{-/-}$ mice (Liu et al., 2013). The latter study also showed pronounced effects of GYY4137 to inhibit expression not only of iNOS but also of P-selectin, lectin-like ox-LDL receptor-1, phosphorylated IκBα, NF-κB, ICAM-1, VCAM-1, and a range of chemokines including CXCL2, CXCR4, CXCL10, and CCL17 indicative of an anti-inflammatory effect which will be discussed in more detail later in this review. GYY4137 has also recently been reported to augment NO-mediated human platelet aggregation in response to collagen and thrombin although the underlying mechanism is not yet clear (Ilkan et al., 2013). Similarly, and equally as important, is a potential role for H_2S in regulating diabetic myocardial injury. In this context, high glucose concentrations promote cellular damage within cardiac tissues *via* the formation of advanced endpoint glycation products, increased reactive oxygen species (ROS) production and the generation of a proinflammatory environment. GYY4137 protects cardiac myoblast cells (H9c2) against high glucose-induced cytotoxicity by activation of the AMPK/mTOR signaling pathway. The activation of AMPK orchestrates the metabolic responses involved in the maintenance of appropriate energy levels needed for cell survival (Beauloye, Bertrand, Horman, & Hue, 2011; Wei, Hu, Zhuang, Liao, & Li, 2014). Such effects of GYY4137 may be related to a generalized action on cell metabolism which will be discussed below.

5.2. Effect of GYY4137 on nonvascular smooth muscle

A similar smooth muscle relaxant effect of GYY4137 has now been described in a range of tissues including human airway smooth muscle cells *via* opening of sarcolemma K_{ATP} channels (Fitzgerald et al., 2014), mouse intrapulmonary airways by an effect on intracellular calcium release (Castro-Piedras & Perez-Zoghbi, 2013), pig bladder neck tissues again by opening K_{ATP} channels (Fernandes et al., 2013), pregnant rat myometrial smooth muscle cells (Robinson & Wray, 2012) and in the bovine ciliary artery (Chitnis et al., 2013). Intriguingly, GYY4137 can occasionally, and in defined conditions, exert the opposite effect on blood vessels. Thus, in animals with experimentally induced endotoxic shock, GYY4137 reverses the hypotensive effect (i.e., increases blood pressure) due to *E coli* lipopolysaccharide (LPS) injection (Li et al., 2009). This likely reflects the anti-inflammatory effect of GYY4137 and, among other mechanisms, its ability to reduce the expression of inducible nitric oxide synthase (iNOS) which generates massive amounts of vasodilator nitric oxide (NO) in this condition.

5.3. Inflammation: Is GYY4137 pro- or anti-inflammatory?

GYY4137 exhibits anti-inflammatory properties in cultured cells such as macrophages *in vitro* and in a variety of animal models of inflammation *in vivo*. The anti-inflammatory properties ascribed to GYY4137 are typically concentration dependent with effects seen *in vitro* at concentrations ranging from 0.05 to 5 m*M* and an effective dose administered intraperitoneally in mice of 50–100 mg/kg. In terms of inflammation, GYY4137 has been most extensively evaluated for its effect on cultured cells *in vitro*. In most cases, GYY4137 reduces the functionality of inflammatory cells by inhibiting NF-κB signaling in the target cells. Thus, GYY4137 pretreatment reduces the secretion of proinflammatory mediators like interleukin-1β (IL-1β), interleukin-6 (IL-6), tumor necrosis factor-α (TNF-α), NO, and prostaglandin E_2 (PGE_2) (Whiteman et al., 2010) in mouse macrophages stimulated with LPS as well as in IL-8-stimulated airway smooth muscle cells (Perry et al., 2011). In human synoviocytes and articular chondrocytes, GYY4137 decreased LPS-evoked production of nitrite (NO_2^-), PGE_2, TNF-α, and IL-6 (Li, Fox, et al., 2013). Similarly, in human chondrocytes isolated from osteoarthritis tissues, GYY4137 reduces NO, PGE_2, IL-6, and MMP-13 synthesis from cells stimulated with IL-1β (Fox et al., 2012). Stimulation of human chondrocytes isolated from osteoarthritis sufferers with IL-1β activates the NF-κB signaling pathway and the nuclear translocation and accumulation of this proinflammatory transcription factor. Induction of the NF-κB pathway is accompanied by increased production of the proinflammatory mediators PGE_2, IL-6, and expression of proinflammatory proteins like MMP-13, cyclooxygenase-2 (COX-2), and iNOS, (Burguera, Vela-Anero, Magalhães, Meijide-Faílde, & Blanco, 2014). In this study, both GYY4137 and NaHS significantly reduced the levels of proinflammatory molecules and proteins in chondrocytes. An inhibitory effect of GYY4137 on NF-κB signaling has also been described in mouse macrophages infected with *Mycoplasma fermentans*. In this study, GYY4137 reduced the production of the proinflammatory cytokine MCP-1 and diminished nuclear accumulation of the NF-κB heterodimer p65/p52 (Benedetti et al., 2014).

GYY4137 also exhibits anti-inflammatory activity in animals *in vivo*. For example, injection of GYY4137 prior to LPS administration in the rat protects against the ensuing endotoxic shock both in terms of partially restoring the depressed blood pressure and with respect to inhibiting the generation of proinflammatory cytokines and NO (Li et al., 2009; Wang, Liu, et al., 2014). In addition, GYY4137 administered 18 h after intra-articular injection of Complete Freund's Adjuvant (CFA) in the mouse reduced paw swelling but when administered 1 h before CFA, GYY4137 increased swelling suggesting that the effect of this donor in this animal model critically depends on the stage of the inflammation. This data implies that H_2S exerts different (perhaps even opposite) effects on the inflammatory response (Li, Fox, et al.,

2013; Li, Yang, Long, Yang, & Shen, 2013). In a recent report, GYY4137 coadministered with histamine into mouse skin inhibited the accompanying pruritis but did not affect the extravasations response (Rodrigues et al., 2013) again suggesting differences in the response to GYY4137 in different systems. Finally, administration of GYY4137 to newborn mouse pups subjected to hyperoxia (85% O_2 for 10 days) augmented lung alveoli development whilst blunting leuokocyte infiltration into the alveolar spaces indicative of an overall anti-inflammatory effect (Madurga et al., 2014). Cumulatively, these studies show that GYY4137 consistently reduces the expression and/or generation of proinflammatory mediators *in vitro* but has a more complex pattern of effects in animal models of inflammation *in vivo*. Inflammation is a multifactorial process which depends on the concerted and time-dependent interplay of numerous cell types each of which is able to generate a range of both pro- and anti-inflammatory mediators. Bearing in mind the somewhat promiscuous effect of H_2S on different cells and transduction mechanisms it is perhaps not surprising that the *in vivo* effects of GYY4137 are complex. It is clear from *in vitro*, and from some studies *in vivo*, that the anti-inflammatory effect of GYY4137 is mediated by inhibiting NF-κB signaling. However, this by no means rules out other possible molecular targets for GYY4137 derived H_2S which should continue to be sought to obtain a more complete picture of the mechanism of action of this H_2S donor.

5.4. Effect of GYY4137 in the reproductive system

The effect of GYY4137 has also been examined in the reproductive system. Roles for H_2S in female reproductive biology have only recently come to light. For example, both CBS and CSE are expressed in human intrauterine tissues and placenta (Patel, Vatish, Heptinstall, Wang, & Carson, 2009; You et al., 2011) as well as in epithelial cells of the fallopian tubes (Ning et al., 2014). The first direct evidence suggesting a role for H_2S in reproductive biology was in preeclampsia, a hypertensive condition that affects around 4% of all pregnancies. The administration of DL-propargylglycine (PAG, an inhibitor of the H_2S synthesizing enzyme, CSE) to pregnant mice resulted in hypertension and the promotion of abnormal vascularization within the placenta. This was correlated with reduced placental growth factor production and increased soluble fms-like tyrosine kinase-1 (sFlt-1) and soluble endoglinin (sEng) release from human umbilical vein endothelial cells which are considered to be markers for the onset of preeclampsia in humans. In mice, GYY4137 reversed this effect of PAG treatment, reducing circulating sFlt-1 and sEng levels and restoring fetal growth (Wang et al., 2013).

In the human fallopian tube, epithelium cell-derived H_2S has been postulated to promote relaxation of the oviduct. Interestingly, egg retention within the fallopian tube is associated with ectopic pregnancy and deregulation of fallopian tube contractility may play a part in this process. Indeed, disruption of H_2S signaling has recently been reported to promote embryo retention and cause developmental delay in the mouse whilst pharmacological treatment with either GYY4137 or NaHS reversed these effects and restored the status quo (Ning et al., 2014).

Although more work is clearly needed, these limited data do suggest possible roles for H_2S in protecting against conditions such as preeclampsia and ectopic pregnancy and thereby raise the possibility that an H_2S donor, like GYY4137, may be of value therapeutically in this regard.

5.5. GYY4137: Apoptosis and cell cycle progression

Whilst most reports in the literature have concentrated on the effect of GYY4137 in the cardiovascular system and in inflammation, it is worth noting that a potential use of this drug in the treatment of cancer has also recently begun to be addressed. To date, only a few reports have examined the involvement of H_2S in cancer cell signaling and/or in the induction and promotion of apoptosis in cancer cells. From the available literature, a number of key findings have been made about the anticancer properties of H_2S and donor molecules. Importantly, both pro- and anti-apoptotic effects of H_2S have been reported perhaps related to the choice of H_2S generating system used. For example, as noted earlier, conventional H_2S donors such as NaHS and Na_2S generate H_2S rapidly in aqueous systems and the released H_2S has a very short half life in cell culture media. It might be argued that GYY4137, which releases H_2S slowly and continuously, may therefore be a better "tool" with which to evaluate the longer term effects of the gas in cell culture systems.

With this in mind, GYY4137 has been shown to influence cell cycle progression in a range of different cell types. This effect of GYY4137 appears to be cell type specific since it caused G2/M phase cell cycle arrest in immortalized MCF-7 cells (Lee et al., 2011), G1/S cell cycle transition in HepG2 cells (Lu, Gao, Huang, & Wang, 2014), and inhibits abnormal proliferation in lymphocytes derived from systemic lupus erythematosus patients (Han et al., 2013). Aside from affecting cell cycle progression, GYY4137 also caused a concentration dependent loss of cell viability across a range of immortalized human cancer cell lines including HeLa, HCT-116, Hep G2, HL-60, MCF-7, MV4-11, and U2OS (Lee et al., 2011). Preliminary work has revealed that a number of GYY4137 chemical analogues exhibit similar activity (Liao et al., 2013). Mechanistic studies revealed that these effects were mediated *via* a caspase dependent pathway since PARP-1 and caspase 9 cleavage were evident in GYY4137-treated MCF-7 cells (Lee et al., 2011). The same study also reported that GYY4137 exerted a dose-dependent inhibitory effect on tumor growth in the mouse *in vivo*. Likewise, GYY4137 suppressed cell proliferation in human hepatocellular carcinoma (HCC) cell lines and reduced tumor growth *in vivo* most likely, in this case, by inhibition of the signal transducer and activator of transcription 3 (STAT3) pathway (Lu et al., 2014).

GYY4137 alters cell metabolism in a number of different ways some of which may have relevance in terms of its antitumor activity. For example, treatment of HeLa cells with GYY4137 increased calcium signaling by upregulating expression of the inositol 1,4,5-trisphosphate (IP3) receptors 1 and 2 (Lencesova et al., 2013). Moreover, depletion of calcium stores from the endoplasmic reticulum (ER) is accompanied by an increase in the expression of X-box, CHOP and ATF4, known biomarkers associated with ER stress (Lencesova et al., 2013). Similarly, upregulation of the sodium calcium exchanger (NCX) by GYY4137 elevates cAMP levels leading to increased expression of β_1 and β_3 adrenoceptors and the induction of apoptosis (Markova et al., 2013). These findings raise the possibility that GYY4137 may promote endoplasmic reticular stress *via* mobilization of calcium stores.

Historically, H_2S gained notoriety due to its ability to inhibit the catalytic activity of cytochrome C oxidase, as H_2S directly reduces the oxygen-reactive a3CuB binuclear center of this protein (Nicholls, Marshall, Cooper, & Wilson, 2013). It is perhaps no great surprise therefore that recent work has indicated subtle effects of GYY4137 on primary metabolic pathways within cancer cells. Aerobic glycolysis provides cancer cells with a growth or survival strategy as it provides valuable metabolic intermediates needed to support rapidly proliferating cells (Gatenby & Gillies, 2004). GYY4137 has recently been reported to inhibit glycolysis across a range of immortalized cancer cell lines with little or no effect on noncancer cells. Interestingly, overproduction of lactate and inhibition of the anion exchanger and sodium/proton exchanger lead to defective pH regulation, increased intracellular acidification, and cell death (Lee et al., 2014).

Clearly more work is needed to determine the role of extrinsic and intrinsic signaling pathways in the apoptotic effects of GYY4137 and to assess in greater detail its antitumor action *in vivo*. Moreover, the precise molecular mechanisms observed in these experimental studies also need to be examined.

5.6. GYY4137 and aging

Aging is intrinsically linked to many of the pathophysiological diseases described in this review. Any description of the biology of the aging process is beyond the scope of this review but readers will undoubtedly be aware of the profound and often deleterious effects that the aging process has on the cardiovascular system, metabolic functions, and immune system of mammalian species. Recently, it has been hypothesized that H_2S plays a fundamental role in the aging process (Zhang et al., 2013). Evidence that H_2S plays a role in this process stems from a number of observations notably, that H_2S can inhibit oxidative stress (Benetti et al., 2013; Whiteman et al., 2004), increase longevity in some model organisms (Miller & Roth, 2007) and promote the induction of genes associated with longevity (Lee, Kennedy, Tolonen, & Ruvkun, 2003; Qabazard et al., 2014). H_2S also reduces cell senescence and promotes cellular cytoprotective systems (Predmore, Alendy, Ahmed, Leeuwenburgh, & Julian, 2010). Indeed, these studies provided a potential approach to treat age-related deterioration in health. The big question is how does H_2S influence the aging process and can GYY4137 be utilized as an antiaging drug in the sense of reducing the incidence or severity of age-related diseases? Recent evidence has indicated that GYY4137 can indeed reduce the deleterious effects of the aging process at least in the model nematode worm *Caenorhabditis elegans* (Qabazard et al., 2013, 2014). For example, GYY4137 treatment increased the expression of several age-related, stress response, and antioxidant genes. Moreover, GYY4137 increases lifespan in short-lived mev-1 mutants with elevated oxidative stress and protected wild-type *C. elegans* against paraquat induced poisoning which is believed to occur *via* the generation of an excess of superoxide anions (Qabazard et al., 2013, 2014).

Taken together the available literature points to a fundamental role for H_2S signaling in the aging process and provides rational avenues of intervention using H_2S donor molecules like GYY4137.

6. THE EFFECT OF GYY4137 IN NONMAMMALIAN SYSTEMS

Whilst the majority of work highlighting a therapeutic role for GYY4137 has been derived from studies in mammalian models of disease less has been reported about the use of GYY4137 in other biological systems. This may be important since it provides further experimental evidence for biological roles for H_2S and likely bolsters additional, albeit nonclinical, translational uses, for example to alleviate stress within agronomic crops.

To date, only a few reports have studied the effect of GYY4137 in nonmammalian species (for summary, see Table 2). For example, GYY4137 prevents Na^+ uptake in larval zebra fish, *Danio rerio* (Kumai, Porteus, Kwong, & Perry, 2014) and has antiaging properties when tested in the nematode worm, *C. elegans* (Qabazard et al., 2013, 2014). The effect of GYY4137 on plant metabolism has also been evaluated. Multiple reports over the years have shown that plants respond to H_2S in a biphasic manner with both positive and negative effects on growth observed. From an environmental context, exposure of plants to high levels of exogenous sources of H_2S gas can have phytotoxic effects with common symptoms associated with but, not restricted to, reduced oxygen production, reducing nutrient uptake, and the promotion of leaf damage (Thompson & Kats, 1978). In contrast, increased rates of photosynthesis, germination rates, and increased plant productivity upon exposure to H_2S have also been documented (Chen et al., 2011; Dooley, Nair, & Ward, 2013). Interestingly, H_2S production in plant tissues is likely derived from the sulfur assimilation pathways and L-cysteine desulfydrase (Ravilious & Jez, 2012; Romero et al., 2014; Takahashi, Kopriva, Giordano, Saito, & Hell, 2011). It is known that in the presence of high concentrations of sulfate (SO_4^{2-}), sulfur dioxide (SO_2) and/or L-cysteine, plant tissues can synthesize, and emit H_2S (Hällgren & Fredriksson, 1982; Sekiya, Schmidt, Wilson, & Filner, 1982; Wilson, Bressan, & Filner, 1978). Surprisingly, it was not until recently that a potential physiological role for H_2S in plants was considered. To date, the reported effects of H_2S on plants are wide ranging and this has led to many researchers postulating that H_2S may function as a potential plant signaling molecule. Sadly, descriptions of potential roles for H_2S releasing donor molecules in plants are rare. The use of H_2S donor molecules should be considered in plant studies since H_2S has the ability to promote plant growth and vigor, reduce the effect of biotic and abiotic stress and increase crop yields (Dooley et al., 2013; Thompson, Kats, & Lennox, 1979).

Table 2 Studies utilizing GYY4137 to demonstrate novel physiological roles for H_2S across a range of model species

Experimental model	Reported effect	Outcome	Reference
Arabidopsis thaliana ecotypes Landsberg erecta and Columbia ecotypes	Stomatal function	GYY4137 promotes stomatal opening in the light and prevented stomatal closure in the dark. Reduces the accumulation of NO within guard cells	Lisjak et al. (2010)
Capsicum anuum	Stomatal function	Reduced accumulation of NO induced by abscisic acid (ABA) treatment of leaf tissues. Promotes stomatal opening	Lisjak et al. (2011)
Zea mays L., Yedan no. 51	Prevention of heat damage	GYY4137 increased endogenous H_2S levels, promoted survival, and reduced electrolyte leakage in seedlings under heat stress	Li, Fox, et al. (2013) and Li, Yang, et al. (2013)
Capsicum anuum	Prevention of salt stress	Accumulation of proline induced by salt treatment	Lisjak et al. (2013)
Caenorhabditis elegans	Aging	GYY4137 increased the expression of several age-related, stress response, and antioxidant genes. Increased lifespan in short-lived mev-1 mutants with elevated oxidative stress. Protected wild-type *C. elegans* against paraquat poisoning	Qabazard et al. (2014)
Caenorhabditis elegans	Survival and H_2S homeostasis	GYY4137, extended median survival and increased tolerance toward oxidative and endoplasmic reticulum (ER) stress. Restored life span in cysl-2 mutant worms.	Qabazard et al. (2013)

Such studies highlighting a fundamental role of H_2S in the regulation of aging, cellular stress pathways, and in plant physiological responses.

GYY4137 has already been used to explore the role of H_2S in regulating stomatal function in plants (Lisjak et al., 2010, 2011). In these studies it was noted that H_2S, derived either from NaHS or GYY4137, caused stomatal opening in the model plant *Arabidopsis thaliana* and in the commercially important crop species, *Capsium anuum*. Further experimentation revealed that both donors reduced abscisic acid induced accumulation of NO within leaf tissues suggestive of a potential cross talk between the two gases. Similarly, García-Mata and Lamattina (2010) also reported that GYY4137 promotes stomatal closure in *Vicia faba (L.) var. major* and in *Impatiens walleriana*.

The involvement of H_2S, derived from GYY4137, in the prevention of plant stress responses has also recently been explored. In plants, exposure to high temperatures increases membrane fluidity and reduces membrane integrity, promotes protein denaturation, and reduces chloroplast and mitochondrial function leading to a state of oxidative and osmotic stress. In this regard, GYY4137 reduces the impact of heat stress and preserves the viability of *Zea Mays* (Li, Yang, et al., 2013). Moreover, preliminary data shows that GYY4137 induces expression of the cytoprotective amino acid proline in plant tissues following salt stress (Lisjak et al., 2013).

7. CONCLUSION

Much progress has been made in the characterization of the pharmacological effects of GYY4137 over the past several years. The biological activity of GYY4137 have now been comprehensively evaluated in a range of *in vitro* cell systems as well as in mammalian and nonmammalian species *in vivo* and also in plants. It is now clear that GYY4137 exhibits a very wide range of biological effects from vasodilator activity in isolated blood vessels through antiaging activity in *C. elegans* to crop preservation in plants. GYY4137 has the advantage that it is not only very water soluble but also easy to synthesize chemically as well as being commercially available. GYY4137 was an early addition to the family of H_2S donors and can therefore be considered as the prototype for this class of compounds. Further progress will undoubtedly lead to new second and third generation compounds with the ability to release their H_2S "payload" at well-defined rates, in defined organs and cells and in the appropriate subcellular location. Experimental "tools" of this type will prove invaluable in unraveling the complex biological roles of H_2S in multiple systems and perhaps also act as the forerunners of a new class of therapeutic agents.

REFERENCES

Beauloye, C., Bertrand, L., Horman, S., & Hue, L. (2011). AMPK activation, a preventive therapeutic target in the transition from cardiac injury to heart failure. *Cardiovascular Research*, *90*, 224–233.

Benedetti, F., Davinelli, S., Krishnan, S., Gallo, R. C., Scapagnini, G., Zella, D., et al. (2014). Sulfur compounds block MCP-1 production by Mycoplasma fermentans-infected macrophages through NF-κB inhibition. *Journal of Translational Medicine*, *12*, 1–11.

Benetti, L. R., Campos, D., Gurgueira, S. A., Vercesi, A. E., Guedes, C. E., Santos, K. L., et al. (2013). Hydrogen sulfide inhibits oxidative stress in lungs from allergic mice in vivo. *European Journal of Pharmacology*, *698*, 463–469.

Bucci, M., Papapetropoulos, A., Vellecco, V., Zhou, Z., Zaid, A., Giannogonas, P., et al. (2012). cGMP-dependent protein kinase contributes to hydrogen sulfide-stimulated vasorelaxation. *PLoS One*, *7*, e53319.

Burguera, E. F., Vela-Anero, A., Magalhães, J., Meijide-Faílde, R., & Blanco, F. J. (2014). Effect of hydrogen sulfide sources on inflammation and catabolic markers on interleukin 1β-stimulated human articular chondrocytes. *Osteoarthritis and Cartilage*, *22*, 1026–1035.

Calderwood, A., & Kopriva, S. (2014). Hydrogen sulfide in plants: From dissipation of excess sulfur to signalling molecule. *Nitric Oxide*. http://dx.doi.org/10.1016/j.niox.2014.02.005.

Castro-Piedras, I., & Perez-Zoghbi, J. F. (2013). Hydrogen sulphide inhibits Ca^{2+} release through InsP3 receptors and relaxes airway smooth muscle. *Journal of Physiology*, *591*, 5999–6015.

Chen, J., Wu, F. H., Wang, W. H., Zheng, C. J., Lin, G. H., Dong, X. J., et al. (2011). Hydrogen sulphide enhances photosynthesis through promoting chloroplast biogenesis, photosynthetic enzyme expression, and thiol redox modification in Spinacia oleracea seedlings. *Journal of Experimental Botany*, *62*, 4481–4493.

Chitnis, M. K., Njie-Mbye, Y. F., Opere, C. A., Wood, M. E., Whiteman, M., & Ohia, S. E. (2013). Pharmacological actions of the slow release hydrogen sulfide donor GYY4137 on phenylephrine-induced tone in isolated bovine ciliary artery. *Experimental Eye Research*, *116*, 350–354.

Clarke, P. H. (1953). Hydrogen sulphide production by bacteria. *Microbiology*, *8*, 397–407.

Devarie-Baez, N. O., Bagdon, P. E., Peng, B., Zhao, Y., Park, C. M., & Xian, M. (2013). Light-induced hydrogen sulfide release from "caged" gem-dithiols. *Organic Letters*, *15*, 2786–2789.

Dooley, F. D., Nair, S. P., & Ward, P. D. (2013). Increased growth and germination success in plants following hydrogen sulfide administration. *PLoS One*, *8*, e62048.

Fernandes, V. S., Ribeiro, A. S., Barahona, M. V., Orensanz, L. M., Martínez-Sáenz, A., Recio, P., et al. (2013). Hydrogen sulfide mediated inhibitory neurotransmission to the pig bladder neck: Role of K_{ATP} channels, sensory nerves and calcium signalling. *Journal of Urology*, *190*, 746–756.

Fitzgerald, R., DeSantiagoa, B., Lee, D. Y., Yang, G., Jae Yeon, K., Foster B, D., et al. (2014). H_2S relaxes isolated human airway smooth muscle cells via the sarcolemmal K_{ATP} channel. *Biochemical and Biophysical Research Communications*, *446*, 393–398.

Foster, J. C., Powell, C. R., Radzinski, S. C., & Matson, J. B. (2014). S-aroylthiooximes: A facile route to hydrogen sulfide releasing compounds with structure-dependent release kinetics. *Organic Letters*, *16*, 1558–1561.

Fox, B., Schantz, J. T., Haigh, R., Wood, M. E., Moore, P. K., Viner, N., et al. (2012). Inducible hydrogen sulfide synthesis in chondrocytes and mesenchymal progenitor cells: Is H_2S a novel cytoprotective mediator in the inflamed joint? *Journal of Cellular and Molecular Medicine*, *16*, 896–910.

García-Mata, C., & Lamattina, L. (2010). Hydrogen sulphide, a novel gasotransmitter involved in guard cell signalling. *New Phytologist, 188*, 977–984.
Gatenby, R. A., & Gillies, R. J. (2004). Why do cancers have high aerobic glycolysis? *Nature Reviews Cancer, 4*, 891–899.
Geng, B., Cai, B., Liao, F., Zheng, Y., Zeng, Q., Fan, X., et al. (2013). Increase or decrease hydrogen sulfide exert opposite lipolysis, but reduce global insulin resistance in high fatty diet induced obese mice. *PLoS One, 8*, e73892.
Grambow, E., Mueller-Graf, F., Deliagina, E., Frank, M., Kuhula, A., & Vollmar, B. (2014). Effect of the hydrogen sulfide donor GYY4137 on platelet activation and microvascular thrombus formation in mice. *Platelets, 25*, 166–174.
Greiner, R., Palinkas, Z., Basell, K., Becher, D., Antelmann, H., Nagy, P., et al. (2013). Polysulfides link H_2S to protein thiol oxidation. *Antioxidants & Redox Signaling, 19*, 1749–1765.
Guo, W., Cheng, Z. Y., & Zhu, Y. Z. (2013). Hydrogen sulfide and translational medicine. *Acta Pharmacologica Sinica, 34*, 1284–1291.
Hällgren, J. E., & Fredriksson, S. A. (1982). Emission of hydrogen sulfide from sulfur dioxide-fumigated pine trees. *Plant Physiolology, 70*, 456–459.
Han, Y., Zeng, F., Tan, G., Yang, C., Tang, H., Luo, Y., et al. (2013). Hydrogen sulfide inhibits abnormal proliferation of lymphocytes via AKT/GSK3β signal pathway in systemic lupus erythematosus patients. *Cellular Physiology and Biochemistry, 31*, 795–804.
Hasegawa, U., & van der Vlies, A. J. (2014). Design and Synthesis of Polymeric Hydrogen Sulfide donors. *Bioconjugate Chemistry, 25*, 1290–1300.
Hosoki, R., Matsuki, N., & Kimura, H. (1997). The possible role of hydrogen sulfide as an endogenous smooth muscle relaxant in synergy with nitric oxide. *Biochemical and Biophysical Research Communications, 237*, 527–531.
Ida, T., Sawa, T., Ihara, H., Tsuchiya, Y., Watanabe, Y., Kumagai, Y., et al. (2014). Reactive cysteine persulfides and S-polythiolation regulate oxidative stress and redox signalling. *Proceedings of the National Academy of Sciences of the United States of America, 111*, 7606–7611.
Ilkan, Z., Mustafa, F., Apostoli, G., Solomon, A., Whiteman, M., & Emerson, M. (2013). OP16 Hydrogen sulfide inhibits human platelet aggregation. *Nitric Oxide, 31*(Suppl 2), S26. http://dx.doi.org/10.1016/j.niox.2013.06.046.
Isenberg, J. S., Jia, Y., Field, L., Ridnour, L. A., Sparatore, A., Del Soldato, P., et al. (2007). Modulation of angiogenesis by dithiolethione-modified NSAIDs and valproic acid. *British Journal of Pharmacology, 151*, 63–72.
Julian, D., Statile, J. L., Wohlgemuth, S. E., & Arp, A. J. (2002). Enzymatic hydrogen sulfide production in marine invertebrate tissues. *Comparative Biochemistry and Physiology Part A, Molecular & Integrative Physiology, 133*, 105–115.
Kashfi, K., & Olson, K. R. (2013). Biology and therapeutic potential of hydrogen sulfide and hydrogen sulphide releasing chimeras. *Biochemical Pharmacology, 85*, 689–703.
Kolluru, G. K., Shen, X., Bir, S. C., & Kevil, C. G. (2013). Hydrogen sulfide chemical biology: Pathophysiological roles and detection. *Nitric Oxide, 35*, 5–20.
Kumai, Y., Porteus, C. S., Kwong, R. W., & Perry, S. F. (2014). Hydrogen sulfide inhibits Na + uptake in larval zebrafish, Danio rerio. *Pflügers Archiv/European Journal of Physiology*. http://dx.doi.org/10.1007/s00424-014-1550-y.
Lee, S. S., Kennedy, S., Tolonen, A. C., & Ruvkun, G. (2003). DAF-16 target genes that control *C. elegans* life-span and metabolism. *Science, 300*, 644–647.
Lee, Z. W., Teo, X. Y., Tay, E. Y., Tan, C. H., Hagen, T., Moore, P. K., et al. (2014). Utilizing hydrogen sulfide as a novel anti-cancer agent by targeting cancer glycolysis and pH imbalance. *British Journal of Pharmacology, 171*, 4322–43226.
Lee, Z. W., Zhou, J., Chen, C. S., Zhao, Y., Tan, C. H., Li, L., et al. (2011). The slow-releasing hydrogen sulfide donor, GYY4137, exhibits novel anti-cancer effects in vitro and in vivo. *PLoS One, 6*, e21077.

Lencesova, L., Hudecova, S., Csaderova, L., Markova, J., Soltysova, A., Pastorek, M., et al. (2013). Sulphide signalling potentiates apoptosis through the up-regulation of IP3 receptor types 1 and 2. *Acta Physiologica*, *208*, 350–361.

Le Trionnaire, S., Perry, A., Szczesny, B., Szabo, C., Winyard, P. G., Whatmore, J. L., et al. (2014). The synthesis and functional evaluation of a mitochondria-targeted hydrogen sulfide donor, (10-oxo-10-(4-(3-thioxo-3H-1,2-dithiol-5-yl)phenoxy) decyl)triphenylphosphonium bromide (AP39). *Medicinal Chemistry Communications*, *5*, 728–736.

Le Trionnaire, S., Perry, A., Whatmore, J. L., Wood, M. E., & Whiteman, M. (2013). P50 Mitochondria-targeted hydrogen sulfide donors: A novel twist to an old "tail"? *Nitric Oxide*, *31*(Suppl 2), S57. http://dx.doi.org/10.1016/j.niox.2013.06.112.

Li, L., Fox, B., Keeble, J., Salto-Tellez, M., Winyard, P. G., Wood, M. E., et al. (2013). The complex effects of the slow releasing hydrogen sulfide donor GYY4137 in a model of acute joint inflammation and in human cartilage cells. *Journal of Cellular and Molecular Medicine*, *17*, 365–376.

Li, L., Rose, P., & Moore, P. K. (2011). Hydrogen sulfide and cell signaling. *Annual Review of Pharmacology and Toxicology*, *51*, 169–187.

Li, L., Salto-Tellez, M., Tan, C. H., Whiteman, M., & Moore, P. K. (2009). GYY4137, a novel hydrogen sulphide releasing molecule, protects against endotoxic shock in the rat. *Free Radical Biology and Medicine*, *47*, 103–113.

Li, L., Whiteman, M., Guan, Y. Y., Neo, K. L., Cheng, Y., Lee, S. W., et al. (2008). Characterization of a novel, water-soluble hydrogen sulfide-releasing molecule (GYY4137): New insights into the biology of hydrogen sulphide. *Circulation*, *117*, 2351–2360.

Li, Z. G., Yang, S. Z., Long, W. B., Yang, G. X., & Shen, Z. Z. (2013). Hydrogen sulphide may be a novel downstream signal molecule in nitric oxide-induced heat tolerance of maize (*Zea mays* L.) seedlings. *Plant, Cell and Environment*, *36*, 1564–1572.

Liao, L. X., Tsai, C. Y., Atan, M. S., Lee, Z. W., Deng, L. W., Dymock, B. W., et al. (2013). Identification of a novel slow releasing hydrogen sulfide donor for cancer therapy. *Nitric Oxide*, *31*(Suppl 2), S47. http://dx.doi.org/10.1016/j.niox.2013.06.090.

Lisjak, M., Srivastava, N., Teklic, T., Civale, L., Lewandowski, K., Wilson, I., et al. (2010). A novel hydrogen sulfide donor causes stomatal opening and reduces nitric oxide accumulation. *Plant Physiology and Biochemistry*, *48*, 931–935.

Lisjak, M., Teklic, T., Wilson, I. D., Whiteman, M., & Hancock, J. T. (2013). Hydrogen sulfide: Environmental factor or signalling molecule? *Plant, Cell and Environment*, *36*, 1607–1616.

Lisjak, M., Teklić, T., Wilson, I. D., Wood, M., Whiteman, M., & Hancock, J. T. (2011). Hydrogen sulfide effects on stomatal apertures. *Plant Signaling and Behaviour*, *6*, 1444–1446.

Liu, Z., Han, Y., Li, L., Lu, H., Meng, G., Li, X., et al. (2013). The hydrogen sulfide donor, GYY4137, exhibits anti-atherosclerotic activity in high fat fed apolipoprotein E(−/−) mice. *British Journal of Pharmacology*, *169*, 1795–1809.

Lu, S., Gao, Y., Huang, X., & Wang, X. (2014). GYY4137, a hydrogen sulfide (H_2S) donor, shows potent anti-hepatocellular carcinoma activity through blocking the STAT3 pathway. *International Journal of Oncology*, *44*, 1259–1267.

Madurga, A., Mižíková, I., Ruiz-Camp, J., Vadász, I., Herold, S., Mayer, K., et al. (2014). Systemic hydrogen sulfide administration partially restores normal alveolarization in an experimental animal model of bronchopulmonary dysplasia. *The American Journal of Physiology - Lung Cellular and Molecular Physiology*, *306*, L684–L697.

Markova, J., Hudecova, S., Soltysova, A., Sirova, M., Csaderova, L., Lencesova, L., et al. (2013). Sodium/calcium exchanger is upregulated by sulfide signaling, forms complex with the β1 and β3 but not β2 adrenergic receptors, and induces apoptosis. *Pflügers Archiv European Journal of Physiology*, *466*, 1329–1342.

Martelli, A., Testai, L., Citi, V., Marino, A., Pugliesi, I., Barresi, E., et al. (2013). Arylthioamides as H_2S donors: l-cysteine-activated releasing properties and vascular effects in vitro and in vivo. *ACS Medicinal Chemistry Letters*, *4*, 904–908.

Miller, D. L., & Roth, M. B. (2007). Hydrogen sulfide increases thermotolerance and lifespan in *Caenorhabditis elegans*. *Proceedings of the National Academy of Sciences of the United States of America*, *104*, 20618–20622.

Miranda, K. M., & Wink, D. A. (2014). Persulfides and the cellular thiol landscape. *Proceedings of the National Academy of Sciences of the United States of America*, *111*, 7505–7506.

Nicholls, P., Marshall, D. C., Cooper, C. E., & Wilson, M. T. (2013). Sulfide inhibition of and metabolism by cytochrome c oxidase. *Biochemical Society Transactions*, *41*, 1312–1316.

Ning, N., Zhu, J., Du, Y., Gao, X., Liu, C., & Li, J. (2014). Dysregulation of hydrogen sulphide metabolism impairs oviductal transport of embryos. *Nature Communications*. http://dx.doi.org/10.1038/ncomms5107.

Olas, B. (2014). Hydrogen sulfide in hemostasis: Friend or foe? *Chemico-Biological Interactions*, *217*, 49–56.

Olson, K. R., Healy, M. J., Qin, Z., Skovgaard, N., Vulesevic, B., Duff, D. W., et al. (2008). Hydrogen sulfide as an oxygen sensor in trout gill chemoreceptors. *The American Journal of Physiology - Regulatory, Integrative and Comparative Physiology*, *295*, R669–R680.

Papenbrock, J., Riemenschneider, A., Kamp, A., Schulz-Vogt, H. N., & Schmidt, A. (2007). Characterization of cysteine degrading and H_2S releasing enzymes of higher plants - from the field to the test tube and back. *Plant Biology*, *9*, 582–588.

Patel, P., Vatish, M., Heptinstall, J., Wang, R., & Carson, R. J. (2009). The endogenous production of hydrogen sulphide in intrauterine tissues. *Reproductive Biology and Endocrinology*, *7*, 1–9.

Perry, M. M., Hui, C. K., Whiteman, M., Wood, M. E., Adcock, I., Kirkham, P., et al. (2011). Hydrogen sulfide inhibits proliferation and release of IL-8 from human airway smooth muscle cells. *The American Journal of Respiratory Cell and Molecular Biology*, *45*, 746–752.

Predmore, B. L., Alendy, M. J., Ahmed, K. I., Leeuwenburgh, C., & Julian, D. (2010). The hydrogen sulfide signaling system: Changes during aging and the benefits of caloric restriction. *Age*, *32*, 467–481.

Qabazard, B., Ahmed, S., Li, L., Arlt, V. M., Moore, P. K., Stürzenbaum, S. R., et al. (2013). *C. elegans* aging is modulated by hydrogen sulfide and the sulfhydrylase/cysteine synthase cysl-2. *PLoS One*, *8*, e80135.

Qabazard, B., Li, L., Gruber, J., Peh, M. T., Ng, L. F., Kumar, S. D., et al. (2014). Hydrogen sulfide is an endogenous regulator of aging in *Caenorhabditis elegans*. *Antioxidant & Redox Signaling*, *20*, 2621–2630.

Ravilious, G. E., & Jez, J. M. (2012). Structural biology of plant sulfur metabolism: From assimilation to biosynthesis. *Natural Products Reports*, *29*, 1138–1152.

Robinson, H., & Wray, S. (2012). A New Slow Releasing, H_2S Generating Compound, GYY4137 Relaxes spontaneous and oxytocin-stimulated contractions of human and rat pregnant myometrium. *PLoS One*, *7*, 1–12.

Rodrigues, L., Santos, B. C., Florenzano, J., Dos Santos, K. T., Lopes, K. C., Lira, F. B., et al. (2013). P44 Does the slow-releasing hydrogen sulfide donor GYY4137 controls inflammation and pruritogen responses in mouse skin? *Nitric Oxide*, *31*(Suppl 2), S54–S55. http://dx.doi.org/10.1016/j.niox.2013.06.106.

Romero, L. C., Aroca, M. Á., Laureano-Marín, A. M., Moreno, I., García, I., & Gotor, C. (2014). Cysteine and cysteine-related signaling pathways in *Arabidopsis thaliana*. *Molecular Plant*, *7*, 264–276.

Sekiya, J., Schmidt, A., Wilson, L. G., & Filner, P. (1982). Emission of hydrogen sulfide by leaf tissue in response to. L-cysteine. *Plant Physiology*, *70*, 430–436.

Shen, X., Peter, E. A., Bir, S., Wang, R., & Kevil, C. G. (2012). Analytical measurement of discrete hydrogen sulfide pools in biological specimens. *Free Radical Biology & Medicine*, *52*, 2276–2283.

Song, Z. J., Ng, M. Y., Lee, Z.-W., Dai, W., Hagen, T., Moore, P. K., et al. (2014). Hydrogen sulfide donors in research and drug development. *Medicinal Chemistry Communications*, *5*, 557–570.

Sparatore, A., Perrino, E., Tazzari, V., Giustarini, D., Rossi, R., Rossoni, G., et al. (2009). Pharmacological profile of a novel $H(_2)S$-releasing aspirin. *Free Radical Biology & Medicine*, *46*, 586–592.

Takahashi, H., Kopriva, S., Giordano, M., Saito, K., & Hell, R. (2011). Sulfur assimilation in photosynthetic organisms: Molecular functions and regulations of transporters and assimilatory enzymes. *Annual Review of Plant Biology*, *62*, 157–184.

Thompson, C. R., & Kats, G. (1978). Effects of continuous hydrogen sulfide fumigation on crop and forest plants. *Environmental Science & Technology*, *12*, 550–553.

Thompson, C. R., Kats, G., & Lennox, R. W. (1979). Effects of fumigating crops with hydrogen sulfide or sulphur dioxide. *Californian Agriculture*, *33*, 9–10.

Wang, R. (2012). Physiological implications of hydrogen sulfide: A whiff exploration that blossomed. *Physiological Reviews*, *92*, 791–896.

Wang, K., Ahmad, S., Cai, M., Rennie, J., Fujisawa, T., Crispi, F., et al. (2013). Dysregulation of hydrogen sulfide producing enzyme cystathionine γ-lyase contributes to maternal hypertension and placental abnormalities in preeclampsia. *Circulation*, *127*, 2514–2522.

Wang, C. N., Liu, Y. J., Duan, G. L., Zhao, W., Li, X. H., Zhu, X. Y., et al. (2014). CBS and CSE Are Critical for Maintenance of Mitochondrial Function and Glucocorticoid Production in Adrenal Cortex. *Antioxidants & Redox Signaling*. http://dx.doi.org/10.1089/ars.2013.5682.

Wang, K., Peng, H., Ni, N., Dai, C., & Wang, B. (2014). 2,6-dansyl azide as a fluorescent probe for hydrogen sulphide. *Journal of Fluorescence*, *24*, 1–5.

Wei, W. B., Hu, X., Zhuang, X. D., Liao, L. Z., & Li, W. D. (2014). GYY4137, a novel hydrogen sulfide-releasing molecule, likely protects against high glucose-induced cytotoxicity by activation of the AMPK/mTOR signal pathway in H9c2 cells. *Molecular Cell Biochemistry*, *389*, 249–256.

Whiteman, M., Armstrong, J. S., Chu, S. H., Jia-Ling, S., Wong, B. S., Cheung, N. S., et al. (2004). The novel neuromodulator hydrogen sulfide: An endogenous peroxynitrite 'scavenger'? *Journal of Neurochemistry*, *90*, 765–768.

Whiteman, M., Li, L., Rose, P., Tan, C. H., Parkinson, D. B., & Moore, P. K. (2010). The effect of hydrogen sulfide donors on lipopolysaccharide-induced formation of inflammatory mediators in macrophages. *Antioxidants & Redox Signaling*, *12*, 1147–1154.

Wilson, L. G., Bressan, R. A., & Filner, P. (1978). Light-dependent Emission of Hydrogen Sulfide from Plants. *Plant Physiology*, *61*, 184–189.

Yong, Q. C., Cheong, J. L., Hua, F., Deng, L. W., Khoo, Y. M., Lee, H. S., et al. (2011). Regulation of heart function by endogenous gaseous mediators-crosstalk between nitric oxide and hydrogen sulfide. *Antioxidant & Redox Signaling*, *14*, 2081–2091.

You, X. J., Xu, C., Lu, J. Q., Zhu, X. Y., Gao, L., Cui, X. R., et al. (2011). Expression of cystathionine β-synthase and cystathionine γ-lyase in human pregnant myometrium and their roles in the control of uterine contractility. *PLoS One*, *6*, e23788.
Zhang, Y., Tang, Z. H., Ren, Z., Qu, S. L., Liu, M. H., Liu, L. S., et al. (2013). Hydrogen sulfide, the next potent preventive and therapeutic agent in aging and age-associated diseases. *Molecular and Cellular Biology*, *33*, 1104–1113.

CHAPTER TEN

Neuroprotective Effects of Hydrogen Sulfide in Parkinson's Disease Animal Models: Methods and Protocols

Xue Xue, Jin-Song Bian[1]
Department of Pharmacology, Yong Loo Lin School of Medicine, National University of Singapore, Singapore
[1]Corresponding author: e-mail address: phcbjs@nus.edu.sg

Contents

Abstract

Hydrogen sulfide (H_2S) was first proposed to be a neuromodulator in 1996. After that, multiple biological functions of H_2S have been revealed. In brain, it regulates intracellular calcium, intracellular pH, and cAMP level in different brain cells via regulation of the functions of different proteins. In pathological situations, H_2S produces

Methods in Enzymology, Volume 554
ISSN 0076-6879
http://dx.doi.org/10.1016/bs.mie.2014.11.015

anti-inflammatory, -oxidant, and -apoptotic effects, and therefore is potentially used to treat neurodegenerative diseases, especially Parkinson's disease (PD). In this chapter, we summarized the methods commonly used to create PD animal models followed by description of evaluations of PD pathology. The PD models described in this chapter included those caused by various neurotoxins like 6-hydroxydopamine, rotenone, and 1-methyl-4-phenyl-1,2,3,6-tetrahydropyridine. Methods for examination of dopaminergic neuron loss and microglial activation in both substantia nigra and striatum are also described. The role of H_2S and its therapeutic potentials are discussed in the last section of this chapter.

1. INTRODUCTION

Parkinson's disease (PD) is the second most common neurodegenerative disease after Alzheimer's disease and is essentially a disease of an aging population. Currently, about 1% of the population at the age of 55 is affected and the prevalence may increase to about 5% by the age of 85 (Andrew Siderowf & Matthew Stern, 2003; Schober, 2004). Prominent clinical features of PD include motor symptoms (e.g., tremor, bradykinesia, postural instability, etc.,) and nonmotor-related symptoms (e.g., autonomic dysfunction, olfactory deficits, cognitive deficitse, etc.,). The key neuropathology involves the loss of dopaminergic neurons and the formation of Lewy bodies (α-synuclein) mainly in the substantia nigra of the midbrain (Braak et al., 2003). The Parkinson symptoms turn to be obvious when 50–60% of dopaminergic neurons are lost in the substantia nigra pars compacta (SNpc; Schwab & Bartholdi, 1996). Apart from loss of dopaminergic neurons, microglial activation also occurs in PD (Imamura et al., 2003; Langston et al., 1999).

Unfortunately, effective drugs to treat PD are still limited. Hydrogen sulfide (H_2S) is a new gasotransmitter alongside nitric oxide and carbon oxide. Recent studies revealed that H_2S may produce anti-inflammatory, -oxidant, and -apoptotic effects in brain and is demonstrated to be useful to treat PD. This chapter summarizes the methods used to study the neuroprotective and therapeutic effects of H_2S in PD. The methods described in this chapter are commonly used previously in our and other laboratories.

2. PD ANIMAL MODELS

The etiologies of most common forms of PD remain poorly understood. Environmental factors, or a combination of both environment and genetic susceptibility, have been proposed to play a role in sporadic PD

(Dauer & Przedborski, 2003). The use of toxin-induced animal models is able to replicate most pathological and phenotypic features of PD. These models are also often used to screen drugs to treat PD. The neurotoxins commonly used include 6-hydroxydopamine hydrobromide (6-OHDA), rotenone, and 1-methyl-4-phenyl-1,2,3,6-tetrahydropyridine (MPTP) (Fig. 1). Stereotaxic surgery is needed for 6-OHDA injection and this will cause unilateral dopaminergic neuron degeneration. Being systemically active, both MPTP and rotenone do not require the skilled stereotaxic injection technique. Moreover, the systemic injection produces a bilateral neurodegeneration of the nigrostriatal tract. This may reflect more pathological process occurred in PD. We will describe PD models induced by three above neurotoxins first. All animal experiments described in this chapter are subject to be approved according to Institutional Animal Care and Use Committee of individual institutes.

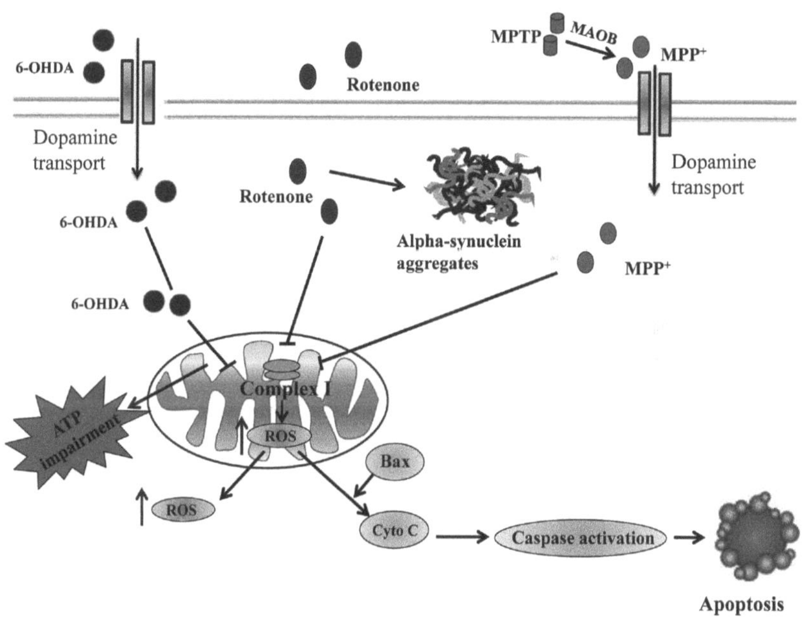

Figure 1 Schematic illustration showing the mechanisms for different neurotoxinsinduced PD models. All neurotoxins-induced neuronal damages are mainly caused by the massive oxidative stress. Rotenone is highly lipophilic and readily enters all cells, where it blocks complex I activity, causing massive formation of reactive ROS. Different from Rotenone, both 6-OHDA and MPP^+ have a similar affinity as dopamine to the dopamine transporter (DPT) and are therefore uptaken into dopaminergic neurons to generate ROS such as superoxide radicals. This explains why and how these two neurotoxins selectively kill dopaminergic neurons. (See the color plate.)

2.1. 6-OHDA-induced PD rat model

The first ever generated PD animal model is 6-OHDA model (Blandini, Armentero, & Martignoni, 2008). This prototypical model is based on local (i.e., intracerebral) injection of a neurotoxin, 6-OHDA. It is widely used to produce lesions in the nigrostriatal dopaminergic system of rodents. 6-OHDA does not efficiently cross the blood–brain barrier, which necessitates its direct injection into the SNpc or the striatum. This model procedure evolves over a period of 4–6 weeks. This may mimic the gradual evolution of the neurodegenerative process of human PD (Ungerstedt, 1968).

Following its injection, 6-OHDA is uptaken into the dopaminergic neurons via the dopamine transporter and gain access to the cytosol where it is auto-oxidized, hence generating an intracellular oxidative stress and mitochondrial respiratory dysfunction (Lee, Sauer, & Bjorklund, 1996). Since 6-OHDA has high affinity to the dopamine transporter, it selectively kills dopaminergic neurons by generating reactive oxygen species (ROS). 6-OHDA is often given unilateral, as the mortality rate of the bilateral injection of this compound into the striatum is higher. The method described below is adopted from Stephen D. Skaper and Giulia Mercanti (Mercanti, Bazzu, & Giusti, 2012) work with modification.

Male Sprague–Dawley (SD) male rats (180–220 g) are often used. The animals need to be housed in standard environmental conditions (22 ± 1 °C, humidity $60 \pm 5\%$, 12 h light:12 h dark cycle) and are allowed to freely access to water and food.

6-OHDA (H116, Sigma) needs to be prepared carefully as it is oxidized rapidly when exposed to light at room temperature. 8 mg of 6-OHDA–HCl is usually dissolved in 4 ml of sterile 0.02% ascorbic acid saline solution (Sigma–Aldrich). The final concentration of 6-OHDA is 2 μg/μl. It can be aliquoted to 6 μl/Eppendorf® tubes and stored in −80 °C freezer. All Eppendorf tubes must be covered with aluminum foil to avoid exposure to light. Repeated thawing/freezing should be avoided. If the solution turns to pinkish color, it suggests that 6-OHDA is oxidized and new fresh 6-OHDA solution should be prepared.

Before surgery, a10 μl Hamilton syringe (0.46 mm in diameter from Hamilton Bonaduz AG, Bonaduz, GR, Switzerland) is needed. It can connect to Harvard compact infusion syringe pump (American Laboratory Trading, Inc. East Lyme, CT, USA). The syringe and needle must be filled with saline solution before the experiment. The needle is placed on the Stereotaxic apparatus (Stoelting Instruments, Wood Dale, IL, USA) and make sure that it is perfectly vertical on the left site. All the surgery instruments need to be autoclaved at 121.6 °C for 20 min.

The heating pad (two Biological Instruments, Besozzo, Italy) should be turned on and be kept warm at 37 °C before anesthesia. Once ready, animals are anesthetized with Ketamine 75 mg/kg + Xylazine 10 mg/kg intraperitoneal (ketamine hydrochloride/xylazine hydrochloride solution (Sigma–Aldrich)). The volume of drug solution is about 1 μl/g body weight. Animals should be put in a cage with bedding until they are fully anesthetized and unconscious. The surgeons should wash hands thoroughly with a bactericidal scrub and wear a clean lab coat, sterile surgical gloves, and a surgical mask.

After the rats are fully anesthetized (negative toe pinch), they are placed in the stereotaxic frame on a flat-skull position. Two ear bars are pushed into the ears to fix the skull. The teeth of animal are fixed on the tooth bar to prevent head movement vertically. This procedure is sophisticated and essential, as it helps to keep the bregma and lambda at the same horizontal level.

The surgical area is disinfected with 70% alcohol. A contact time of at least 15 min should be allowed. The rat's hair in the surgical area should be shaved with an electrical shaver and scrubbed with a scrub of 70% alcohol cotton pad. The surgical area should be draped with sterile drapes. A midline incision on the skin of about 1 cm in length will be performed between two eyes. The connective tissue above the bone should be removed so that the bregma can be easily located. After exposure of the skull, a burr hole is drilled to introduce a syringe for a single injection of the 6-OHDA solution. The position of left striatum is: AP = + 3.0 mm; ML = + 1.0 mm; DV = − 4.5 mm (4 μl). The needle is put right on the bregma and then retracted and the location of injection can be calculated by subtracting the coordinates for left striatum.

A hole can be drilled with a drill (Dremel 300 Series drill, Dremel Italia) in the skull bone above the injection coordinate. Special care is needed to avoid rupturing the meninges during drilling. A small hole can be made in the meninges with an injection needle. After rinse with saline containing 0.02% ascorbic acid, the syringe is filled with the above 6-OHDA solution. The Hamilton syringe is placed on the pump, and connected with the needle. After lowering the injection needle down to the desired position, the desired volume of 6-OHDA solution is given at a maximum rate of 0.5 μl/min. Sham-lesioned rats are infused with 4 μl saline containing 0.02% ascorbic acid into the left striatum and served as controls. The needle is left in the place for 3 min before slowly retracting back.

The syringe is then cleaned by rinse with ethanol and saline containing 0.02% ascorbic acid.

After surgery, the wound area is rehydrated with sterile water and closed by suturing. 2 ml of saline is injected to prevent dehydration during recovery from anesthesia. For ethical reasons, analgesic treatment should be administered to the animals before they regain consciousness following anesthesia. All surgical tools and the table should be disinfected. The postoperative analgesia consists of the following mixture: Glucose solution (glucose 5%): 3 ml intraperitoneal injection; Antibiotic (ceftriaxone, Rocefin® 1, Roche): 100 μl subcutaneously. One animal should be placed in one clean cage to avoid they bite each other after regains consciousness and recovers.

During surgery, the body temperature of the rat is often decreased. To prevent the temperature loss, a pad needs to be placed underneath the animal's body. When surgery is finished, the rats are sent back into their cages and covered with cloth. On the days following surgery, it is necessary to check whether the animals recover to their normal activities, the wound healing situation, and the body weight.

To test whether the model is successfully established, apomorphine (Tocris, USA) is used. It is dissolved in sterile 0.02% ascorbic acid saline solution at a concentration of 0.5 mg/ml. It is aliquoted to 0.4 ml/Eppendorf tube and stored in 4 °C. Apomorphine is rapidly oxidized when exposed to light at room temperature, so it needs to be prepared in a dark environment and covered with aluminum foil.

At the end of the third week after the lesion, 6-OHDA treated rats are placed in rotometer bowls and secured to count rotation. Once they are acclimated for at least 10 min, apomorphine (0.5 mg/kg) is given subcutaneously at the back of neck. After 10 min, the number of net rotations (360° contralateral turns) is continuously recorded for 30 min. Only those animals showing at least seven turns per min in both tests are considered to be successful model creation (Fig. 2). Repeat the Contralateral rotations again at the end of the fourth week.

2.2. Rotenone-induced PD rat model

Rotenone is an insecticide and piscicide extracted from *Leguminosa* plants. It is highly lipophilic and readily crosses the blood–brain barrier and enters all cells freely independent of a specific transporter. Once in the cell, rotenone blocks complex I activity, causing massive formation of reactive ROS (Pan-Montojo et al., 2010), and inhibits proteasome activity, thereby generating proteolytic stress. Rotenone can replicate most the hallmarks of PD.

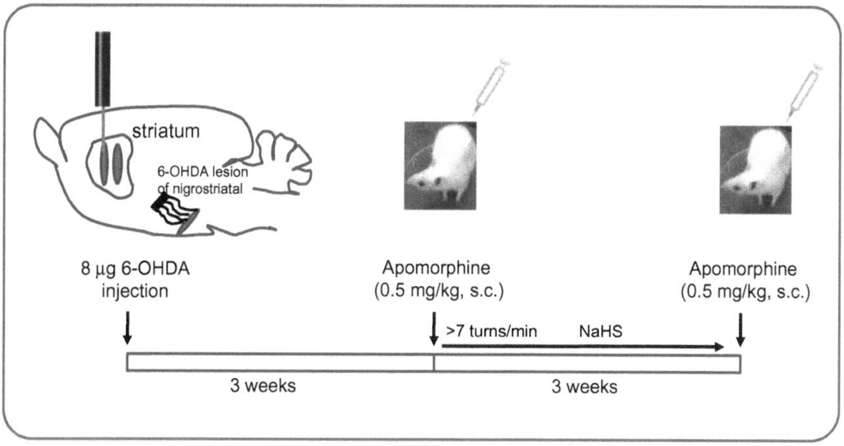

Figure 2 A diagram showing 6-OHDA-induced rat PD model. Unilateral injection of 6-OHDA to the left striatum is performed 3 weeks before behavioral test. The tendency of animals to rotate is induced by application of apomorphine (0.5 mg/kg, s.c.). Contralateral rotations are measured 2 in 2 weeks time. Only those rats showing greater than seven turns per min/30 min can be considered to be successful. After that, rats are treated with H2S. The treatment period lasts for 3 weeks. After H2S/vehicle treatment, contralateral rotations are then measured.

These may include behavioral alterations, complex I blockade, synuclein aggregation, inflammation, and oxidative stress.

Male SD or wistar rats weighing 200–230 g are often used. The animals are housed in groups of two in cages with bedding. They are maintained in a temperature-controlled room (22 ± 2 °C) on a 12-h light–dark cycle. Rats are allowed to freely access to water and food throughout the experiment. Dimethylsulfoxide (DMSO)/ polyethylene glycol (1:1) is used to dissolve rotenone (Sigma). Rats are injected with rotenone (2.5 mg/kg) subcutaneously. It is given t.i.d. daily at 8 AM, 2 PM, and 8 PM for 4 weeks. The injection volume is no more than 0.1 ml/100 g body weight. All rats are tested twice, before and after 4 weeks NaHS/vehicle treatment.

2.3. MPTP-induced subacute PD mice model

MPTP is selective nigral toxic. It brings a new impetus to animal models of PD. Like rotenone, MPTP is a lipophilic protoxin that rapidly crosses the blood–brain barrier following systemic injection (usually i.p.). Once inside the brain, MPTP is taken up by the astrocytes and is converted to the intermediary, 1-methyl-4-phenyl-2, 3, dihydropyridinium ($MPDP^+$) by monoamine oxidase-B (MAO-B). $MPDP^+$ is then rapidly and spontaneously oxidized to the toxic moiety, 1-methyl-4-phenylpyridinium (MPP^+;

Chiba, Trevor, & Castagnoli, 1984). After that, MPP^+ is then uptaken via dopamine transporter into dopaminergic neurons. Cytoplasmic MPP^+ in dopaminergic neurons inhibits complex I of the mitochondrial electron transport chain and results in ATP depletion and increased oxidative stress. For this reason, MPP^+ also selectively kills dopaminergic neurons.

Precautionary measures should be taken to ensure safe handling, detoxification, and disposal. MAO inhibitors may be utilized in case of accidental human intoxication. Prior to MPTP injection, appropriate personal protective equipment (PPE) should be used. It includes disposable lab coat, double nitrile gloves, surgical mask, hair cover, and so on. It is highly recommended that all MPTP experiments should be performed in a dedicated procedure room.

Eight- to ten-week-old male C57BL/6 mice weighing 22–30 g are often used. They are housed of five animals per cage. The animals are allowed to freely access food pellets and water. The room is maintained on a 12-h light–dark cycle at constant temperature and humidity. Probenecid is often used to inhibit the rapid clearance and excretion of MPTP from the brain and kidney after each injection. Probenecid is dissolved in dimethyl sulfoxide (DMSO) at 250 mg/ml just 30 min before injection. MPTP (20 mg/kg) is injected subcutaneously 1 h before injection of probenecid (250 mg/kg, i.p.) for 5 days.

Decontamination of work surface should be performed with bleach after each experiment. Leftover MPTP solution should be disposed by addition of an equivalent volume of 10% bleach, and the contents should be discarded as biohazardous liquid waste. All used PPE should be sprayed with 10% bleach before discarding into a labeled bin.

2.3.1 Postoperative care

Animals received MPTP treatment will be housed in open disposable cages. Since mice experience transient hypothermia up to 12 h after MPTP injection, the room temperature should be between 22.2 and 24.4 °C. Both the rooms and cages with MPTP-treated mice should be clearly labeled. Animals are sacrificed 3.5 days after the last injection of MPTP. All cages and liners should be disposed in appropriately labeled biohazard bags.

3. H_2S AND ITS RELEASING COMPOUND TREATMENT

3.1. NaHS

The successful model rats tested by apomorphine above are further treated with H_2S releasing compounds. Rats are grouping into three groups:

6-OHDA model saline vehicle group, sham-operated saline vehicle group, and NaHS (i.p.) treatment groups. At the end of treatment with either saline or NaHS treatment for 3 weeks, apomorphine at 0.5 mg/kg is given subcutaneously to induce the rotational behavior.

For rotenone model, NaHS is injected with NaHS (i.p.) for three consecutive days prior to rotenone administration. From day 4 onward, NaHS is given 1 h before the rotenone injection on a daily basis for at least 4 weeks.

For MPTP model, there are three groups: vehicle and MPTP model mice treated with or without NaHS. NaHS is given (i.p.) for 7 days and MPTP is administrated (i.p.) 30 min after application of NaHS on day 3 to day 7. Saline is used as vehicle. Animals are sacrificed 3.5 days after the last injection of MPTP.

3.2. ACS84

4-(3-thioxo-3*H*-1,2-dithiol-4-yl)-benzoic acid (ACS84) is a hybrid compound derived from Levodopa (L-Dopa) methyl ester and ACS50 (a H_2S-releasing moiety). This compound can penetrate blood–brain barrier and release H_2S in cells. It is synthesized by Sparatore's group (Lee et al., 2010).

The successful 6-OHDA model rats tested by apomorphine above are used to test the therapeutic effect of ASC84 compounds. These animals are divided into two groups: vehicle treatment group, ACS84 administered intravenously (40 mg/kg) group. An additional sham-operated rats received vehicle treatment is also needed. Animals are treated with ACS84 for three more weeks. The rotational behavior is recorded once a week until the end of treatment. After 3 weeks of treatment, the rats are used for behavioral and biochemical experiments. The behavioral tests are usually performed before drug treatment to avoid any disturbance.

4. BEHAVIOR TESTS

4.1. Rotational behavior test

After treatment with H_2S for 3 weeks, apomorphine at 0.5 mg/kg (s.c.) is used again to test the rotational behavior. The number of net rotations (360° contralateral turns) is continuously recorded for 30 min after 10 min injection of apomorphine. The therapeutic effect of H_2S can be tested by whether it can reduce the number of turns.

4.2. Locomotor activity test

Locomotor activity is detected by measuring animal travel distance in an open-field arena. Which is thoroughly cleaned with a 75% alcohol solution before each test. The movement activity of each animal is automatically recorded in the arena and detected with a computer-based video tracking system during a 5-min test session. Traveling distance information is processed and analyzed online with EthoVision software (Noldus, Wageningen, The Netherlands). Animals are often tested at two time points, before and after 4 weeks treatment with H_2S.

4.3. Rearing activity test

Rearing activity is to record the number of times the animals stood on their hind paws. All animals are tested and videotaped. Animals are usually placed in a clear Plexiglas cylinder (17.6 cm inside diameter, 34 cm height). The typical rearing is that the animals place their forelimbs along the wall to show a kind of exploratory behavior. Because rotenone affects motor behavior bilaterally, activity is measured by counting the number of rears made by each animal in a 5-min period without recording specific limb use. The number of rears in a 5-min period is counted as rearing activity. Videotape recordings are analyzed by an experimenter blind to experimental condition.

5. IMMUNOHISTOCHEMICAL ASSAY

After the experiments, immunohistochemistry is often performed to demonstrate the loss of dopaminergic neurons, glial activation, and the expression of H_2S generating enzymes.

The method described below is adopted from Jae-Joon Park (Park & Cunningham, 2007) and Gregory J. Gage (Gage, Kipke, & Shain, 2012) with modification. The animals are anesthetized with Ketamine 75 mg/kg + Xylazine 10 mg/kg intraperitoneal. The toe pinch-response method is used to test whether the animals are deeply anesthetized. After animals are unresponsive, the chest is opened and the heart is exposed. A 15-G blunt perfusion needle is inserted into the ascending aorta through the cut ventricle. The tip should not reach the aortic arch where the brachial and carotid arteries diverge. A clamp is needed to secure the needle and prevent leakage. An incision is made to the animal's right atrium to create an outlet. After perfusion with saline till the liver turning to pale, 4% paraformaldehyde (50 ml) is given. Fixation tremors are usually observed within seconds.

The brain tissues need to be carefully collected. The olfactory bulbs and nervous connections are cut off with a scalpel blade. The brain is gently peeling away from the head. The brain is placed in a tube of fixative solution. The volume of this solution is at least 10 × of the that of the brain itself. It is kept in 4% paraformaldehyde for 24 h at 4 °C. After that, the brain is washed with phosphate-buffered saline (PBS) by exchanging the media. They are transferred to graded sucrose solutions (20% sucrose in PBS overnight and then to 30% sucrose overnight till the brain sunk to the bottom of the tube). Tissues are cryopreserved 30% glycerol in PBS to remove excess sucrose solution, frozen on dry ice, and stored at −80 °C until use. Tissue samples are placed in optimum cutting temperature compound (OCT, Sakura Finetek, Torrance, CA, USA). Freezing column is prepared with outer ring projecting about 5 mm above top of column forming well for OCT. Tissues and OCT are completely frozen within 1 min onto center of freezing column surface with crushed dry ice. Coronal sections are collected at 30 μm thickness and washed twice in 0.01 *M* PBS (pH 7.4). The sections are stored in freeze-stored liquid in 1.5 ml microcentrifuge tubes at −20 °C.

5.1. Tyrosine-hydroxylase positive neurons

Tyrosine-hydroxylase positive (TH^+) neuron staining is a very important parameter to show the neuroprotective effect of H_2S in PD. The sectioned midbrain slides are placed in polystyrene inserts with polyester mesh bottoms for free-floating immunochemistry. After rinse in PBS buffer for 3–5 ×, sections are transferred to 96-well plate with 200 μl 3% H_2O_2 to quench endogenous peroxidases and incubated for 15 min in the dark. After rinse 3 × in 0.01 *M* PBS, samples are treated with permeabilized solution (PBS + 0.1% Triton X-100) for 30 min and blocked with 10% goat serum in PBS for another 30 min. All steps are performed with light agitation on shaker. Sections are then incubated with primary antibody (200 μl/well, mouse monoclonal anti-TH, 1:2000, Sigma) diluted in PBS with 1% BSA and 0.3% TX-100, for 24 h at 4 °C. IgG from the primary species is often used as the negative immunostaining control. Samples are then washed in PBS buffer for 3 × (5 min/time) at room temperature. Sections are then incubated in biotinylated secondary antibodies against mouse IgG (horse polyclonal, Vector Laboratories) at room temperature for 1 h followed by rinse in PBS for 3× (5 min/time). Samples are then incubated in the ABC-complex (1:50 for solutions A and B, Vector Laboratories) for 3×. After washing in 0.1 *M* PBS for 3×, the TH^+ staining can be detected with 0.02% DAB (Sigma) for 8 min at room temperature.

5.2. Glia activation

Apart from loss of dopaminergic neurons, microglial activation (MAC-1 (CD11b)) and an increase in astroglia activation (glial fibrillary acidic protein, GFAP) also occur in PD. An increase in glial cells in postmortem tissue from the brains of PD patients has been reported (Damier, Hirsch, Zhang, Agid, & Javoy-Agid, 1993). Similar method as the determination of TH^+ can be used except that different primary antibodies are employed. Sections are incubated with primary antibody GFAP (mouse monoclonal, 1:400, from Sigma) and Mac-1 (mouse anti rat monoclonal antibody, 1:50 dilution, CD11b; AbD Serotec, Oxford, UK) which diluted in PBS with 1% BSA for 24 h (48 h for MAC-1) at 4 °C. IgG from the primary species is often used as the negative immunostaining control. All other steps are similar as above in Section 4.1.

6. BRAIN H_2S ACTIVITY TESTS

6.1. CSE, CBS, and 3-MST expression

H_2S can be formed from cysteine by pyridoxal-5′-phosphate (PLP)-dependent enzymes, including cystathionine β-synthase (CBS), cystathionine γ-lyase (CSE), and PLP-independent enzyme, 3-mercaptopyruvate sulfurtransferase (3-MST) in combination with cysteine aminotransferase (CAT; Enokido et al., 2005). The synthesis of H_2S is regulated by alterations in the expressions and activities of these enzymes. The up- or downregulation of protein expression can be determined with immunohistochemistry or Western blots in the brain tissues of different animal models. Similar methods mentioned above will be used except the primary antibodies. Anti-CSE (rabbit polyclonal antibody, 1:3000, Protein Tech, Manchester, UK), anti-CBS (rabbit polyclonal antibody, 1:5000, Ab-nova, Taipei, Taiwan), anti-CAT (rabbit polyclonal antibody, 1:2000,

Novus Biologicals, Littleton, CO, USA), anti-3-MST (rabbit polyclonal antibody, 1:200, Sigma–Aldrich) are used.

To study the colocalization of these enzymes in different cells, tissues are incubated with one of the above H_2S synthesis antibody and one of the following three antibodies, a postmitotic neuronal nuclei marker: anti-NeuN (mouse monoclonal, 1:200, Millipore, MAB377), an astrocyte marker: anti-GFAP (mouse monoclonal,1:200, Sigma–Aldrich), or a microglial cell marker: anti-MAC-1(CD11b, mouse monoclonal,1:200, AbD Serotec). After incubation for 24–48 h and rinse for 5 min for 3 ×. The slides are then incubated with FITC conjugated goat anti-rabbit IgG used against CBS, CSE, 3-MST antibody, and Cy3 conjugated goat anti-mouse IgG used against anti-NeuN, anti-GFAP, or anti-MAC-1 for 1 h for fluorescent labeling. Sections are then incubated in 4,6-diamidino-2-phenylindole dihydrochloride (DAPI, Sigma–Aldrich, 0.5 μg/ml in PBS) for 5 min, mounted and observed under a confocal microscope (Olympus, Tokyo, Japan).

6.2. Brain H_2S generating enzyme activity test

6.2.1 Brain tissue H_2S-producing capacity

The brain tissue H_2S production rate is measured according to our previous publication (Webb et al., 2008; Yong et al., 2008). Brain tissues are homogenized in potassium phosphate buffer (100 m*M*, pH 7.4). The homogenates (460 μl) are then treated and incubated with 10 m*M*L-cysteine (20 μl), 2 m*M* pyridoxal 5′-phosphate (20 μl) in the tightly sealed Eppendorf vials for 30 min at 37 °C. 250 μl Zinc acetate (1% wt/vol) is added to trap the generated H_2S and trichloroacetic acid (10% wt/vol, 250 μl) is used to stop the reaction. After this treatment, equal amount (133 μl each) of *N*,*N*-dimethyl-*p*-phenylenediamine sulfate (20 m*M*,) and FeCl3 (30 m*M*) are added into the reaction solution followed by centrifugation at 14,000 g for 10 min. After that, supernatants are collected and the absorbance at 670 n*M* of 200 μl of the supernatant is determined using a 96-well microplate reader. The H_2S concentration of samples is calculated using a calibration curve of NaHS (3.125–200 μ*M*), and the results are expressed as nanomoles of H_2S produced per milligram of soluble protein per 30 min. Protein is determined with NanoDrop Spectrophotometer (ND-1000, NanoDrop technology) or other commercial available kits.

Before the discovery of 3-MST, CBS was thought to be the main enzyme to produce H_2S in brain (Shibuya et al., 2009). To study the activity of CBS, aminooxyacetic acid (AOAA), a CBS inhibitor, can be given together with L-cysteine as mentioned above. The decreased H_2S generation capacity caused by AOAA can be defined as the CBS activity. Kimura's group recently found another important enzyme, namely 3-MST, is also able to produce H_2S in brain (Shibuya et al., 2009), as they found that brain homogenates from CBS-knockout mice are still able to produce H_2S. To study the activity of 3-MST, brain homogenate (300 μl, 14.3% w/v) is mixed with 3-mercaptopyruvate (2 m*M*, sodium salt, Sigma–Aldrich) and 2-mercaptoethnol (10 m*M*, Sigma–Aldrich) in 50 m*M* potassium phosphate buffer (pH 6.8). 2-Ketobutyric acid (40 m*M*), an uncompetitive inhibitor of 3-MST, can be given in the reaction tube in an ice bath. The difference between treatment with and without 2-ketobutyric acid is considered to be the activity of 3-MST. The detailed method for measurement of 3-MST activity is described in the previous publications (Shibuya et al., 2009; Zhao, Chan, Ng, & Wong, 2013).

7. PROSPECTS OF H_2S THERAPY ON PD AND CONCLUSIONS

Currently, L-Dopa is still the most common drug to treat PD. It compensates for the dopamine deficiency. However, L-Dopa cannot reverse/stop the progression of PD. Moreover, long term treatment with L-Dopa may induce side effects like dyskinesia (Jankovic & Stacy, 2007) and accelerates the neuron degeneration due to oxidative stress (Hattoria et al., 2009; Mytilineou, Han, & Cohen, 1993). Our group has demonstrated that H_2S is able to suppress oxidative stress, inhibit inflammation (Hu, Pan, Neo, Yong, & Bian, 2008; Hu et al., 2005), and protect cells against apoptosis (Hu, Lu, Wu, Wong, & Bian, 2009; Fig. 3). These beneficial effects are mediated by either suppression or activation of different protein kinases including PKC, PI3K/Akt, p38, JNK, and ERK–MAPKs (Tiong, Lu, & Bian, 2010). H_2S can also stimulate glutamate transporter function and induce direct sulfhydration of p66SHC which prevents phosphorylation of this protein and therefore inhibits ROS production in mitochondria (Xie et al., 2014). In addition, H_2S may also stimulate glutamate transporter activity by suppression of ERK/MAPK pathway during oxidative stress (Lu, Hu, Hu, & Bian, 2008; Fig. 3). Based on these reports, it is speculated that the combination of L-Dopa and H_2S may have a better therapeutic value. By using ACS84, a compound releasing both H_2S and

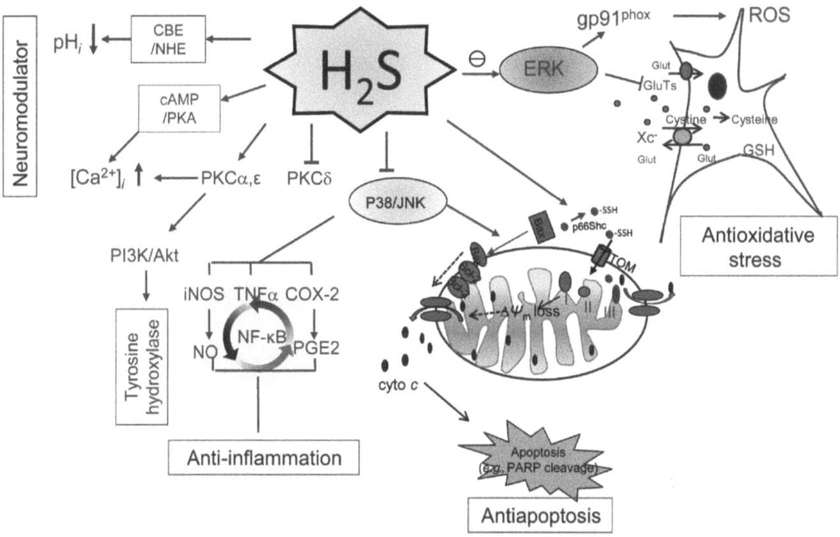

Figure 3 Schematic illustration showing the neuromodulatory and -protective effects of H_2S. Physiologically, H_2S regulates intracellular calcium, pHi, and cAMP level in brain cells. In pathological situations, H_2S produces anti-inflammatory, -oxidant, and -apoptotic effects via multiple mechanisms. This can be mediated by suppression or activation of different protein kinases including PKC, PI3K/Akt, p38, JNK, and ERK–MAPKs. Direct effect is mediated by S-sulfhydration of p66SHC at Cys-59, which prevents oxidative stress-induced phosphorylation and translocation of this protein to mitochondria. This effect further contributes to the inhibitory effect of H_2S on ROS production in mitochondria. (See the color plate.)

L-Dopa, we demonstrated that ACS84 produces stronger stimulatory effect on SOD activity and scavenges more ROS when compared with either NaHS or L-Dopa alone. For this reason, ACS84 is able to produce stronger protective effects against 6-OHDA-induced neuronal cell death and injury (Xie et al., 2013). However, even though the better protective effects of H_2S releasing L-Dopa is demonstrated at cellular level, it is still too early to get any conclusion that this compound may produce better clinical outcomes. Clinical uses of any H_2S releasing compound to treat neurodegenerative diseases should be very careful. One should bear in mind that high concentration of H_2S is toxic to brain function. Accumulation of H_2S in long term treatment may produce harmful/toxic effects on brain. How to control brain H_2S concentration at the therapeutic level is still a challenge. To this end, more studies are warranted to test the effect of ACS84 and other H_2S-releasing L-dopa in different animal models in the future.

ACKNOWLEDGMENT

This work was supported by NUHS B2B research grant (NUHSRO/2011/012/STB/B2B-08).

REFERENCES

Andrew Siderowf, M. D., & Matthew Stern, M. D. (2003). Update on Parkinson disease. *Annals of Internal Medicine*, *138*, 651–658.

Blandini, F., Armentero, M. T., & Martignoni, E. (2008). The 6-hydroxydopamine model: News from the past. *Parkinsonism & Related Disorders*, *14*(Suppl. 2), S124–S129. http://dx.doi.org/10.1016/j.parkreldis.2008.04.015.

Braak, H., Del Tredici, K., Rub, U., de Vos, R. A., Jansen Steur, E. N., & Braak, E. (2003). Staging of brain pathology related to sporadic Parkinson's disease. *Neurobiology of Aging*, *24*(2), 197–211.

Chiba, K., Trevor, A., & Castagnoli, N., Jr. (1984). Metabolism of the neurotoxic tertiary amine, MPTP, by brain monoamine oxidase. *Biochemical and Biophysical Research Communications*, *120*(2), 574–578.

Damier, P., Hirsch, E. C., Zhang, P., Agid, Y., & Javoy-Agid, F. (1993). Glutathione peroxidase, glial cells and Parkinson's disease. *Neuroscience*, *52*(1), 1–6.

Dauer, W., & Przedborski, S. (2003). Parkinson's disease: Mechanisms and models. *Neuron*, *39*(6), 889–909.

Enokido, Y., Suzuki, E., Iwasawa, K., Namekata, K., Okazawa, H., & Kimura, H. (2005). Cystathionine beta-synthase, a key enzyme for homocysteine metabolism, is preferentially expressed in the radial glia/astrocyte lineage of developing mouse CNS. *FASEB Journal*, *19*(13), 1854–1856.

Gage, G. J., Kipke, D. R., & Shain, W. (2012). Whole animal perfusion fixation for rodents. *Journal of Visual Experiments*, (65), 3564. http://dx.doi.org/10.3791/3564.

Hattoria, N., Wanga, M., Taka, H., Fujimura, T., Yoritaka, A., Kubo, S., et al. (2009). Toxic effects of dopamine metabolism in Parkinson's disease. *Parkinsonism & Related Disorders*, *15*(Suppl. 1), S35–S38. http://dx.doi.org/10.1016/S1353-8020(09)70010-0.

Hu, L. F., Lu, M., Wu, Z. Y., Wong, P. T., & Bian, J. S. (2009). Hydrogen sulfide inhibits rotenone-induced apoptosis via preservation of mitochondrial function. *Molecular Pharmacology*, *75*(1), 27–34. http://dx.doi.org/10.1124/mol.108.047985, mol.108.047985 [pii].

Hu, L. F., Pan, T. T., Neo, K. L., Yong, Q. C., & Bian, J. S. (2008). Cyclooxygenase-2 mediates the delayed cardioprotection induced by hydrogen sulfide preconditioning in isolated rat cardiomyocytes. *Pflügers Archiv*, *455*(6), 971–978. http://dx.doi.org/10.1007/s00424-007-0346-8.

Hu, L. F., Wang, S., Shi, X. R., Yao, H. H., Sun, Y. H., Ding, J. H., et al. (2005). ATP-sensitive potassium channel opener iptakalim protected against the cytotoxicity of MPP+ on SH-SY5Y cells by decreasing extracellular glutamate level. *Journal of Neurochemistry*, *94*(6), 1570–1579. http://dx.doi.org/10.1111/j.1471-4159.2005.03306.x, JNC3306 [pii].

Imamura, K., Hishikawa, N., Sawada, M., Nagatsu, T., Yoshida, M., & Hashizume, Y. (2003). Distribution of major histocompatibility complex class II-positive microglia and cytokine profile of Parkinson's disease brains. *Acta Neuropathologica*, *106*(6), 518–526. http://dx.doi.org/10.1007/s00401-003-0766-2.

Jankovic, J., & Stacy, M. (2007). Medical management of levodopa-associated motor complications in patients with Parkinson's disease. *CNS Drugs*, *21*(8), 677–692.

Langston, J. W., Forno, L. S., Tetrud, J., Reeves, A. G., Kaplan, J. A., & Karluk, D. (1999). Evidence of active nerve cell degeneration in the substantia nigra of humans years after 1-methyl-4-phenyl-1,2,3,6-tetrahydropyridine exposure. *Annals of Neurology*, *46*(4), 598–605.

Lee, C. S., Sauer, H., & Bjorklund, A. (1996). Dopaminergic neuronal degeneration and motor impairments following axon terminal lesion by instrastriatal 6-hydroxydopamine in the rat. *Neuroscience*, *72*(3), 641–653.

Lee, M., Tazzari, V., Giustarini, D., Rossi, R., Sparatore, A., Del Soldato, P., et al. (2010). Effects of hydrogen sulfide-releasing L-DOPA derivatives on glial activation: Potential for treating Parkinson disease. *The Journal of Biological Chemistry*, *285*(23), 17318–17328. http://dx.doi.org/10.1074/jbc.M110.115261.

Lu, M., Hu, L. F., Hu, G., & Bian, J. S. (2008). Hydrogen sulfide protects astrocytes against H (2)O(2)-induced neural injury via enhancing glutamate uptake. *Free Radical Biology and Medicine*, *45*(12), 1705–1713. http://dx.doi.org/10.1016/j.freeradbiomed.2008.09.014, S08915849(08)00517-0 [pii].

Mercanti, G., Bazzu, G., & Giusti, P. (2012). A 6-hydroxydopamine in vivo model of Parkinson's disease. *Methods in Molecular Biology*, *846*, 355–364. http://dx.doi.org/10.1007/978-1-61779-536-7_30.

Mytilineou, C., Han, S. K., & Cohen, G. (1993). Toxic and protective effects of L-dopa on mesencephalic cell cultures. *Journal of Neurochemistry*, *61*(4), 1470–1478.

Pan-Montojo, F., Anichtchik, O., Dening, Y., Knels, L., Pursche, S., Jung, R., et al. (2010). Progression of Parkinson's disease pathology is reproduced by intragastric administration of rotenone in mice. *PLoS One*, *5*(1), e8762. http://dx.doi.org/10.1371/journal.pone.0008762.

Park, J. J., & Cunningham, M. G. (2007). Thin sectioning of slice preparations for immunohistochemistry. *Journal of Visual Experiments*, (3), 194. http://dx.doi.org/10.3791/194.

Schober, A. (2004). Classic toxin-induced animal models of Parkinson's disease: 6-OHDA and MPTP. *Cell and Tissue Research*, *318*(1), 215–224. http://dx.doi.org/10.1007/s00441-004-0938-y.

Schwab, M. E., & Bartholdi, D. (1996). Degeneration and regeneration of axons in the lesioned spinal cord. *Physiological Reviews*, *76*(2), 319–370.

Shibuya, N., Tanaka, M., Yoshida, M., Ogasawara, Y., Togawa, T., Ishii, K., et al. (2009). 3-Mercaptopyruvate sulfurtransferase produces hydrogen sulfide and bound sulfane sulfur in the brain. *Antioxidants & Redox Signaling*, *11*(4), 703–714. http://dx.doi.org/10.1089/ARS.2008.2253.

Tiong, C. X., Lu, M., & Bian, J. S. (2010). Protective effect of hydrogen sulphide against 6-OHDA-induced cell injury in SH-SY5Y cells involves PKC/PI3K/Akt pathway. *British Journal of Pharmacology*, *161*(2), 467–480. http://dx.doi.org/10.1111/j.1476-5381.2010.00887.x.

Ungerstedt, U. (1968). 6-Hydroxy-dopamine induced degeneration of central monoamine neurons. *European Journal of Pharmacology*, *5*(1), 107–110.

Webb, G. D., Lim, L. H., Oh, V. M., Yeo, S. B., Cheong, Y. P., Ali, M. Y., et al. (2008). Contractile and vasorelaxant effects of hydrogen sulfide and its biosynthesis in the human internal mammary artery. *Journal of Pharmacology and Experimental Therapeutics*, *324*(2), 876–882. http://dx.doi.org/10.1124/jpet.107.133538, jpet.107.133538 [pii].

Xie, L., Hu, L. F., Teo, X. Q., Tiong, C. X., Tazzari, V., Sparatore, A., et al. (2013). Therapeutic effect of hydrogen sulfide-releasing L-Dopa derivative ACS84 on 6-OHDA-induced Parkinson's disease rat model. *PLoS One*, *8*(4), e60200. http://dx.doi.org/10.1371/journal.pone.0060200.

Xie, Z. Z., Shi, M. M., Xie, L., Wu, Z. Y., Li, G., Hua, F., et al. (2014). Sulfhydration of p66Shc at cysteine59 mediates the antioxidant effect of hydrogen sulfide. *Antioxidants & Redox Signaling*. http://dx.doi.org/10.1089/ars.2013.5604.

Yong, Q. C., Lee, S. W., Foo, C. S., Neo, K. L., Chen, X., & Bian, J. S. (2008). Endogenous hydrogen sulphide mediates the cardioprotection induced by ischemic postconditioning. *American Journal of Physiology. Heart and Circulatory Physiology*, *295*(3), H1330–H1340. http://dx.doi.org/10.1152/ajpheart.00244.2008, 00244.2008 [pii].

Zhao, H., Chan, S. J., Ng, Y. K., & Wong, P. T. (2013). Brain 3-mercaptopyruvate sulfurtransferase (3MST): Cellular localization and downregulation after acute stroke. *PLoS One*, *8*(6), e67322. http://dx.doi.org/10.1371/journal.pone.0067322.

SECTION III

Hydrogen Sulfide Metabolism in Mammalian Tissues

CHAPTER ELEVEN

Assay Methods for H_2S Biogenesis and Catabolism Enzymes

Ruma Banerjee[1], Taurai Chiku, Omer Kabil, Marouane Libiad, Nicole Motl, Pramod K. Yadav

Department of Biological Chemistry, University of Michigan Medical School, Ann Arbor, Michigan, USA

[1]Corresponding author: e-mail address: rbanerje@umich.edu

Contents

Abstract

H_2S is produced from sulfur-containing amino acids, cysteine and homocysteine, or a catabolite, 3-mercaptopyruvate, by three known enzymes: cystathionine β-synthase, γ-cystathionase, and 3-mercaptopyruvate sulfurtransferase. Of these, the first two enzymes reside in the cytoplasm and comprise the transsulfuration pathway, while the third enzyme is found both in the cytoplasm and in the mitochondrion. The following mitochondrial enzymes oxidize H_2S: sulfide quinone oxidoreductase, sulfur dioxygenase, rhodanese, and sulfite oxidase. The products of the sulfide oxidation pathway are thiosulfate and sulfate. Assays for enzymes involved in the production and oxidative clearance of sulfide to thiosulfate are described in this chapter.

1. INTRODUCTION

The steady-state intracellular concentrations of H_2S are the product of its enzymatic generation and clearance rates (Kabil, Motl, & Banerjee, 2014; Vitvitsky, Kabil, & Banerjee, 2012). H_2S is synthesized by at least three

Methods in Enzymology, Volume 554
ISSN 0076-6879
http://dx.doi.org/10.1016/bs.mie.2014.11.016

known enzymes: cystathionine β-synthase (CBS), γ-cystathionase (CSE), and mercaptopyruvate sulfurtransferase (MST) (Kabil & Banerjee, 2014). Of these, the first two constitute the cytoplasmic transsulfuration pathway, while the third is found in the cytoplasm and in mitochondria and is part of the cysteine catabolic pathway (Fig. 1). The contribution of each of these enzymes to net H_2S production is dictated by its presence and relative tissue concentration, which varies in a cell-specific manner (Chiku et al., 2009; Kabil, Vitvitsky, Xie, & Banerjee, 2011; Singh, Padovani, Leslie, Chiku, & Banerjee, 2009; Yadav, Yamada, Chiku, Koutmos, & Banerjee, 2013). The enzymes in the mitochondrial sulfide oxidation pathway convert H_2S to thiosulfate and sulfate, which are subsequently eliminated. The component enzymes include sulfide quinone oxidoreductase (SQR), sulfur dioxygenase (also known as ETHE1 or persulfide dioxygenase), rhodanese (or thiol sulfurtransferase), and sulfite oxidase (Fig. 1). Of these, SQR is anchored in the inner mitochondrial membrane, rhodanese and sulfur dioxygenase are in the mitochondrial matrix, and sulfite oxidase is in the intermitochondrial membrane space. The overlapping substrate specificities combined with the substrate ambiguity of CBS and CSE have confounded

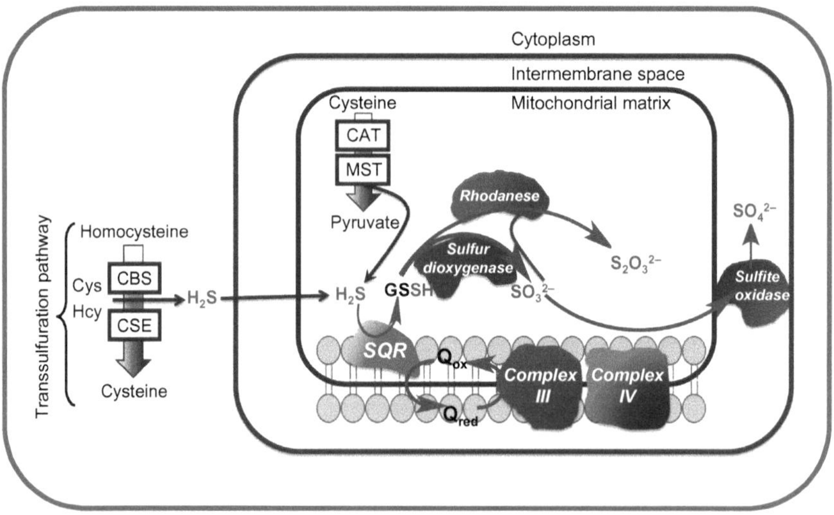

Figure 1 Scheme showing enzymes involved in H_2S biogenesis and oxidation. For simplicity, H_2S production by MST is shown in the mitochondrion only although the enzyme can be present in the cytoplasm and mitochondrion. CBS, CSE, CAT, MST, and SQR denote cystathionine β-synthase, γ-cystathionase, cysteine aminotransferase, mercaptopyruvate sulfur transferase, and sulfide quinone oxidoreductase, respectively. Electrons from the mitochondrial sulfide oxidation pathway are transferred via ubiquinone to complex III in the electron transfer chain.

assessment of their contributions to H_2S production in cell and tissue samples. Similarly, multiple acceptors of the persulfide moiety generated in the SQR-, rhodanese-, and MST-catalyzed reactions complicate their enzymatic assays and interpretation of the organizational logic of the sulfide oxidation pathway (Hildebrandt & Grieshaber, 2008; Jackson, Melideo, & Jorns, 2012; Libiad, Yadav, Vitvitsky, Martinov, & Banerjee, 2014). Methods for assaying the individual enzymes are discussed in this chapter.

2. ASSAYS FOR H_2S BIOGENESIS

CBS and CSE catalyze the successive conversion of homocysteine to cystathionine and then cysteine in the transsulfuration pathway (Eqs. 1 and 2). The substrate ambiguity of each enzyme supports multiple H_2S-generating reactions in which the amino acids cysteine and homocysteine are utilized as the sulfur source. CBS catalyzes the β-replacement of cysteine by homocysteine (Eq. 3), cysteine by water (Eq. 4) or cysteine by a second mole of cysteine (Eq. 5) generating H_2S and the corresponding products. CSE catalyzes the α,β-cleavage of cysteine (Eq. 6), the α,γ-cleavage of homocysteine (Eq. 7), and the γ-replacement of homocysteine by a second mole of homocysteine (Eq. 8) generating H_2S. In addition, CSE, like CBS, catalyzes the condensation of two moles of cysteine (Eq. 5) and of cysteine and homocysteine (Eq. 3), producing H_2S:

$$\text{Serine} + \text{Homocysteine} \rightarrow \text{Cystathionine} + H_2O \quad (1)$$

$$\text{Cystathionine} \rightarrow \text{Cysteine} + \alpha\text{-Ketobutyrate} + NH_3{}^{+} \quad (2)$$

$$\text{Cysteine} + \text{Homocysteine} \rightarrow \text{Cystathionine} + H_2S \quad (3)$$

$$\text{Cysteine} + H_2O \rightarrow \text{Serine} + H_2S \quad (4)$$

$$\text{Cysteine} + \text{Cysteine} \rightarrow \text{Lanthionine} + H_2S \quad (5)$$

$$\text{Cysteine} \rightarrow \text{Pyruvate} + NH_3{}^{+} + H_2S \quad (6)$$

$$\text{Homocysteine} \rightarrow \alpha\text{-Ketobutyrate} + NH_3{}^{+} + H_2S \quad (7)$$

$$\text{Homocysteine} + \text{Homocysteine} \rightarrow \text{Homolanthionine} + H_2S \quad (8)$$

Under V_{max} conditions, the k_{cat} for the CBS-catalyzed reactions decrease in the following order: (3) $[61\times] \gg$ (4) $[1.6\times] >$ (5) $[1\times]$ where the round and square brackets denote the reaction number and the relative rate constants with the slowest reaction designated as $1\times$. Under V_{max} conditions, the k_{cat} for the CSE-catalyzed H_2S-generating reactions show the following order: (6) $[270{,}000\times] >$ (7) $[111{,}500\times] \gg$ (8) $[1400\times] \gg$ (3) $[50\times] \gg$ (5) $[1\times]$.

MST is a sulfurtransferase that transfers the sulfur from mercaptopyruvate to an active site cysteine to form a persulfide intermediate, MST-SSH (Eq. 9). A variety of small molecules in addition to thioredoxin can accept the persulfide group and in the presence of a reductant, H_2S is released (Eq. 10):

$$\text{Mercaptopyruvate} + \text{MST-SH} \rightarrow \text{Pyruvate} + \text{MST-SSH} \quad (9)$$

$$\text{MST-SSH} + 2\text{RSH} \rightarrow \text{MST-S-H} + \text{R-S-S-R} + H_2S \quad (10)$$

2.1. Assays for CBS and CSE

2.1.1 H_2S formation from cysteine or cysteine + homocysteine

Reagents

- 200 m*M* HEPES buffer, pH 7.4
- 400 m*M* D,L-homocysteine (pH adjusted to 7.4 with 10 *M* NaOH)
- 1 *M* L-cysteine
- 40 m*M* lead acetate
- 2.0 mg ml^{-1} purified human CBS or purified human CSE in 100 m*M* HEPES buffer, pH 7.4
- Deionized water

Method: In a polystyrene cuvette, add 500 μl HEPES buffer, 125 μl homocysteine, and 25 μl of cysteine or only 25 μl cysteine, 10 μl lead acetate, and water to bring the reaction volume to 990 μl. Place the cuvette in a spectrophotometer for 5 min at 37 °C, maintained using a circulating water bath. Initiate the reaction with 10 μl of CBS or 10 μl of CSE and monitor the increase in absorbance at 390 nm due to formation of lead sulfide. From the slope of the absorbance change, calculate the specific activity of CBS or CSE using a molar extinction coefficient of 5500 M^{-1} cm^{-1} for lead sulfide (Singh et al., 2009).

2.1.2 Methanethiol formation from methylcysteine

Reagents

- 200 m*M* HEPES buffer, pH 7.4
- 500 m*M* methylcysteine
- 4.0 mg ml^{-1} purified human CBS or purified human CSE in 100 m*M* HEPES, pH 7.4
- 50 m*M* DTNB (dithiobisnitrobenzoic acid)
- Deionized water

Method: In a polystyrene cuvette, add 500 μl HEPES buffer, 20 μl methylcysteine, 10 μl DTNB, and water to bring the reaction volume to 990 μl. Place the cuvette in a spectrophotometer connected to a water bath maintained at 37 °C for 5 min. Add 10 μl CBS or CSE and monitor the increase in absorbance at 412 nm due to generation of the nitrobenzene thiolate anion (which forms due to the reaction of the methanethiol product with DTNB). From the slope of the absorbance change, calculate the specific activity of CBS or CSE using a molar extinction coefficient of 13,600 M^{-1} cm^{-1}.

2.1.3 Assessing H_2S production by CBS versus CSE in tissue samples

Reagents

- 300 m*M* propargylglycine
- 200 m*M* HEPES buffer, pH 7.4
- 400 m*M* D,L-homocysteine
- 400 m*M* L-cysteine
- 40 m*M* lead acetate
- 100 mg ml^{-1} tissue extract in 100 m*M* HEPES buffer, pH 7.4
- Deionized water

Method: Using a glass homogenizer, disrupt frozen tissue in 200 m*M* HEPES, pH 7.4, to obtain a concentration of 100 mg tissue wet weight ml^{-1}. Incubate on ice for 20 min. In a polystyrene cuvette, add 815 μl HEPES buffer, 10 μl lead acetate, and 100 μl tissue extract (total volume 925 μl). Place the cuvette in a cuvette holder for 5 min, at 37 °C maintained using a circulating water bath. Initiate the reaction by adding a mixture of 50 μl homocysteine and 25 μl cysteine (to obtain 10 m*M* final concentration each of the L-form of the amino acid) and monitor the increase in absorbance at 390 nm due to formation of lead sulfide. From the slope of the absorbance change, calculate the specific activity using a molar extinction coefficient of 5500 M^{-1} cm^{-1} for lead sulfide.

To determine the contribution of CBS activity only, preincubate 100 μl tissue extract with 10 μl of propargylglycine, a suicide inactivator of CSE (Abeles & Walsh, 1973), for 5 min on ice before using it in the above reaction. The total H_2S production rate, contributed by CBS and CSE, is obtained from the reaction rate in the absence of propargylglycine. Calculate the contribution of CBS to total H_2S production from the assay in the presence of propargylglycine. Subtract the contribution of CBS from the total H_2S production rate to determine the contribution of CSE.

2.2. Assays for MST

2.2.1 MST assay using small molecule acceptors

Reagents

- 400 m*M* HEPES buffer, pH 7.4
- 30 m*M* 3-mercaptopyruvate
- 400 m*M* D,L-homocysteine
- 400 m*M* dihydrolipoic acid, pH adjusted to ~7.4 by using 10 *M* NaOH
- 400 m*M* glutathione
- 1 *M* L-cysteine
- 10 mg ml^{-1} bovine serum albumin
- 40 m*M* lead acetate
- 1.0 mg ml^{-1} purified MST in 100 m*M* Tris, pH 8.0
- Deionized water

Method: In a polystyrene cuvette, add 500 μl HEPES buffer, 10 μl 2-mercaptopyruvate, 10 μl bovine serum albumin, 10 μl lead acetate, and 50 μl dihydrolipoic acid, or 100 μl homocysteine, or 25 μl cysteine or 125 μl GSH. Adjust the reaction volume to 990 μl using deionized water and place the cuvette in spectrophotometer with a water-jacketed cuvette holder maintained at 37 °C. Add 10 μl of MST and monitor the increase in absorbance at 390 nm due to formation of lead sulfide. From the linear portion of the reaction curve, calculate the specific activity of MST using a molar extinction coefficient of 5500 M^{-1} cm^{-1} for lead sulfide (Yadav et al., 2013).

2.2.2 MST assay using thioredoxin

Reagents

- 400 m*M* HEPES buffer, pH 7.4
- 300 m*M* sodium mercaptopyruvate
- 1 m*M* purified human thioredoxin
- 180 μ*M* purified human thioredoxin reductase
- 20 m*M* NADPH
- 40 m*M* lead acetate
- 10 mg ml^{-1} bovine serum albumin
- 1.0 mg ml^{-1} purified human MST in 100 m*M* Tris, pH 8.0
- Deionized water

Method: In a polystyrene cuvette, add 500 μl of HEPES buffer, 10 μl 3-mercaptopyruvate, 20 μl thioredoxin, 20 μl thioredoxin reductase, 10 μl NADPH, 10 μl bovine serum albumin, and 10 μl of lead acetate.

Adjust the volume of the reaction mixture to 990 μl with deionized water and place the cuvette for 5 min in a spectrophotometer with a water-jacketed cuvette holder maintained at 37 °C. Add 10 μl of purified MST and monitor the increase in absorbance at 390 nm due to formation of lead sulfide. From the linear portion the reaction curve, calculate the specific activity of MST using a molar extinction coefficient of 5500 M^{-1} cm^{-1} for lead sulfide (Yadav et al., 2013).

3. ASSAYS FOR ENZYMES INVOLVED IN H_2S CATABOLISM

The mitochondrial sulfide oxidation pathway converts sulfide to thiosulfate and sulfate, which are eliminated. SQR catalyzes the first step in the pathway, which involves a two-electron oxidation of sulfide to persulfide and proceeds via an enzyme-bound sulfane sulfur intermediate. The physiologically relevant sulfide acceptor of the reaction is controversial, and sulfide, thiols (Eq. 11), and sulfite all function as acceptors in the *in vitro* assay. The electrons are transferred from SQR, a flavoprotein, to coenzyme Q (Eq. 12) and thence, to the electron transfer chain:

$$H_2S + RSH + SQR_{ox} \rightarrow RS\text{-}SH + SQR_{red} + 2H^+ \quad (11)$$

$$SQR_{red} + CoQ_{ox} \rightarrow SQR_{ox} + CoQ_{red} \quad (12)$$

Sulfur dioxygenase catalyzes the oxygen-dependent four-electron oxidation of glutathione persulfide (GSSH) to sulfite (Eq. 13). The enzyme belongs to the family of 3His-1Asp mononuclear iron oxygenases (Kabil & Banerjee, 2012).

$$GSSH + O_2 + H_2O \rightarrow GSH + SO_3^{2-} + 2H^+ \quad (13)$$

Rhodanese or thiosulfate sulfurtransferase catalyzes a transsulfuration reaction using a variety of sulfur donors and acceptors. The common method for assaying rhodanese was described by Sörbo (1955) and monitors the transfer of the sulfane sulfur from thiosulfate to cyanide and yields sulfite and thiocyanate (Eq. 14). Other sulfur donors in the rhodanese-catalyzed reaction include thiosulfonates and GSSH (Eq. 15), while compounds such as sulfite (Eq. 15), glutathione, cysteine, and homocysteine can serve as sulfur acceptors:

$$S_2O_3^{2-} + KCN \rightarrow SO_3^{2-} + KSCN \quad (14)$$

$$GSSH + SO_3^{2-} \rightarrow S_2O_3^{2-} + GSH \quad (15)$$

3.1. Assay for SQR

Reagents

- 200 m*M* sodium phosphate buffer, pH 7.4, containing 0.06% 1,2-diheptanoyl-*sn*-glycero-3-phosphocholine
- 15 m*M* sodium sulfide
- 100 m*M* sodium sulfite
- 40 m*M* CoQ_1 in DMSO
- 10 mg ml^{-1} bovine serum albumin
- 400 m*M* glutathione
- 1 *M* L-cysteine
- 400 m*M* D,L-homocysteine, pH adjusted to ~7.4 by using 10 *M* NaOH
- 0.01 mg ml^{-1} purified SQR in 100 m*M* Tris, pH 8.0, containing 0.03% 1,2-diheptanoyl-*sn*-glycero-3-phosphocholine
- Deionized water

Method: In a polystyrene cuvette, add 500 μl of sodium phosphate buffer, 1.5 μl CoQ_1, and 10 μl bovine serum albumin and mix by pipetting. Add 10 μl of sulfite or 125 μl glutathione or 50 μl cysteine or 250 μl homocysteine (Libiad et al., 2014). Adjust the reaction volume to 985 μl using deionized water. Incubate the reaction mixture in a water-jacketed cuvette holder maintained at 25 °C for 3 min. Add 5 μl of SQR and 10 μl of sulfide simultaneously using two pipettes and monitor the decrease in absorbance at 278 nm due to reduction of CoQ_1. From the slope of the reaction progress, calculate the specific activity using the molar extinction coefficient ($\Delta\varepsilon_{ox-red} = 12{,}000\,M^{-1}\,cm^{-1}$) for reduction of CoQ_1 at 278 nm (Jackson et al., 2012).

3.2. Assay for sulfur dioxygenase (or persulfide dioxygenase or ETHE1)

3.2.1 Preparation of GSSH

Reagents

- 350 m*M* sodium phosphate, pH 7.4
- Solid sodium sulfide
- 50 m*M* GSSG dissolved in anaerobic 350 m*M* sodium phosphate buffer, pH 7.4

Method: In an anaerobic glass vial, add 5 ml GSSG and add 240 mg of solid sodium sulfide (to obtain a final concentration of 200 m*M*). Seal the reaction vial to prevent loss of H_2S, remove the vial from the anaerobic chamber and incubate at 37 °C for 20 min. Measure the concentration of the GSSH product using the cold cyanolysis method as described (Wood, 1987).

3.2.2 Oxygen consumption assay

Reagents

- 100 m*M* sodium phosphate, pH 7.4
- 50 m*M* GSSH
- 0.5–2 μg sulfur dioxygenase in 50 m*M* Tris, pH 8.0, containing 0.5 *M* NaCl
- Deionized water

Method: In a Gilson-type chamber containing a Clark oxygen electrode and a magnetic stir bar washed with deionized water, add 1.5 ml of sodium phosphate. Cap the chamber to prevent additional O_2 from dissolving into the buffer and inject 0.5–2 μg of sulfur dioxygenase. Initiate the reaction by addition of 30 μl GSSH. Record oxygen consumption using a Kipp and Zonen BD single channel chart recorder. From the linear portion of the curve, calculate the rate of oxygen consumption. The concentration of dissolved O_2 at room temperature is ~280 μ*M*. The rate of oxygen consumption is reported in units of μmol O_2 consumed min^{-1} mg of $enzyme^{-1}$.

3.3. Assays for rhodanese

3.3.1 Assay for thiocyanate formation by rhodanese

Reagents

- 1 *M* sodium thiosulfate
- 1 *M* potassium cyanide
- 15% (w/v) formaldehyde solution
- Ferric nitrate nonahydrate solution (6.6 g of $Fe(NO_3)_3 \cdot 9H_2O$ dissolved in 3.4 ml nitric acid. The final volume is adjusted to 50 ml).
- 300 m*M* HEPES buffer, pH 7.4, containing 150 m*M* NaCl.

Method: In a polystyrene cuvette add 12.5 μl of sodium thiosulfate, 12.5 μl of potassium cyanide, and 225 μl of HEPES buffer at 25 °C. The reaction is initiated by addition of 0.5 μg of rhodanese. After 5 min, the reaction is terminated by addition of 250 μl of 15% (w/v) formaldehyde and the reaction mixture is centrifuged for 5 min at 10,000 × *g* to remove the protein. Addition of 0.5 ml of ferric nitrate to the supernatant results in the development of a red color due to ferric thiocyanate formation, which is monitored at 460 nm. A control reaction lacking rhodanese is run in parallel. The amount of thiocyanate formed is determined using a standard curve generated with thiocyanate ranging from 0.1 to 5 μmol. One unit of enzyme activity catalyzes the formation of 1 μmol of thiocyanate min^{-1} at 25 °C. The specific activity is expressed as units mg $protein^{-1}$.

3.3.2 Assay for thiosulfate production by rhodanese

Reagents

- 1 *M* sodium thiosulfate
- 100 m*M* sodium sulfite
- 50 m*M* GSSH (see Section 3.2.1 for preparation)
- 100 m*M* monobromobimane dissolved in DMSO
- 100% methanol
- 100% acetic acid
- 0.2 m*M* sodium citrate, pH 2.0
- 100 m*M* HEPES buffer, pH 7.4, containing 150 m*M* NaCl

Method: The assay mixture is made by mixing 2 μl of sodium sulfite, 8 μl GSSH in 200 μl of HEPES buffer, and 1 μg of rhodanese. The reaction is initiated by addition of 1 μg of rhodanese and incubated for 5 min at 25 ° C followed by derivatization of samples with 2 μl of monobromobimane. Incubation is continued for 10 min prior to acidification with 100 μl of sodium citrate. A control reaction lacking rhodanese was prepared in parallel. The derivatized samples are centrifuged at 10,000 × *g* for 10 min at 4 °C and 50 μl of the supernatant is injected onto a C8 reverse phase HPLC column (4.6 × 150 mm, 3 μm packing, Phenomenex) pre-equilibrated with 80% solvent A (10% methanol and 0.25% acetic acid) and 20% of solvent B (90% methanol and 0.25% acetic acid). The sample is eluted using the following gradient: solvent B: 20% from 0 to 10 min, 20–40% from 10 to 25 min, 40–90% from 25 to 30 min, 90–100% from 30 to 32 min, 100% from 32 to 35 min, 100–20% from 35 to 37 min, and 20% from 37 to 40 min. The flow rate is 0.75 ml min^{-1}. The bimane adduct of thiosulfate elutes at ~22 min under these conditions and is detected by excitation at 340 nm and emission at 450 nm. The concentration of thiosulfate is determined using thiosulfate standards of known concentration. One unit of enzyme activity catalyzes the formation of 1 μmol of thiosulfate min^{-1} at 25 °C. Specific activity is expressed as units mg^{-1} protein.

3.3.3 Assay for H_2S production by rhodanese

Reagents

- 1 *M* sodium thiosulfate
- 200 m*M* glutathione
- 200 m*M* cysteine
- 200 m*M* homocysteine
- 100 m*M* lead acetate
- 100 m*M* HEPES buffer, pH 7.4, containing 150 m*M* NaCl

Method: Prepare the reaction mixture in a polystyrene cuvette by adding 1.5 μl of thiosulfate, 50 μl glutathione (or 50 μl cysteine or 100 μl homocysteine), 2 μl lead acetate, and 350 μl HEPES buffer in a final volume of 500 μl. The cuvette in placed in a spectrophotometer maintained at 37 ° C for 4 min. The reaction is initiated by addition of 1–10 μg of rhodanese. The increase in absorbance at 390 nm due to formation of lead sulfide is monitored. The specific activity is calculated using a molar extinction coefficient of 5500 M^{-1} cm^{-1} for lead sulfide (Singh et al., 2009).

ACKNOWLEDGMENTS

This work was supported in part by grants from the NIH (HL58984 and GM112455 to R. B.) and by the American Heart Association (14POST18760003 to P. K. Y. and 13SDG17070096 to O. K.).

REFERENCES

Abeles, R. H., & Walsh, C. (1973). Acetylenic enzyme inactivators. Inactivation of γ-cystathionase, in vitro and in vivo, by propargylglycine. *Journal of the American Chemical Society*, *95*, 6124–6125.

Chiku, T., Padovani, D., Zhu, W., Singh, S., Vitvitsky, V., & Banerjee, R. (2009). H_2S biogenesis by human cystathionine gamma-lyase leads to the novel sulfur metabolites, lanthionine and homolanthionine, and is responsive to the grade of hyperhomocysteinemia. *Journal of Biological Chemistry*, *284*, 11601–11612.

Hildebrandt, T. M., & Grieshaber, M. K. (2008). Three enzymatic activities catalyze the oxidation of sulfide to thiosulfate in mammalian and invertebrate mitochondria. *The FEBS Journal*, *275*(13), 3352–3361.

Jackson, M. R., Melideo, S. L., & Jorns, M. S. (2012). Human sulfide:quinone oxidoreductase catalyzes the first step in hydrogen sulfide metabolism and produces a sulfane sulfur metabolite. *Biochemistry*, *51*(34), 6804–6815.

Kabil, O., & Banerjee, R. (2012). Characterization of patient mutations in human persulfide dioxygenase (ETHE1) involved in H_2S catabolism. *Journal of Biological Chemistry*, *287*, 44561–44567.

Kabil, O., & Banerjee, R. (2014). Enzymology of H_2S biogenesis, decay and signaling. *Antioxidants & Redox Signaling*, *20*(5), 770–782.

Kabil, O., Motl, N., & Banerjee, R. (2014). H_2S and its role in redox signaling. *Biochimica et Biophysica Acta*, *1844*(8), 1355–1366.

Kabil, O., Vitvitsky, V., Xie, P., & Banerjee, R. (2011). The quantitative significance of the transsulfuration enzymes for H_2S production in murine tissues. *Antioxidants & Redox Signaling*, *15*, 363–372.

Libiad, M., Yadav, P. K., Vitvitsky, V., Martinov, M., & Banerjee, R. (2014). Organization of the human mitochondrial H_2S oxidation pathway. *Journal of Biological Chemistry*, *289*, 30901–30910.

Singh, S., Padovani, D., Leslie, R. A., Chiku, T., & Banerjee, R. (2009). Relative contributions of cystathionine beta-synthase and gamma-cystathionase to H_2S biogenesis via alternative trans-sulfuration reactions. *Journal of Biological Chemistry*, *284*(33), 22457–22466.

Sörbo, B. H. (1955). Rhodanese. *Methods in Enzymology*, *2*, 334–337.

Vitvitsky, V., Kabil, O., & Banerjee, R. (2012). High turnover rates for hydrogen sulfide allow for rapid regulation of its tissue concentrations. *Antioxidants & Redox Signaling*, *17*(1), 22–31.

Wood, J. L. (1987). Sulfane sulfur. *Methods in Enzymology*, *143*, 25–29.

Yadav, P. K., Yamada, K., Chiku, T., Koutmos, M., & Banerjee, R. (2013). Structure and kinetic analysis of H_2S production by human mercaptopyruvate sulfurtransferase. *Journal of Biological Chemistry*, *288*, 20002–20013.

CHAPTER TWELVE

Oxidation of H_2S in Mammalian Cells and Mitochondria

Abbas Abou-Hamdan*,†,‡, Hala Guedouari-Bounihi*,†,‡, Véronique Lenoir*,†,‡, Mireille Andriamihaja§, François Blachier§, Frédéric Bouillaud*,†,‡,1

*Inserm U1016, Institut Cochin, Paris, France
†CNRS UM8104, Institut Cochin, Paris, France
‡Université Paris Descartes UMR-S1016, Institut Cochin, Paris, France
§INRA-CRNH-IdF-AgroParisTech, UMR 914 Nutrition Physiology and Ingestive Behavior, Paris, France
[1]Corresponding author: e-mail address: frederic.bouillaud@inserm.fr

Contents

Methods in Enzymology, Volume 554
ISSN 0076-6879
http://dx.doi.org/10.1016/bs.mie.2014.11.042

Abstract

Hydrogen sulfide (H_2S) is the third gasotransmitter described in mammals. These gasotransmitters (H_2S, CO, and NO) are small molecules able to diffuse freely across membranes and thus susceptible to reach easily intracellular targets, one of which is the respiratory enzyme cytochrome oxidase subject to complete inhibition by low micromolar concentrations of these gases. However in contrast to NO or CO, H_2S can be metabolized by a sulfide quinone reductase feeding the mitochondrial respiratory chain with the hydrogen atoms of sulfide. Sulfide is thus a two-sided molecule: substrate or poison according to the concentration. The aim of this chapter is to present a mean to monitor sulfide oxidation by isolated mitochondria or cells and to summarize how the properties of this amazing couple (mitochondria and sulfide) translate into practical and conceptual consequences.

1. INTRODUCTION

1.1. Sulfide gasotransmitters and mitochondria

The gasotransmitters nitric oxide (NO), carbon monoxide (CO), and hydrogen sulfide (H_2S) are small molecules able to diffuse freely across membrane and thus susceptible to reach easily intracellular targets. Actually, the possible cellular targets are multiple and when concentration of these three gases rises above a certain level a toxic effect based on the inhibition of mitochondrial respiration takes place. The gaseous signaling is expected to occur with lower concentrations and to imply different pathways than toxicity. However, the pharmacological use of gaseous transmitters should take care not to cause a partial intoxication, or to take it into consideration with its possible deleterious but also positive consequences linked to stress-induced adaptations (hormesis). In contrast to NO or CO, H_2S shows the specific property to act as a hydrogen donor for the mitochondrial respiratory chain. The hydrogen atoms are extracted by a specific enzyme: a sulfide quinone reductase (SQR). Consequently, H_2S is a substrate for mitochondrial oxidation where/when SQR is present. This surprising activity is likely to be explained both by the ancestral presence of the SQR in the bacteria at the origin of mitochondria and by conservation during evolution due to the need to eliminate any endogenous H_2S resulting from cellular metabolism. The final outcome be that the same molecule H_2S is a poison, a food, and a signal in the eukaryotic cell (Bouillaud & Blachier, 2011).

1.2. Issues treated in and audience of this chapter

Our aim, based on our past experience, is to present a convenient mean to monitor sulfide oxidation by isolated mitochondria or cells and to summarize how the properties of this amazing couple (mitochondria and sulfide) translate into practical and conceptual consequences. This concerns primarily people interested in the study the interaction of sulfide with mitochondria but should not be ignored by those interested in the signaling properties of H_2S. Moreover, the issue of the interaction with the complex IV of mitochondrial respiratory chain is shared by the three gasotransmitters (NO, CO, and H_2S). The terms H_2S and sulfide should be considered as synonymous here.

2. SULFIDE IN THE CONTEXT OF MITOCHONDRIAL BIOENERGETICS

2.1. Cellular bioenergetics and mitochondria

Living cells collect in their environment what is necessary to feed their permanent energy needs. While some could gather directly energy in the form of light, many other import molecules engaged into redox reactions leading to new products (considered as waste) and to chemical energy. Both light and chemical energy need to be converted into the forms directly usable by the cell: such as "high energy." Phosphate bond (ATP) or electrochemical gradients, with the important property that these different forms of energy are made mutually convertible through the action of "bioenergetically competent" enzymes. The situation in mammalian mitochondria is schematized in Figs. 1 and 2.

The animal cells import organic compounds which are processed by metabolism for a first round of redox reactions leading to oxidation of carbon into CO_2 (waste) and to the reduction of intermediates coenzymes (NAD, FAD). The reduced forms (NADH, $FADH_2$) of these intermediates behave as hydrogen donors for the mitochondrial respiratory chain that drives oxidation of these hydrogen atoms by oxygen to form water (Fig. 1). The energy released during this reaction is converted into a proton gradient used to maintain a high membrane potential and to regenerate ATP from ADP (Fig. 2). This sequence of events from hydrogen donors to ATP formation is central to mitochondrial bioenergetics. The mitochondrial respiratory chain is composed of different complexes (numbered I–V) catalyzing different steps of the entire process. In addition to bioenergetics, mitochondria are the site where a large part of the metabolism

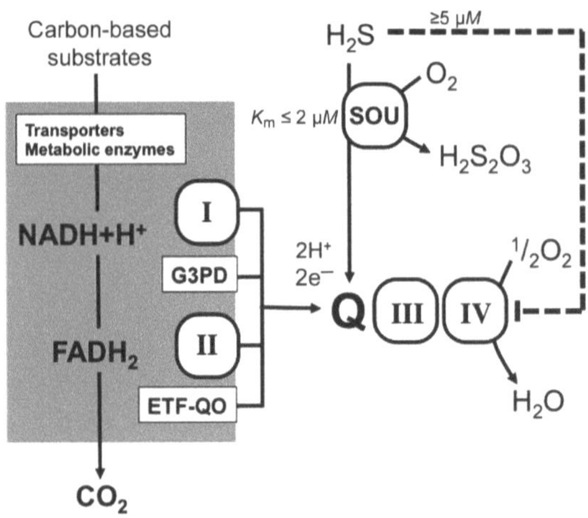

Figure 1 Hydrogen donors and the mitochondrial respiratory chain. The different mitochondrial respiratory complexes (I–IV) as known in animal (mammalian) mitochondria are schematized as boxes with roman numerals. The carbon metabolism is schematized by a gray box with indications of the main partners: transporters, enzymes, and coenzymes as well as the enzymatic activities that provide hydrogen atoms to reduce quinone, complex I; glyceraldehyde 3 phosphate dehydrogenase (G3PD), complex II and the ETF quinone oxidase (ETF-QO). The sulfide oxidation pathway (SOU) associates a dioxygenase and a sulfur transferase to the sulfide quinone reductase. The inhibition of complex IV by sulfide is indicated with a thick dotted line.

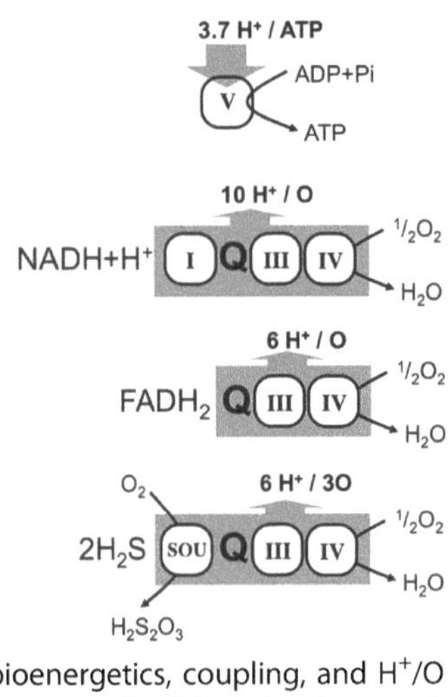

Figure 2 Mitochondrial bioenergetics, coupling, and H^+/O (P/O) ratio. The mitochondrial complex V uses proton transfer driven by the membrane potential to generate the energy necessary for ATP formation, the generation of one ATP needs the passage of 3.7 protons. This process (ATP generation) is the main controller of the assembly of proton motive force generators that are shown in the figure and work according to different regimes: If NADH is reoxidized by complexes I, III, and IV, 10 protons are pumped with consumption of one atom of oxygen. When FADH2 is involved (complex II, G3PD, ETF-QO), only six protons are pumped per oxygen atoms. When SOU is involved, the presence of the dioxygenase increases three times the oxygen usage for this same number of six protons pumped.

(release of CO_2) takes place. Mitochondria are therefore the major and critical site for energy extraction from the nutrients in the animal cell. Cells seem to have a bit more mitochondria than usually needed, which is supposed to guarantee fitting/survival if a challenging situation needs extra energy supply to the cell. Moreover, control analysis explains how in a linear chain of reactions partial impairment of one step could be partially compensated by the properties of the whole chain. This explains why limited damage/dysfunction of a respiratory complex could be well tolerated (remain silent) and a partial inhibition of a respiratory complex, including of course by H_2S, NO, or CO, might be well tolerated by cells. However, above a threshold, further deterioration of the mitochondrial function has consequences that evolve from deleterious to lethal within a short range of variation (Rossignol et al., 2003). A noticeable exception is provided by the cell cultures grown in the presence of glucose that could survive from glycolytic ATP for extended period of time and therefore could survive to massive exposure to H_2S (Leschelle et al., 2005). It should be recalled that these cultured cells have limited bioenergetic performance when compared to the cells present in the tissues of an animal and performing intensively highly specialized tasks that are absolutely dependent on mitochondrial bioenergetics.

2.2. Multiple hydrogen donors to the mitochondrial coenzyme Q

The acceptor of hydrogen atoms in the mitochondrial respiratory chain is the coenzyme Q (redox couple Q/QH_2). Complexes II and I are two parallel pathways able to reduce coenzyme Q: Complex I oxidizes NADH and complex II oxidizes succinate. There are other enzymatic steps yielding hydrogen atoms to coenzyme Q, for example, the reoxidation of the FADH2 generated during fatty acid oxidation, or the oxidation of L-alpha glycerophosphate well developed in glycolytic muscle (Fig. 1). Finally, H_2S itself is a hydrogen donor to coenzyme Q through the action of the SQR. For this reason, H_2S was presented as the first mineral substrate for mammalian cells (Goubern, Andriamihaja, Nubel, Blachier, & Bouillaud, 2007).

Therefore, H_2S oxidation operates in parallel (in fact in competition) with the others pathways for coenzyme Q reduction involving carbon-containing substrates and linked to a complex metabolism releasing CO_2 as waste (Fig. 1). In contrast, H_2S oxidation and its waste-generating process need a very limited set of enzymes (Hildebrandt & Grieshaber, 2008): two other enzymes, including a dioxygenase present in the same compartment, release the oxidized primary waste: thiosulfate. We introduced the term sulfide-oxidizing unit (SOU) to describe this assembly of three enzymes performing sulfide oxidation in thiosulfate and yielding hydrogen atoms to the mitochondrial respiratory chain (Lagoutte et al., 2010). Moreover, H_2S is a gas, diffuses freely across lipidic membranes, and does not need any transporter to reach SQR. Consequently, and in contrast with any other extracellular substrate, there is no preliminary metabolic preparation before the mitochondrial respiratory chain uses hydrogen atoms from H_2S. Moreover, the affinity of the mitochondria for sulfide is in the low micromolar range and therefore orders of magnitude higher than that for equivalent carbon-containing substrates; for a direct comparison between sulfide and succinate, see Szabo et al. (2014). This affinity is even higher than the estimated affinity of mitochondrial complex I for NADH (Birrell & Hirst, 2013), raising the point that H_2S could be better compared to true hydrogen donors (coenzymes) than to the other substrates of mitochondrial respiration.

2.3. Sulfide and gaseous transmitters are toxic to mitochondria

This description of H_2S as a hydrogen donor and the preferred substrate for mammalian mitochondria contrasts sharply with what was known before. Actually, H_2S together with NO, CO, and cyanide is highly toxic and micromolar concentrations are able to inhibit completely mitochondrial respiration (Cooper & Brown, 2008). Moreover, this toxicity is based on their property of interacting with heme-containing proteins so that the possible targets within a mammalian organism are multiple and hemoglobin, for example, is a first binding site (Haouzi et al., 2014), with the consequence that the actual molecular basis of their toxicity might be difficult to define on the basis of "a target" (Goldbaum, Ramirez, & Absalon, 1975). A relatively intense redox turnover of sulfide is thought to take place (Vitvitsky, Kabil, & Banerjee, 2012). SOU appears in the first line to limit endogenous sulfide levels and is able to siphon off any incoming sulfide and to maintain its concentration in the low nanomolar range within cells. Of course, SQR might be overwhelmed and poisoning occurs if a sudden and massive increase of sulfide concentration takes place, with all the possible consequences linked to mitochondrial failure.

2.4. Positive feedback loops for sulfide oxidation/inhibition

Consideration of the reaction network in Fig. 1 indicates that sulfide would inhibit its own oxidation. At the opposite, starting from a sulfide-inhibited state, a decrease in sulfide concentration would allow sulfide oxidation to take place and to accelerate. These positive feedback loops (Fig. 3) create instability and hardly predictable evolution when the different actors SOU and respiratory chain (and notably complex IV) activities together

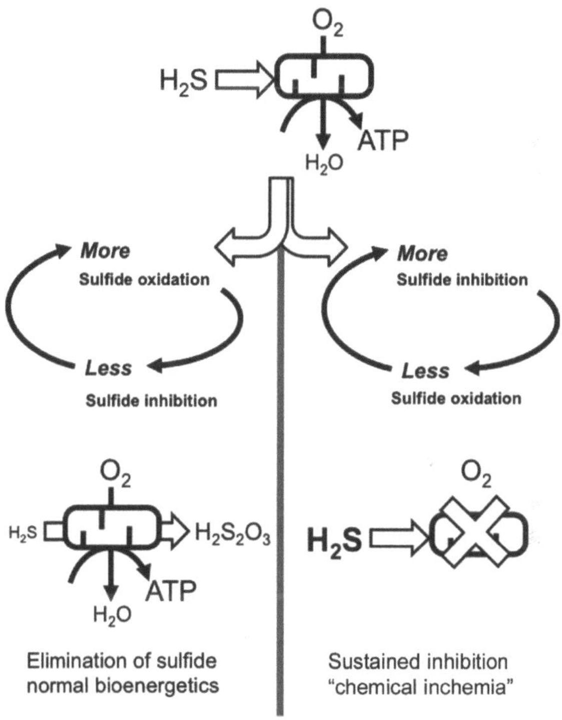

Figure 3 Positive feedback for sulfide oxidation/inhibition. From top to bottom: the response to a sulfide challenge able to inhibit partially mitochondrial respiration. Left: the mitochondria are able to decrease the sulfide concentration by mean of the SOU activity making elimination even faster, ending with a final state in which sulfide has little if any influence on the normal bioenergetics circuit. Right: SOU activity is insufficient to neutralize all the incoming sulfides therefore hydrogen sulfide accumulates and leads with time to complete inhibition of mitochondrial bioenergetics. Single additions (injection) of sulfide eventually end as shown on the left after a variable period of time necessary for the sulfide concentration to be decreased to values authorizing significant oxidation.

with sulfide delivery rate realize any critical combination in which a minor variation would result in radically opposed outcomes.

3. PRACTICAL ISSUES

3.1. Oxygen consumption

Measurement of oxygen consumption is a convenient mean to follow mitochondrial respiratory chain activity as the formation of water, the final product of oxidation, consumes oxygen. Consequently, most of our work on sulfide oxidation in mammals was based on the use of this methodology and the description of practical issues briefly mentioned in these past articles will be detailed hereafter. We use the O2k apparatus (Oroboros Instrument Inc.) equipped with the TIP2k injection pump (Helmy et al., 2014; Lagoutte et al., 2010). Of course, it is possible to arrange alternate settings and, for example, experiments were performed with another oxygen electrode (Hansatech) with (Goubern et al., 2007) or without an infusion pump (Mimoun et al., 2012). The advantages of using an infusion pump are the following: (1) slow infusion of sulfide is necessary to establish and control steady states of sulfide oxidation. (2) The dilution of the sulfide solution increases further the risk of sulfide loss, which means that the smaller the volume injected, the better an injection pump allows to inject/infuse few nanoliters. (3) Introduction of the working solution in the injection syringe results in confinement in a tight container with no air interface, and this prevents sulfide loss by evaporation and limits auto-oxidation by oxygen. For these reasons, the use of the Seahorse technology to monitor a significant sulfide oxidation rate through the SQR needs a few specific adaptations and we present hereafter elements for a possible strategy.

It is of the highest importance to bear in mind that the mitochondrial oxidation of sulfide by SOU involves a dioxygenase that constitutes a second site of oxygen consumption in addition to the mitochondrial complex IV (Figs. 1 and 2). Otherwise, the increase in oxygen consumption noticed in the presence of sulfide would lead to the conclusion that sulfide and/or its oxidation uncouples mitochondrial respiratory chain, which in fact is not the case (Helmy et al., 2014; Lagoutte et al., 2010). However, when compared with other mitochondrial substrates, sulfide oxidation takes place with a deeply modified relationship between proton pumping (ATP generation) and oxygen consumption (Fig. 2).

3.2. Use of inhibitors of mitochondrial respiration

The site of injection of electrons by SQR allows prediction with regard to the effects of inhibitors of mitochondrial respiration. Inhibition of mitochondrial complex I (made usually with rotenone) allows mitochondrial sulfide oxidation to proceed. In contrast, inhibition of mitochondrial complex III (for example, by antimycin) or of complex IV (by cyanide but also NO, CO, or sulfide itself) is expected to block sulfide oxidation, as reoxidation of coenzyme Q by the mitochondrial respiratory chain is made impossible. Inhibition of complex I by rotenone is a convenient mean to generate a situation in which sulfide is the sole/main substrate usable by cells or mitochondria. The comparison of the effect of sulfide in the presence of rotenone versus that in the presence of rotenone + antimycin/cyanide allows to determine the oxygen consumption resulting from a direct supply of electrons to quinone (hence SQR activity) and to subtract all other components including nonmitochondrial cellular oxygen consumption, spontaneous chemical reactions as well as other possible artifacts (Helmy et al., 2014; Fig. 4). Importantly, this choice of inhibitors is made on the ground of their known effect on mitochondrial respiratory complexes, and since the inhibition of the entire process of sulfide oxidation is observed, their possible effects on SQR/SOU itself remain out of reach.

3.3. Other measurements

A possibility would be to measure directly sulfide disappearance from the solution. The use of a commercially available sulfide electrode revealed difficulties because this electrode appeared less reliable (stability, lifetime) than the Clark oxygen electrode and, moreover, at physiological pH, the sensitivity of the sulfide electrode appeared hardly compatible with precise quantitation of the decrease of a concentration of sulfide in the low micromolar. However, the use of a sulfide electrode remains of interest particularly when the effect of sulfide donors is to be considered (Roger et al., 2013). Finally, measurement of other indirect consequences of sulfide oxidation could provide useful information, for example, impact of sulfide presence/absence on mitochondrial membrane potential (Goubern et al., 2007) or NADH oxidation state (Lagoutte et al., 2010). The question of mitochondrial reactive oxygen species (ROS) release in the presence of sulfide needs to be addressed. Unfortunately, our results are still too preliminary to deserve publication.

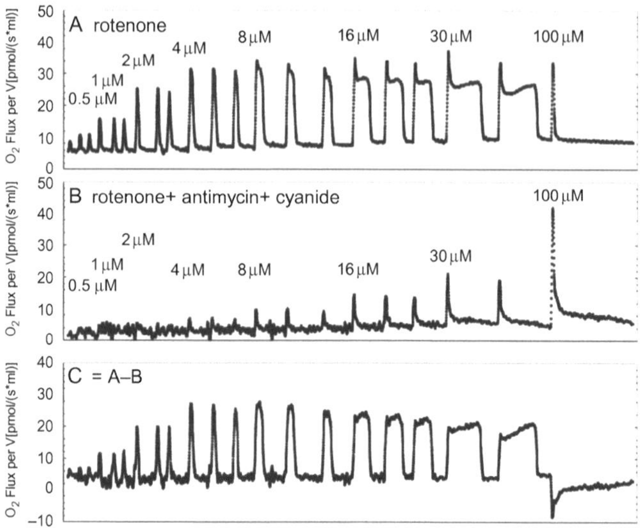

Figure 4 Use of inhibitors to evidence the sulfide oxidation. The same pattern of injection is applied simultaneously to the same suspension of rat liver mitochondria distributed in the two chambers of the O2k Oxygraph with in **(A) the presence of ADP and rotenone and in (B) the presence of ADP, rotenone, antimycin, and cyanide. The default settings were modified to use 10 data points (20 s) for rate calculation. On the bottom (C) is shown the graph resulting from the subtraction A – B which report the oxygen consumption triggered by sulfide injections dependent on the activity of complex III (=SOU, complexes III and IV pathway for oxidation, see Figs. 1 and 2). Notably this subtraction eliminates the sharp peaks present shortly after injection that reflect a dilution artifact of the solubilized oxygen (see text). However, in (B), the drift of the baseline between 16 and 100 μ*M* probably reflects a slow auto-oxidation process of the sulfide accumulated in chamber B (#200 μ*M*), while in (A), this accumulation did not occur. Consequently, this auto-oxidation process is unduly subtracted to A and biases the evaluation of the oxygen consumption rate to negative values at the end of the experiment.

3.4. Sulfide solutions

To increase the free sulfide concentration, the simplest way is to add a sulfide salt in solution that will instantaneously equilibrate into different ionic forms (HS^- S^-) and H_2S according to the pH. At physiological pH, H_2S ranges from 20% to 30% and the amount of S^- is negligible. Because sodium hemisulfide appeared closer to the ionic repartition at physiological pH, it was used in our initial experiments. Now, Na_2S is preferred for two reasons: first, the commercial preparation is usually of higher purity, and second, the stock solution (1 *M*) of Na_2S in water is highly basic and this minimizes the amount of H_2S and loss of sulfide by H_2S evaporation. For the same reason, the working solution (5 m*M*) is prepared in water with no buffering. A slow auto-oxidation of this 5 m*M* solution occurs, which is accompanied by oxygen depletion, and within several tenth of minutes, the 5 m*M* solution becomes anoxic (#250 μ*M* oxygen consumed). To protect this working 5 m*M* solution from the effect of evaporation and oxidation, it is stored exclusively in the tightly closed compartment of the injection syringes. Our observations suggest that the stock (1 *M*) solution could be stored for several days. However, as we cannot exclude a partial loss of sulfide, if the concentration/quantity of sulfide is a critical issue, we still recommend preparing a 1 *M* stock solution each day of experiment. The preparation is then as follows: a sodium disulfide crystal is weighted, the volume of water to reach a 1 *M* solution is added, after complete dissolution, the 5 m*M* working dilution is prepared, and immediately after preparation, it is loaded into the two injection syringes (0.5 ml each) of the Tip2K injection pump. Both stock and working solutions are prepared with ultrapure (milliQ) water.

3.5. Cellular and mitochondrial models

The presence of SOU is a property shared by numerous cell types or mitochondrial preparations (Table 1). However, some cell types are unable to oxidize sulfide. For example, mouse or rat brain mitochondria and the established neuronal cell lines that we could evaluate showed no SOU activity, with the consequence that only the toxic effect of sulfide could be observed (Fig. 5; Lagoutte et al., 2010). First, this is relevant information with regard to sulfide signaling and pharmacology. Second, these "SQR-free" preparations are best suited to evaluate the partial inhibition of mitochondrial respiration by micromolar concentrations of sulfide without the complications brought by SOU activity, which simultaneously increases oxygen consumption and decreases sulfide concentration. At the opposite, colonocytes, which are exposed to the sulfide released by gut microbiota as well as macrophages, show extremely high sulfide oxidation capacities, which in fact might be limited by the rest of the mitochondrial oxidative machinery (complexes I–V) and not by the SOU itself (Mimoun et al., 2012). The high activity found in macrophages might be related to the proposal that sulfide release might well be a strategy of defense used by bacteria to counteract against oxidative stress (Shatalin, Shatalina, Mironov, & Nudler, 2011). Finally, for a given organ, the ratio between SQR activity and that of other respiratory enzymes (complexes III–V) is highly influenced by the species and, for example, human liver mitochondria show maximal sulfide oxidation rates that appear sometimes to be

Table 1 SOU activity in different cell lines or mitochondrial preparations

Cell type	**Cell line**	**Sulfide oxidation**
Colonic carcinoma	HT29	High
Monocytes	THP1	High
Myoblasts	p.c.	High
Colonic carcinoma	CaCo2	High
Mammary carcinoma	ZR75-1	Med
Myotubes	Primary culture	Med
Hepatoma	Hep G2	Med
Mouse macrophages	Primary culture	Med
T cell	Immortalized locally	Med
Fibroblasts	NIH 3T3	Med
Ovarian	HeLa	Med
Ovarian	CHO	Med
Neuroblastoma	SHS-Y5Y	No
Placenta	Jeg3	No
Neuroblastoma	N2A	No
Embryonic Kidney	HEK293	No
Mitochondria from	**Species**	
Liver	Rat/mouse	Low
Liver	Man/Pig	Med
Kidney	Rat/mouse	Low
Heart	Rat/mouse	Low
Brain	Rat/mouse	No

The presentation of data for cells grown in culture (top panel) is a simplified version of that shown in Lagoutte et al. (2010). The sulfide oxidation intensity data should be interpreted as follows: "No," SOU activity could not be detected on the basis of oxygen consumption; "Low," maximal RSE 0.5 or less; "Med," maximal RSE around one; and "High," maximal RSE above 1.

one order of magnitude higher than observed with rodents (mouse, rat) when the sulfide oxidation rate is compared to the normal state 3 rate in the presence of substrate and ADP (Helmy et al., 2014). Apparently, this is also true with pig liver (see below).

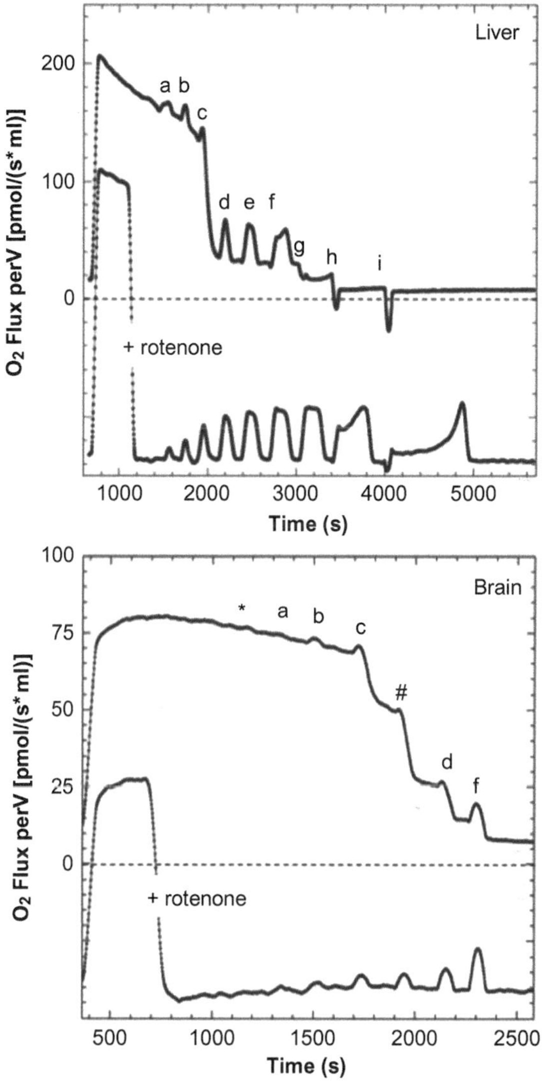

Figure 5 Injection of sulfide in the presence of a preexisting respiratory rate. The same injection pattern is applied to mitochondria in the presence of glutamate malate and ADP or to the same mitochondria but after the inhibition of complex I with rotenone (indicated on the graph). For clarity, the two traces have been shifted vertically and the numerical values of the *Y*-axis are not correct for the bottom trace but the scale is the same. The default settings: data rate 2 s and 40 data points for rate calculation were used. Top: rat liver mitochondria, the injections are as follows (in μ*M*): (a) 1, (b) 2, (c) 4, (d) 8, (e) 12, (f) 16, (g) 20, (h) 30, and (i) 50. Bottom: rat brain mitochondria, the injections are coded as above except: *0.5 μ*M* and #5 μ*M*.

The conclusion is that the SQR activity to be expected in a given experimental model is hardly predictable. Moreover, it is likely to be influenced by yet undetermined factors as our experience suggests that according to the individuals (Helmy et al., 2014) or to growth conditions (F. Bouillaud & A. Abou-Hamdan, unpublished results), the activity of SOU could vary. Finally, the association between SOU components mRNA expression and sulfide oxidation capacity may be loose enough (Mimoun et al., 2012). Therefore, it appears at least wise to evaluate experimentally the sulfide oxidation capacities of the model of interest and to do so in conditions as close as possible to the physiological or pharmacological conditions in which sulfide exposure is supposed to take place.

4. SULFIDE OXIDATION EXPERIMENTS

4.1. Addition of defined concentrations of sulfide

The simplest experiment to evaluate sulfide oxidation presence is the addition of a sulfide salt (Na_2S, for example) to cells or mitochondria. Of course, other sulfide donors might be used, and although the exact concentration of free sulfide would be unknown, the effects of the sulfide donor could be compared to that of sulfide salts providing clues about the possible efficient sulfide concentration resulting from the donor decomposition.

For the reasons mentioned before, the sulfide solution added is likely to be anoxic. Thus, the injection of the sulfide solution into an oxygen-containing medium results in a dilution effect that decreases marginally oxygen concentration but does it in a very short time (injection). The calculator interprets that as an abrupt increase in the oxygen consumption rate (Helmy et al., 2014; Figs. 4–7). Consequently, shortly after injection, the calculation of the oxygen consumption rate does not report faithfully the true oxygen consumption by the preparation studied. If the electrode responds immediately to the abrupt change in oxygen concentration and with the default setting of the O2k (2 s between data points and 40 data points used for rate calculation), this means 80 s, reducing the number of data points increases time resolution (Fig. 4) with the risk of increasing notably the noise (compare Fig. 4 with Figs. 5–8) and this becomes problematic when low oxygen consumption rates are recorded. Notably, the dilution artifact contrasts sharply with the subsequent inhibition when high concentrations of sulfide are injected (Figs. 4–7). Although calculation of this artifact appears feasible, if rate determination in the short term is necessary, then it is safer to run a

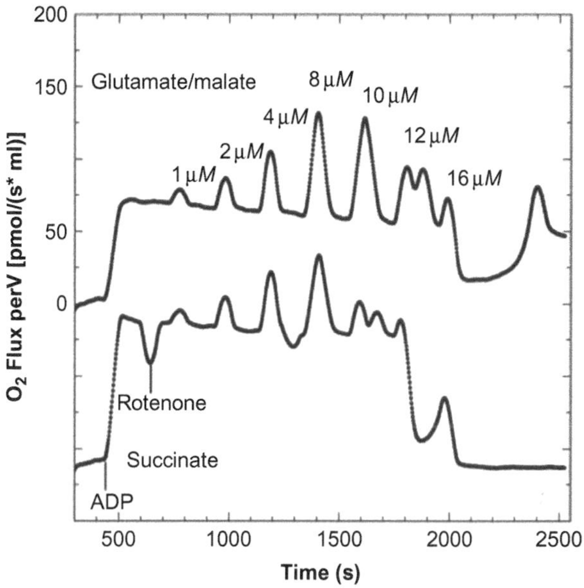

Figure 6 Pig liver mitochondria. Pig liver mitochondria are provided with glutamate malate (top trace) or succinate (bottom trace) as substrates to feed, respectively, mitochondrial complex I or complex II (see Fig. 1). The same remark as in Fig. 5 applies to the *Y*-axis. After establishment of a fast respiratory rate in state 3 by ADP addition (ADP: +1.25 m*M* of ADP) followed by rotenone (bottom trace only), the same sequence of sulfide injections was used. These mitochondria show a high SQR activity and are able to neutralize quickly relatively high amounts (concentrations) of sulfide, the addition of which causes fast and large increase in the oxygen consumption rate. However, above 8 μ*M* no further increase is seen and net inhibition of oxygen consumption takes place for 16 μ*M* (top) or 12 μ*M* (bottom). This earlier inhibition in the presence of succinate is likely to be associated to the higher respiratory rate observed in the presence of this substrate that means a stronger recruitment of complex IV activity. Moreover, the high activity of complex II makes it a potentially better competitor than complex I against SQR for quinone reduction.

parallel experiment in which SOU activity cannot take place (see above) allowing direct subtraction of the injection artifact (Fig. 4C).

4.2. Safe and toxic range for free sulfide concentration

First information brought by this type of experiment is the concentration of sulfide (donor) susceptible to inhibit respiration (toxic effect). As said before, a good model is a cell line or mitochondrial preparation where functional SQR is absent (Table 1); in this situation, the conditions for inhibition can be observed with no other complication than sulfide instability. The

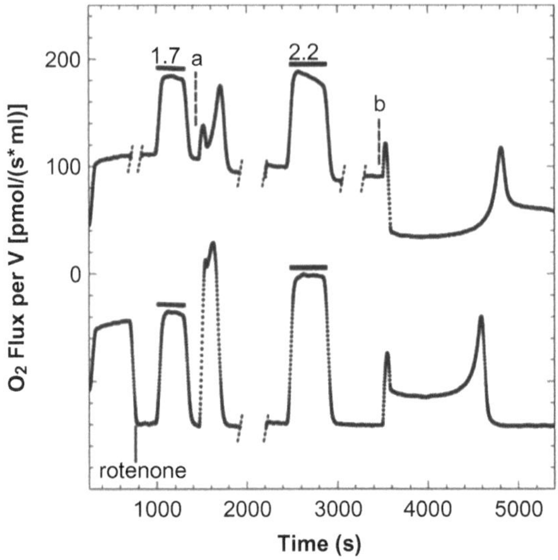

Figure 7 Comparison of infusion and injection protocol. Experiment is made CHO cells transfected with human SQR expression vector (CHO-J in Lagoutte et al., 2010) resuspended in their culture medium, in conditions of normal (routine) cellular respiration (top trace) or in the presence of rotenone (bottom trace). The same remark as in Fig. 5 applies to the *Y*-axis. The sulfide (Na_2S solution 5 m*M*) is added by infusion to for the period indicated by the horizontal black bar. Interruptions in the traces correspond to periods of reoxygenation of the medium. The rate value is given in the relative units (RSE) which is the ratio between the sulfide flux and the reference oxygen consumption flux value (expressed in the same units) during normal cellular respiration as observed at the beginning of the experiment (before the first sulfide infusion (top) or before rotenone addition (value is around 100 pmol/(s ml))). The concentration of cells is 1.10×10^{-6} cells/ml. Each infusion step is followed by the injection of the same quantity of sulfide leading to the final concentrations of 50 μ*M* (a) and 88 μ*M* (b).

respiration must take place and could be either based on the use of the endogenous respiratory rate (cells) or stimulated by the presence of substrate (and possibly ATP or uncoupler) as with isolated mitochondria or permeabilized cells. With this approach, the concentration dependence was determined with cells and sodium sulfide (Lagoutte et al., 2010): inhibition could be detected with 5 and 20 μ*M* caused a 50% decrease in cellular oxygen consumption. Since not all the cellular oxygen consumption is caused by mitochondrial respiration, this 20 μ*M* overestimates somehow the actual IC_{50}. Figure 5 shows similar experiments made with mitochondria from rat brain (no SQR activity) and liver (low SQR activity). In these experiments, a 4 μ*M* concentration was enough to record a significant inhibition

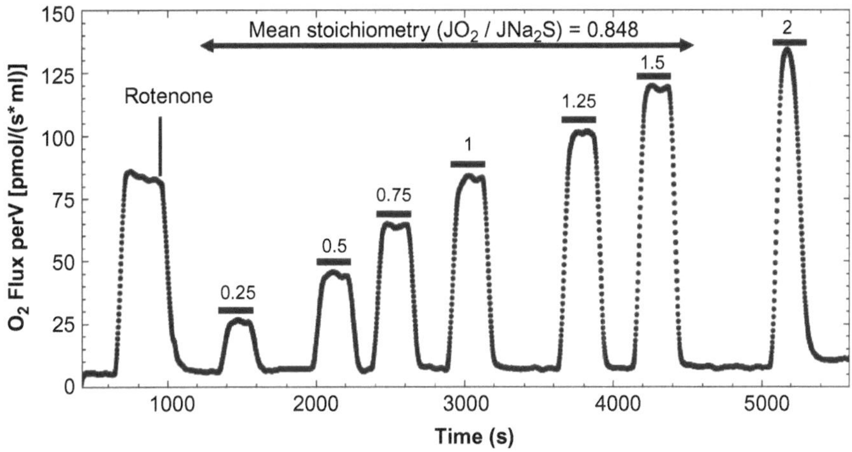

Figure 8 Challenging pig liver mitochondria with increasing RSE. Pig liver mitochondria respiration is settled to state 3 by addition of glutamate malate (5 m*M* each) followed by ADP (1.25 m*M*). This allows the measurement of the reference state 3 rate used for the calculation of the RSE used (see text and legend of Fig. 7). The horizontal black bars indicate infusions time. The last infusion step (RSE = 2) caused a decline of the oxygen consumption before the infusion ceased. This indicates the rise in sulfide concentration to values in the inhibiting range and thus that sulfide oxidation rate by mitochondria is significantly lower than the infusion rate. Accordingly, the maximal RSE value that could be supported by these mitochondria is 1.5 hence 120 pmol/s ml (in the present experiment) or 200 pmol/s mg protein.

of mitochondrial respiration. It should be noticed that the IC_{50} values obtained with respiring preparations (cells/mitochondria) are more than one order of magnitude higher than the values obtained when the effect of sulfide on the enzymatic reaction catalyzed by cytochrome oxidase is studied (Cooper & Brown, 2008; Leschelle et al., 2005). The reasons for this large difference remain unclear. The degree of inhibition to be considered as critical depends on how flexible is cellular metabolism with regard to energy supply/consumption (Leschelle et al., 2005).

A practical conclusion is that the interpretation of experiments using hundreds of micromolar or even millimolar concentrations of sulfide salts should consider the occurrence of a complete but transient inhibition of mitochondrial respiration and its possible consequences. The time necessary to restore respiration depends on the stability of sulfide with the possibility (if sulfide is highly unstable in the protocol used) that the concentration declared was actually never present, thus not meaningful and the inhibition less severe than it should be anticipated.

At this point, it is important to recognize that this transient inhibition of mitochondrial respiration triggered by a single sulfide addition recalls the sequence of events taking place during ischemic preconditioning.

4.3. Concentration dependence of SOU activity

SOU activity could be best evidenced in cells/mitochondria in conditions where no other substrate than sulfide is usable. This is the case when the mitochondrial complex I is inhibited (usually with rotenone), and this blockade of the reoxidation of NADH slows down cellular respiration to a value close to that obtained with complete inhibition of mitochondrial respiratory chain. This occurs probably because carbon metabolism stalls and pathways other than complex I able to yield hydrogen to coenzyme Q run out of substrates. Another possibility is to administrate sulfide in the presence of a preexisting mitochondrial/cellular respiratory rate supported by endogenous or exogenous carbon-containing substrate(s). This situation is closer to physiology as sulfide oxidation is supposed to occur simultaneously with all the other "normal" organic substrates available, but the interpretation is more complicate. When SOU is present, the addition of sulfide in the high nanomolar and low micromolar range causes a sharp increase in the oxygen consumption rate to be distinguished from the injection artifact (Fig. 4). With this approach, the estimation of the apparent K_m for sulfide oxidation by isolated mitochondria led to a value around 2 μM (Szabo et al., 2014), thus lower but close enough to the concentration causing a detectable inhibition of complex IV (4–5 μM and more, see above). A net inhibition of sulfide oxidation by a too high sulfide concentration is detected when a further increase in sulfide concentration in fact decreases the SOU reaction rate (above 8 μM in Fig. 4). The mechanism for this inhibition lies in the dependence of SOU activity from the activity of complex IV. In fact sulfide inhibits the complex IV in a non-competitive manner (Cooper & Brown, 2008) and therefore decreases the maximal activity without affecting the affinity for oxygen. Accordingly, if SOU activity is modest with regard to that of complex IV, a considerable inhibition of the latter (hence high concentration of sulfide) might be tolerated before the sulfide oxidation rate declines (Fig. 5). When other mitochondrial oxidations (normal carbon metabolism) take place at the same time than sulfide oxidation a net decrease in oxygen consumption rate after sulfide injection directly evidences the inhibition by sulfide (Figs. 5 and 6). However, the presence of the dioxygenase in SOU complicates the situation because when sulfide is oxidized, two-thirds of this oxygen consumption is explained by the dioxygenase of SOU and one-third by complex IV (Fig. 2). Schematically, if SOU is not limiting, more than a 67% inhibition of the electron flux in the mitochondrial respiratory chain would be necessary to result a net decrease in oxygen consumption. It exists therefore a range of concentrations of sulfide with which no decrease or even an increase in oxygen consumption is observed although a net decrease in electron transfer in mitochondrial respiratory chain (and deterioration of the ATP production) is likely to occur. Concentrations in the 10–100 μ*M* range are usually required to observe a net decrease in cellular/mitochondrial oxygen consumption (Figs. 5–7), and this is heavily influenced by the maximal activity of SOU that is present. However, even in conditions of severe inhibition, sulfide-consuming reactions continue at a low rate and moreover non-biological processes (evaporation auto-oxidation) participate to an unavoidable sulfide concentration decline. Because of the positive feedback loop, this recovery phase shows an acceleration of oxygen consumption that culminates short before returning to the initial rate.

Therefore, to observe SQR activity with a minimal interference of the inhibiting effects of sulfide, it is advisable to maintain the sulfide concentration in the 10 μ*M* range or below. First, it makes impossible to establish steady states of sulfide addition with a single saturating addition of sulfide solution like it is currently made with the other mitochondrial substrates (5–20 m*M* succinate, for example). Second, this 10 μ*M* concentration is not largely above the K_m with the risk that the SQR oxidation rate declines with time (sulfide concentration). Third, the oxygen consumption rates that could be quantified with a Clark electrode would not allow the quantity of sulfide available in the 10 μ*M* solution to feed oxidation for more than tenth of seconds or few minutes.

4.4. Establishment of steady states by infusion

Infusion of a solution of sulfide is a simple mean to circumvent this toxicity problem (Fig. 7): a sulfide solution is infused by a pump at a fixed rate. As long as this rate of infusion is lower than the maximal possible rate of sulfide oxidation in the studied preparation, different steady states can be established with the sulfide infusion rate compensated by the sulfide oxidation rate (Fig. 8). Accordingly, cessation of infusion results immediately into a decrease in oxygen consumption and return to the rate observed before infusion started. At the opposite, if the infusion rate overmatches, the oxidation rate sulfide accumulates, which is revealed by continuation of the sulfide-stimulated oxygen consumption after cessation of infusion or inhibition of respiration that develops during infusion (last infusion in Fig. 8, see also examples in Goubern et al., 2007; Helmy et al., 2014; Lagoutte et al., 2010; Szabo et al., 2014). An important consequence of this infusion protocol is that when delivery and oxidation are matched, the amount of sulfide oxidized is directly given by the product (infusion rate) × (concentration) and the flux of oxidized sulfide could be expressed in units of flux (mol/s) relatively to volume, mg of proteins, or number of cells. With the Oroboros equipment, we use a 5 m*M* Na_2S working solution that is infused at rates ranging from few to several tenth of nanoliters per second.

4.5. Seahorse

The "Seahorse" technology has become very popular to measure cellular respiration and to evaluate lactate release. The users of the Seahorse interested to evaluate sulfide oxidation should be aware of several issues. The difficulties stem from the fact that the measuring Seahorse plate remains essentially open to the outer atmosphere. A further complicating factor is the fact that the time needed to get one value of respiratory rate is relatively long and particularly because equilibration and mixing times are also to be considered. A low micrometer solution of sulfide at physiological pH would hardly be stable during equilibration and mixing time. It is therefore unrealistic to envisage that sulfide would be added from the beginning. It should then be added during the run with the injection ports. Accordingly, to record possible sulfide oxidation rate of cells in the Seahorse, we used the following modified procedure. First, we load the injection ports with a 10× solution of sulfide leading to final concentration of 0.1–1 m*M*, far too concentrated when compared to the classical experiments and if actually present expected to inhibit severely the cellular respiration. Second, we record for extended period of time the evolution of oxygen concentration (30 min) instead of the 3–5 min used in the several successive cycle of reading used in standard "Seahorse" protocols. The treatment of data would need to consider the evolution of oxygen concentration in individual wells (Fig. 9). First, injection of the sulfide solution produces a large artifact (oxygen decrease) followed by a variable period of inhibition of respiration. Finally, respiration resumes indicating that sulfide has been consumed. In the presence of rotenone, a sharp increase in oxygen consumption followed by return to a zero rate is observed. The interpretation is then the following:

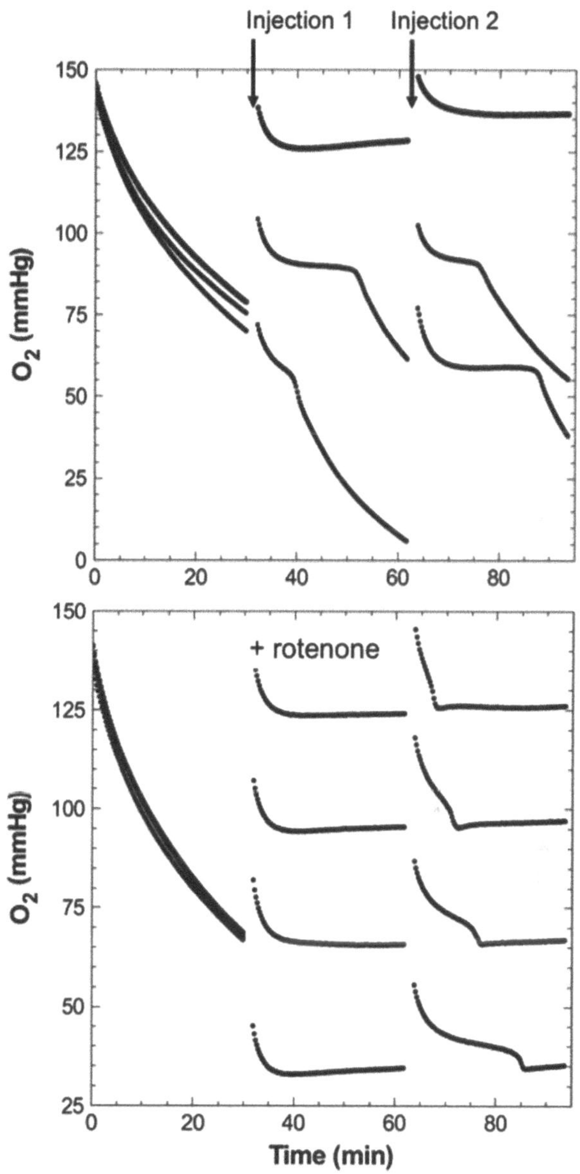

Figure 9 Cellular sulfide oxidation with the Seahorse. The CHO cells used in this "Seahorse" experiment are the same as in Fig. 7 (CHO-J). Three single steps of 30 min of measuring time are used: one before any injection (normal cellular respiration) and the two other after the first and the second injection of a sulfide solution (injections 1 and 2) leading to a theoretical final concentration of 250 μ*M* Na_2S. The graph shows the oxygen concentration traces, in three individual wells. These are unusual protocol and representation of results with the Seahorse. In the bottom graph, the first injection step added rotenone (1 μ*M* final) in addition to sulfide. First, the sulfide

(Continued)

the storage period in the injection port and the mixing step after injection strongly promote H_2S evaporation. Therefore, when the apparatus closes the chamber for oxygen and pH detection, the cells are facing an unknown sulfide concentration, which is lower than that obtained from the dilution factor but still able to inhibit respiration. As said before (in Section 4.3) even in conditions of extreme inhibition, the sulfide concentration declines. Therefore, when the concentration is made low enough to allow respiration, and sulfide oxidation to resume, the positive feedback sequence of events (Fig. 3) takes place with higher and higher respiratory rates (and sulfide oxidation) ending abruptly when sulfide is exhausted. This sequence of events starts at various times after the sulfide injection because the overall sulfide decay is expected to differ slightly from one well to the other and the positive feedback amplifies the consequences of small differences in the initial state. The extraction of an absolute rate of oxygen (and sulfide) consumption appears somehow complicate because the sulfide content of the measuring chambers does not allow more than a few data points to represent the sulfide oxidation phase, which moreover is expected to take place with a highly variable rate (consider the similar sequence of events as shown after the second injection in the top trace of Fig. 7). However, the Seahorse approach could reveal if a given cell type is able or not able to oxidize sulfide and would allow at least semiquantitative studies as the fastest rate observed is likely to approach the maximal rate of oxygen consumption possible in the presence of sulfide. A possible alternative/evolution would be to use sulfide donors able to release and maintain low micromolar sulfide for extended period of time, while it is consumed by the cells.

5. TREATMENT, EXPRESSION, AND INTERPRETATION OF RESULTS

5.1. Steady-state experiments

In the case steady states are established in which the infusion rate is compensated by the oxidation rate, the latter can be directly calculated from the

Figure 9—Cont'd injection caused a large dilution artifact with an immediate drop in the oxygen concentration reported by the detector. Second, when cellular respiration resumes after sulfide injection (top graph) this occurs at different time in the different wells. Third, in the presence of rotenone (bottom graph), a short sequence of accelerating oxygen consumption rate followed by an abrupt return to zero (flat line for oxygen value) indicates a period of a sulfide-based respiration, and the comparison with the injection steps in Fig. 7 is useful.

infusion parameters in mol/time, which should then be expressed relatively to a dimension related to the recipient material (volume, number of cells, mg protein). Actually, both the sulfide oxidation rate and the oxygen consumption rate (calculated from oxygen concentration decrease) can be expressed in the same unit. The ratio between theses two numbers has no dimension and reflects different realities according to the oxygen consumption rate considered as the reference.

If the oxygen consumption rate during sulfide infusion is considered as the reference, then this ratio is related to the experimental stoichiometry between sulfide and oxygen during the oxidation process (Fig. 8). Of course, the sulfide-independent rate has to be taken into account and this is usually the increase in oxygen consumption that is used for this calculation of oxygen to sulfide stoichiometry. This calculation allows predictions with regard to the reaction scheme for sulfide oxidation (Goubern et al., 2007; Lagoutte et al., 2010), with the possible reserves about the purity of the Na_2S solution. At the opposite, if the stoichiometry is supposed to be constant, then a variation in this ratio according to the mitochondrial/cellular respiration regime should reveal the interaction between the sulfide oxidation and the other oxygen-consuming reactions (Lagoutte et al., 2010).

If the oxygen consumption rate before initiating the sulfide infusion is considered as a reference, this ratio defines a relative sulfide exposure (RSE) imposed to the respiring preparation; this relative quantitation is meaningful and convenient.

Meaningful: The oxidation of sulfide is dependent on two functional blocks working sequentially (Fig. 1): the first is SOU that generates the waste thiosulfate and yields electrons and protons to quinone. The second block is the mitochondrial respiratory chain, in fact complexes III–V when the mitochondrial respiration is controlled by ATP demand, which is the physiological state. The respiratory rate in the absence of sulfide provides therefore a good estimation of the possible electron flux inside complexes III and IV (as long as in the absence of sulfide mitochondrial respiration is not limited by substrate supply). This possible electron flux limits the maximal rate of sulfide oxidation to the rate possible when all electrons traveling through complexes III and IV originate from sulfide. In other words, the rate of endogenous respiration in the absence of sulfide allows a calculation of the maximal possible sulfide oxidation when SOU activity is made high enough and not anymore the controlling factor. Consideration of the reaction scheme proposed in Hildebrandt and Grieshaber (2008) leads to the conclusion that the maximal RSE would be $(2H_2S/1/2O_2)=4$, which (because of the major contribution of the dioxygenase; Figs. 1 and 2) would cause a threefold increase in the oxygen consumption rate of a respiring preparation (Lagoutte et al., 2010). Within the same theoretical frame, other equations for the sulfide oxidation reaction would produce different numbers. In coupled mitochondria and because sulfide oxidation feeds only proton pumping by complexes III and IV, the model would need further refinement as the maximal SOU-driven electron flux is expected to be higher than that in the presence of a complex I substrate to compensate for the decrease in the H^+/O ratio associated with the bypass of complex I (Fig. 2).

Convenient: It allows to compare different experiments made with different preparations (mitochondria, homogenates, cells, ...) on the basis of a common internal standard. During experiments, the observation of the endogenous (cells) or possible respiratory rate (in the presence of carbon-containing substrates) allows immediate determination of realistic values for the flux of sulfide that could be administrated to the preparation.

5.2. Injection experiments

A difficulty with infusion experiments is that the actual concentration of sulfide present remains unknown (unless a sulfide electrode is integrated in the measuring chain). The fast injection of a given amount of sulfide is expected to raise abruptly the sulfide concentration to the value calculated by the dilution factor. However, in most case, this concentration would start to decline shortly after injection because of cellular/mitochondrial consumption of sulfide. The interpretation of the results obtained with injections of sulfide has therefore to face the difficulty that the injections induce a sequence of events during which the oxygen/sulfide consumption rate changes continuously. The ideal situation is encountered when following the injection artifact; there is a sufficient period of stable or marginally changing oxygen consumption rate supposed to represent a situation in which SOU is operating close to its maximal possible rate (Figs. 4 and 5). The more complex situation combines the dilution artifact (sharp increase) followed by inhibition (prolonged decrease) and restoration of respiratory activity and if SOU is present by a marked stimulation before return to the original rate. The consideration of the overall duration of these perturbations provides an estimation of an average sulfide oxidation rate with the interest that this evaluation is independent on the actual values of oxygen consumption and of the mechanism involved. It allows, for example, comparison of this average sulfide oxidation rate in the presence/absence of rotenone, which, in Fig. 7, appears shorter in the presence of rotenone after the second injection. This indicates a negative interaction (competition) between complex I and SOU (Fig. 1), but at the opposite in other models, a synergy between the two might occur (Lagoutte et al., 2010; Szabo et al., 2014). Finally, it might be advantageous not to consider the oxygen consumption rate but the oxygen concentration and to deduce the oxygen quantity used to oxidize a given amount of sulfide. This strategy is particularly suited to evidence oxidation in the low range of concentration, and moreover, it provides another estimation of the oxygen to sulfide stoichiometry (Szabo et al., 2014).

6. ORIGINALITY AND INTEREST WITH REGARD TO BIOENERGETICS

Although the determination of the sulfide oxidation capacity of a given respiring preparation would be the main usefulness of the information given in this chapter, we would like to stress out few further characteristics of sulfide oxidation that makes it attractive as mean to illustrate cellular activity/behavior and notably with regard to important basic features of mitochondrial bioenergetics. Firstly, sulfide effect on mitochondria cells show an example of biphasic effect with concentration-dependent opposite effects according to the mode of administration for the same amount engaged (Lagoutte et al., 2010; Szabo et al., 2014; Fig. 7). Second, sulfide effect on cellular respiration provides an example of positive feedback phenomenon causing instability within a critical zone with opposed consequences depending on minor variations in the initial conditions.

6.1. Stoichiometric calculations

The high affinity of SQR means that mitochondria consume quickly all sulfide available. Therefore, a direct calculation of stoichiometry oxygen/substrate is feasible, something uneasy with other substrates for the mammalian mitochondria. The stoichiometry could be obtained both from experiments involving single additions (Szabo et al., 2014) or using infusion protocols (Lagoutte et al., 2010). In our hands, when sulfide is the only available substrate, the experimental values for the stoichiometry (sulfide/O_2) were close enough to the theoretical value of 0.75 derived from the oxidation mechanism as explained in Hildebrandt and Grieshaber (2008). However, this reaction scheme is challenged (Jackson, Melideo, & Jorns, 2012), and experimental values could be criticized on the ground of the purity of sulfide. An important observation is that the stoichiometry is changing according to the conditions in which sulfide oxidation takes place and these changes bring information about the constraints of mitochondrial bioenergetics (Lagoutte et al., 2010).

6.2. Reduction of coenzyme Q and competition between electron donors

In conditions allowing oxidation of other substrates to take place, the addition of sulfide in the micromolar range to respiring cells or mitochondria leads to exhaustion of the incoming sulfide although the concentration of the other substrates (glucose, glutamate/malate, succinate) is 1000 times greater (millimolar range). This is explained by the high affinity of the SQR for sulfide. Then the question is whether sulfide oxidation takes place in addition to or if it replaces partly these other oxidation processes. The answer to this question determines whether or not in the presence of sulfide mitochondrial oxidation remains controlled by respiratory coupling. The change in the apparent stoichiometry observed in the absence/presence of rotenone indicates that when other cellular oxidation processes take place, the dioxygenase reaction only could explain the increase in oxygen consumption (Lagoutte et al., 2010). Therefore, at the level of the respiratory chain, the sulfide oxidation replaces other oxidation processes. This reveals that the precedence of sulfide oxidation over other oxidative substrates together with the fact that cellular respiration, including when sulfide has to be oxidized, is subject to respiratory coupling. With some cell types, the situation could be approached where calculation predicts that almost all the electrons entering the mitochondrial respiratory chain come from sulfide although other substrates are present and could feed to the same extend the respiratory chain. Accordingly, evaluation of cellular NADH reduction shows burst of reduction when sulfide is injected (Lagoutte et al., 2010).

6.3. Reverse flux in complex I

The study of sulfide metabolism in colonocytes, which are cells highly adapted to sulfide exposure allowed to evidence for a quantitatively significant reverse flux in complex I which causes uncoupling of respiration. This led to the surprising observation that in these cells, under conditions allowing sulfide oxidation, rotenone is a coupling factor that reduces the capacity of these cells to neutralize sulfide (Lagoutte et al., 2010).

For all these reasons: importance of administration mode, calculation of stoichiometry, and reverse electron transfer, the study of sulfide oxidation constitutes a remarkable example of training in bioenergetics with the aim of driving the trainees into the use of their knowledge in a somehow unconventional, and thus challenging, situation. Finally, most experiments are feasible with living cells in suspension therefore with no need for preparation of mitochondria.

ACKNOWLEDGMENTS

The authors wish to thank the staff of the surgical center "Marie Lannelongue" (CCML), Le Plessis Robinson, France, for giving us the opportunity to access to pig liver samples. This work was performed within the "Département Hospitalo-Universitaire" (DHU) AUToimmune and HORmonal diseaseS and supported by grants from the "Fondation ARC," and the "Université Sorbonne Paris Cité."

REFERENCES

Birrell, J. A., & Hirst, J. (2013). Investigation of NADH binding, hydride transfer, and NAD (+) dissociation during NADH oxidation by mitochondrial complex I using modified nicotinamide nucleotides. *Biochemistry, 52*(23), 4048–4055. http://dx.doi.org/10.1021/bi3016873.

Bouillaud, F., & Blachier, F. (2011). Mitochondria and sulfide: A very old story of poisoning, feeding, and signaling? *Antioxidants & Redox Signaling, 15*(2), 379–391. http://dx.doi.org/10.1089/ars.2010.3678.

Cooper, C. E., & Brown, G. C. (2008). The inhibition of mitochondrial cytochrome oxidase by the gases carbon monoxide, nitric oxide, hydrogen cyanide and hydrogen sulfide: Chemical mechanism and physiological significance. *Journal of Bioenergetics and Biomembranes, 40*(5), 533–539.

Goldbaum, L. R., Ramirez, R. G., & Absalon, K. B. (1975). What is the mechanism of carbon monoxide toxicity? *Aviation, Space, and Environmental Medicine, 46*(10), 1289–1291.

Goubern, M., Andriamihaja, M., Nubel, T., Blachier, F., & Bouillaud, F. (2007). Sulfide, the first inorganic substrate for human cells. *FASEB Journal, 21*(8), 1699–1706.

Haouzi, P., Sonobe, T., Torsell-Tubbs, N., Prokopczyk, B., Chenuel, B., & Klingerman, C. M. (2014). In vivo interactions between cobalt or ferric compounds and the pools of sulphide in the blood during and after H_2S poisoning. *Toxicological Sciences, 141*(2), 493–504. http://dx.doi.org/10.1093/toxsci/kfu140.

Helmy, N., Prip-Buus, C., Vons, C., Lenoir, V., Abou-Hamdan, A., Guedouari-Bounihi, H., et al. (2014). Oxidation of hydrogen sulfide by human liver mitochondria. *Nitric Oxide, 41*, 105–112. http://dx.doi.org/10.1016/j.niox.2014.05.011.

Hildebrandt, T. M., & Grieshaber, M. K. (2008). Three enzymatic activities catalyze the oxidation of sulfide to thiosulfate in mammalian and invertebrate mitochondria. *The FEBS Journal, 275*(13), 3352–3361.

Jackson, M. R., Melideo, S. L., & Jorns, M. S. (2012). Human sulfide:quinone oxidoreductase catalyzes the first step in hydrogen sulfide metabolism and produces a sulfane sulfur metabolite. *Biochemistry, 51*(34), 6804–6815. http://dx.doi.org/10.1021/bi300778t.

Lagoutte, E., Mimoun, S., Andriamihaja, M., Chaumontet, C., Blachier, F., & Bouillaud, F. (2010). Oxidation of hydrogen sulfide remains a priority in mammalian cells and causes reverse electron transfer in colonocytes. *Biochimica et Biophysica Acta, 1797*(8), 1500–1511. http://dx.doi.org/10.1016/j.bbabio.2010.04.004.

Leschelle, X., Goubern, M., Andriamihaja, M., Blottiere, H. M., Couplan, E., Gonzalez-Barroso, M. D., et al. (2005). Adaptative metabolic response of human colonic epithelial

cells to the adverse effects of the luminal compound sulfide. *Biochimica et Biophysica Acta, 1725*(2), 201–212.
Mimoun, S., Andriamihaja, M., Chaumontet, C., Atanasiu, C., Benamouzig, R., Blouin, J. M., et al. (2012). Detoxification of H(2)S by differentiated colonic epithelial cells: Implication of the sulfide oxidizing unit and of the cell respiratory capacity. *Antioxidants & Redox Signaling, 17*(1), 1–10. http://dx.doi.org/10.1089/ars.2011.4186.
Roger, T., Raynaud, F., Bouillaud, F., Ransy, C., Simonet, S., Crespo, C., et al. (2013). New biologically active hydrogen sulfide donors. *Chembiochem, 14*(17), 2268–2271. http://dx.doi.org/10.1002/cbic.201300552.
Rossignol, R., Faustin, B., Rocher, C., Malgat, M., Mazat, J. P., & Letellier, T. (2003). Mitochondrial threshold effects. *The Biochemical Journal, 370*(Pt 3), 751–762. http://dx.doi.org/10.1042/BJ20021594.
Shatalin, K., Shatalina, E., Mironov, A., & Nudler, E. (2011). H_2S: A universal defense against antibiotics in bacteria. *Science, 334*(6058), 986–990. http://dx.doi.org/10.1126/science.1209855.
Szabo, C., Ransy, C., Modis, K., Andriamihaja, M., Murghes, B., Coletta, C., et al. (2014). Regulation of mitochondrial bioenergetic function by hydrogen sulfide. Part I. Biochemical and physiological mechanisms. *British Journal of Pharmacology, 171*(8), 2099–2122. http://dx.doi.org/10.1111/bph.12369.
Vitvitsky, V., Kabil, O., & Banerjee, R. (2012). High turnover rates for hydrogen sulfide allow for rapid regulation of its tissue concentrations. *Antioxidants & Redox Signaling, 17*(1), 22–31. http://dx.doi.org/10.1089/ars.2011.4310.

CHAPTER THIRTEEN

Redox Regulation of Mammalian 3-Mercaptopyruvate Sulfurtransferase

Noriyuki Nagahara*,[1], Masatoshi Nagano[†], Takaaki Ito[‡], Hidenori Suzuki[†]

*Isotope Research Center, Nippon Medical School, Tokyo, Japan
[†]Department of Pharmacology, Nippon Medical School, Tokyo, Japan
[‡]Department of Pathology and Experimental Medicine, Graduate School of Medical Science, Kumamoto University, Kumamoto, Japan
[1]Corresponding author: e-mail address: noriyuki@nms.ac.jp

Contents

Abstract

A cystine-catabolizing enzyme, 3-mercaptopyruvate sulfurtransferase catalyzes the trans-sulfuration reaction of mercaptopyruvate or thiosulfate to thiol-containing compounds or cyanide. During the catalytic process, persulfide is formed at the catalytic site cysteine residue and a sulfur-acceptor substrate donates the outer sulfur of the persulfide to form a new persulfide molecule. Subsequently, the molecule can be reduced by thioredoxin to form hydrogen sulfide. The enzyme is regulated by redox changes via two redox-sensing molecular switches consisting redox-sensitive cysteine residues. One switch is the catalytic cysteine in itself, which is oxidized to form a cysteine-sulfenate resulting in inhibition of catalytic activity. The sulfenate can be reduced by thioredoxin resulting in restoration of the activity. The redox potential

Methods in Enzymology, Volume 554
ISSN 0076-6879
http://dx.doi.org/10.1016/bs.mie.2014.11.017

of sulfenate is lower than that of glutathione and greater than that of thioredoxin. The other switch involves cysteine residues positioned on the surface of the enzyme. The oxidation the intermolecular disulfide linkage at these cysteine residues, leading to dimer formation, inhibits enzyme activity. On the other hand, reduction-associated monomer formation increases catalytic activity. Thioredoxin reduces the disulfide bond more effectively than dithiothreitol, although the specificity mechanism has not been identified. Congenital defects in this enzyme result in, mercaptolactate-cysteine disulfiduria associated with or without mental retardation. However, the pathogenesis has not been identified. Because 3-mercaptopyruvate sulfurtransferase serves as a cellular antioxidative protein, the other biological functions related to the inhabitant disease are being investigated.

1. INTRODUCTION

1.1. History and molecular properties

3-Mercaptopyruvate sulfurtransferase (MST, EC 2.8.1.2) was discovered in the rat liver (Meister, 1953; Wood & Fiedler, 1953), as a cystine-catabolizing enzyme involved in the mercaptopyruvate pathway (Nagahara & Sawada, 2006). MST is distributed throughout prokaryotes and eukaryotes (Jarabak & Westley, 1978; Meister, 1953; Wood & Fiedler, 1953). The enzyme was first obtained in its highly purified form from rat liver (Nagahara, Okazaki, & Nishino, 1995) and was first cloned by Nagahara and Nishino (1996). The tertiary structures of MSTs isolated from *Escherichia coli* (Spallarossa et al., 2004) and *Leishmania major* (Alphey, Williams, Mottram, Coombs, & Hunter, 2003) are known but that of mammalian MSTs have not been determined.

Rat MST consists of 296 amino acids (molecular mass, 32,808 Da) and is a simple protein enzyme (Nagahara & Nishino, 1996). The enzyme contains three exposed cysteines: one catalytic site cysteine, Cys^{247}, and two cysteines on the surface of the enzyme, Cys^{154} and Cys^{263} (Fig. 1; Nagahara & Nishino, 1996). The cysteine corresponding to position Cys^{263} is conserved among mammalian MSTs; on the other hand, Cys^{154} is unique to murine MST.

The finding that MST is evolutionarily related to mitochondrial rhodanese is of importance; sequence identity in protein and cDNA between rat MST and rhodanese are 60% and 65%, respectively (Nagahara & Nishino, 1996; Nagahara et al., 1995). An MST motif sequence "GSG" follows the catalytic site cysteine of MST, and a rhodanese motif sequence "RKG" follows the catalytic site cysteine of rhodanese (Fig. 1; Nagahara & Nishino, 1996; Nagahara et al., 1995). Using site-directed mutagenesis of these amino

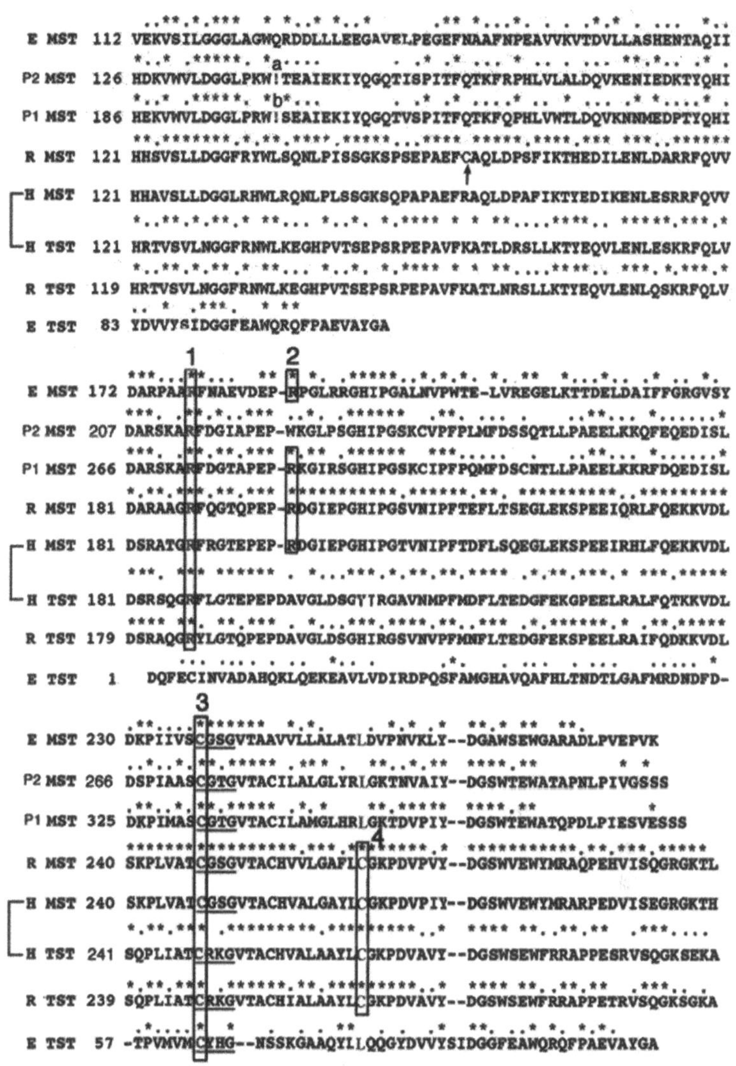

Figure 1 Amino acid sequences of 3-mercaptopyruvate sulfurtransferases (MST) and rhodaneses (TST). The key to binomial names is as follows: E, *Escherichia coli* (D10496 for MST, NP_417883 for TST); H, *Homo sapiens* (BC009450 for MST, D87292 for TST); P1 and P2, *Arabidopsis thaliana* (AB032864 and AB032865 for MSTs); and R, *Rattus norvegicus* (D50564 for MST, BC088449 for TST). Symbols used to highlight specific amino acid residues are as follows: Box 1, a substrate binding site; Box 2, another substrate binding site; Box 3, catalytic site; Box 4, redox-sensing switch; underlined, consensus sequence; arrow, Cys[154] of rat MST; ·, related amino acid residues; *, identical amino acid residues; a!, amino acid sequence "RASGYDVESSASGDAILKASAA" omitted; and b!, amino acid sequence "RASGYDVESSVSNDAILKA" omitted. Sequence identity was analyzed using the GENETYX analysis program. *Modified fig. 1 from Nagahara (2007).*

acids, we succeeded in switching the catalytic activity of MST to that of rhodanese and vice versa (Nagahara & Nishino, 1996; Nagahara et al., 1995).

Structural features of the gene encoding the enzyme have also been identified (Nagahara & Nishino, 1996; Nagahara et al., 2004). Promoter activity assay and cap site hunting analysis for human MST (Nagahara et al., 2004) revealed that the promoter region has features of a typical promoter of a housekeeping gene, containing GC box and silencer element (Nagahara et al., 2004). We also showed that each point mutation in the silencer module (−394G→T or C, −393C→T, or −392T→G; nucleotide A of initiation ATG is designated as +1) markedly enhanced the silencing function and reduced translation of MST (Nagahara et al., 2004). It is interesting to note that, MST expressions in the mouse brain and lung were increaed on embryonic Day 14 and were identical or slightly decreased in postnatal period (Fig. 2).

1.2. Catalytic properties

MST catalyzes a trans-sulfuration reaction in which the sulfur of mercaptopyruvate or thiosulfate is transferred to thiol-containing compounds or cyanide (Nagahara & Nishino, 1996; Nagahara et al., 1995). K_m values for mercaptopyruvate and thiosulfate are 1.2 and 73 mM, respectively (Nagahara et al., 1995). During the catalytic process on the rat enzyme, Arg^{187} at the orifice of the active center serves as a binding site of the substrate, mercaptopyruvate, and a critical residue for catalysis. The positively charged nitrogen of the side chain of Arg^{187} interacts with the oxygen of the carbonyl group of mercaptopyruvate (referred as to nucleophilic assistance; Nagahara & Nishino, 1996).

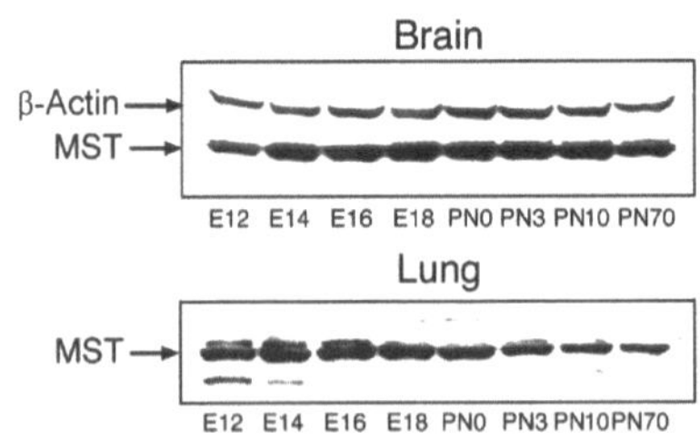

Figure 2 Expression of 3-mercaptopyruvate sulfurtransferase (MST) in embryonic and postnatal stages in mice MST expressions in the mouse brain and lung increased on embryonic Day 14. Western blot analyses were performed on the mouse brain and lung samples at embryonic days (E) 12, 14, 16, and 18 and postnatal days (PN) 3, 10, and 70. β-Actin was used as the loading control.

The catalytic site cysteine is redox active and the pK_a (approximately 6.0 in rat MST) is lower than that of free cysteine. After substrate binding, a stable persulfide is formed at the catalytic site cysteine as a reaction intermediate. It is noteworthy that the persulfide is stable at 4 °C, for at least 1 month. In rat MST, hydrogen bonds between the outer sulfur of the persulfide at Cys^{247} and Ser^{248} and between the outer sulfur of the persulfide at Cys^{247} and Thr^{252} may stabilize the persulfide (deduced by the results in *Leishmania* MST, Alphey et al., 2003). We recently recognized that the stable persulfide could serve as a reservoir of sulfur atom for subsequent formation of hydrogen sulfide.

The reactions involved in MST catalysis are shown in Fig. 3 (Nagahara, Ito, & Minami, 1999). When cyanide attacks the persulfide, less toxic thiocyanate is formed. This reaction is thought to detoxify cyanide. When thiol-containing compounds attack the persulfide, a new persulfide molecule is formed which can be subsequently reduced to hydrogen sulfide by

MST catalysis

$$HSCH_2COCOO^- + RSH \rightarrow CH_3COCOO^- + RSSH$$
$$HSCH_2COCOO^- + RSSH \rightarrow CH_3COCOO^- + RSSSH$$
$$HSCH_2COCOO^- + CN^- \rightarrow CH_3COCOO^- + SCN^-$$
$$HSCH_2COCOO^- + Trx \rightarrow CH_3COCOO^- + Trx\text{-}SH$$
$$HSCH_2COCOO^- \rightarrow CH_3COCOO^- + HSSCH_2COCOO^-$$
$$SSO_3^{2-} + RSH \rightarrow SO_3^{2-} + RSSH$$
$$SSO_3^{2-} + CN^- \rightarrow SO_3^{2-} + SCN^-$$

Trx reduction

$$RSSH \rightarrow RSH + H_2S$$
$$RSSSH \rightarrow RSSH + H_2S$$
$$Trx\text{-}SH \rightarrow Trx + H_2S$$
$$HSSCH_2COCOO^- \rightarrow HSCH_2COCOO^- + H_2S$$

Figure 3 Catalysis by 3-mercaptopyruvate sulfurtransferase (MST) and hydrogen sulfide production by thioredoxin-specific reduction. MST catalysis, a single and two-substrates reactions catalyzed by MST using mercaptopyruvate ($HSCH_2COCOO^-$) or thiosulfate (SSO_3^{2-}) as the sulfur-donor substrate and thiol-containing compound (RSH), persulfurated thiol-containing compound (RSSH), cyanide (CN^-), thioredoxin (Trx), or mercaptopyruvate as the sulfur-acceptor substrate are shown. Corresponding to the acceptor substrate used, persulfurated thiol-containing compound, polysulfide (RSSSH), thiocyanate (SCN^-), persulfurated thioredoxin (Trx-SH), or persulfurated mercaptopyruvate ($HSSCH_2COCOO^-$) is produced. Trx reduction, reduction of persulfurated thiol-containing compound, polysulfide, persulfurated thioredoxin, and persulfurated mercaptopyruvate by thioredoxin to produce hydrogen sulfide.

thioredoxin (Fig. 3; Shibuya et al., 2009; Yadav, Yamada, Chiku, Koutmos, & Banerjee, 2013). Polysulfides are usually formed during the sulfuration reaction (Fig. 3; Kimura, 2013). If thioredoxin serves as a sulfur, thioredoxin persulfide is formed which can be subsequently reduced to hydrogen sulfide by thioredoxin (Fig. 3; Yadav et al., 2013). A single substrate reaction occurs and hydrogen sulfide can be produced (Fig. 3).

1.3. Distribution

MST is found in all the tissues in rat; however, its activity differs in each tissue (Nagahara, Ito, Kitamura, & Nishino, 1998). The specific activity is highest in the kidney, followed by that in liver and the heart, findings supported by northern blot analysis (Nagahara et al., 1998). Immunohistochemical analysis using a laser confocal microscopy have shown that the MST is localized in the proximal renal tubular cells in the kidney, pericentral hepatic cells in the liver, myocardial cells in the heart, perivascular glial cells in the brain, and bronchiolar epithelial cells in the lung (Nagahara et al., 1998).

Subcellular fractionation analysis revealed that eukaryotic MST activity was observed in both the cytoplasm and mitochondria with the specific activities of 1.39 ± 0.05 and 5.62 ± 0.15 units/mg, respectively (Nagahara et al., 1998; Nakamura, Yamaguchi, & Sano, 2000). The findings were supported by Western blot analysis and immunoelectron microscopy study (Nagahara et al., 1998). Cytoplasmic and mitochondrial enzyme molecules are identical; however, a mitochondrial transport mechanism for MST has not yet been identified.

1.4. Biological function

MST activity is regulated by redox change (Nagahara, 2008, 2013; Nagahara & Katayama, 2005; Nagahara, Yoshii, Abe, & Matsumura, 2007). Under oxidizing conditions, MST is reversely inhibited to increase the cysteine pool, resulting in increase of cellular reductants thioredoxin, glutathione, or glutaredoxin. Thus, MST partly contributes to maintaining redox homeostasis and serves as an antioxidative protein.

Mercaptolactate-cysteine disulfiduria is a genetic disease known to be caused by abnormal MST activity. These patients show insufficiency or deficiency of MST. The symptoms include oversecretion of mercaptolactate-cysteine disulfide in the urine, with or without mental retardation (Crawhall et al., 1968). Due to MST deficiency, mercaptopyruvate is catalyzed by lactate dehydrogenase (EC 1.1.1.27) to produce mercaptolactate. Mental retardation might be due to incomplete development of the brain. The function of antioxidative agents, hydrogen sulfide production, and/or sulfur oxides production (described below) might also be impaired at specific stages of embryonic development. However, the mechanism underlying functional abnormality in the brain has not been identified.

2. REDOX REGULATION OF CYSTEINE METABOLISM AND MST

The cysteine pool in cells is increased under oxidizing conditions mainly because of post-translational inhibition of methionine synthase (EC 2.1.1.13; Mosharov, Cranford, & Banerjee, 2000; Nagahara & Sawada, 2006; Taokam, Ohja, Shan, Kruger, & Banerjee, 1998) and post-translational activation of cystathionine β-synthase (EC 4.4.1.1; Fig. 4; Chen, Chakraborty, & Banerjee, 1995; Nagahara & Sawada, 2006). As a result, *de novo* synthesis of reductants, such as glutathione, thioredoxin, and/or glutaredoxin, is facilitated. Further, the transcription of glutamate-cysteine ligase (EC 6.3.2.2), which is a key enzyme for glutathione synthesis, is increased under oxidative stress (Dormer, Westwater, Stephen, & Jamieson, 2002). MST also contributes to increasing the cysteine pool under oxidizing conditions (Fig. 4).

MST activity is regulated by redox change; it is inhibited by oxidation, resulting in suppression of cysteine degradation, and activated by reduction, resulting in facilitation of cysteine degradation (Fig. 4; Nagahara, 2008, 2013; Nagahara & Katayama, 2005; Nagahara et al., 2007). The enzyme activity is regulated at post-transcriptional level via redox-sensing molecular switches.

3. REGULATION OF MST ACTIVITY VIA REDOX-SENSING MOLECULAR SWITCHES

3.1. Redox-sensing molecular switches

Redox-sensitive cysteines in a switch-carrying protein such as enzymes, transcriptional factors, transcriptional factor modulators, receptor proteins, and sensor proteins serve as redox-sensing molecular switches. These switches can be classified into intermolecular and intramolecular types (Nagahara, 2011a, 2011b, 2012, 2013). Further, these switches can be classified into three subtypes: thioredoxin-, glutathione-, and

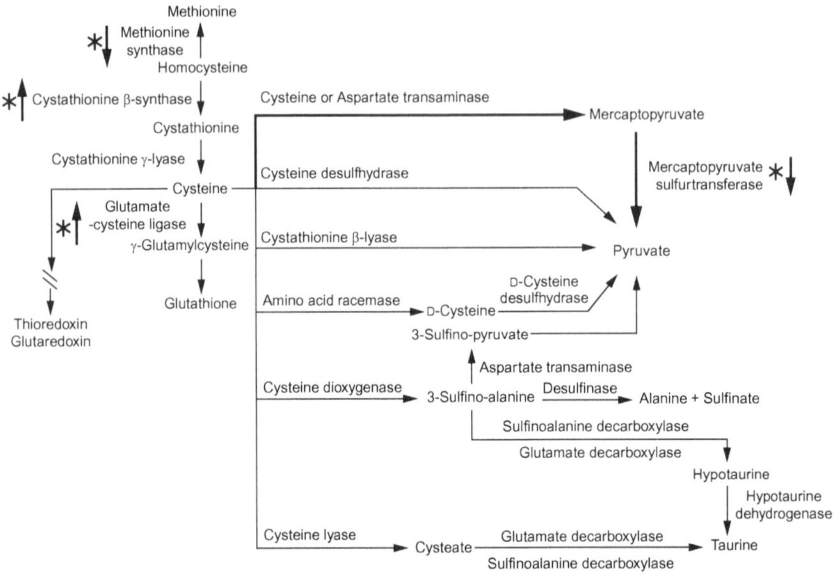

Figure 4 Map of cysteine metabolism and the mercaptopyruvate pathway. The thick line traces the mercaptopyruvate pathway. The catabolism of cysteine involves six pathways, producing pyruvate, alanine, or taurine as metabolites: (1) cysteine is metabolized to form mercaptopyruvate via oxidative deamination by cysteine transaminase (EC 2.6.1.3) or aspartate transaminase (EC 2.6.1.1). Mercaptopyruvate is then converted to pyruvate via trans-sulfuration by MST. (2) Cysteine is directly converted to pyruvate by cysteine desulfhydrase (EC 4.4.1.15). (3) Cysteine is also directly converted to pyruvate by cystathionine β-lyase (EC 4.4.1.8). (4) L-Cysteine is transformed to D-cysteine by amino acid racemase (EC 5.1.1.10). (5) Cysteine is converted to 3-sulfinoalanine by cysteine dioxygenase via oxidation of a sulfhydryl group (EC 1.13.11.20). (6) Cysteine is converted to cysteate by cysteine lyase (EC 4.4.1.10). Additionally, in the "alanine pathway," 3-sulfinoalanine is converted to alanine by desulfinase (EC 4.1.1.12). In the "taurine pathway," 3-sulfinoalanine or cysteate is converted to taurine. Glutamate-cysteine ligase (EC 6.3.2.2) is a key enzyme for glutathione synthesis and its transcription is increased under oxidative stress. Further, metabolic conversion between methionine and cysteine is regulated by cystathionine γ-lyase, cystathionine β-synthase (EC 4.2.1.22), and methionine synthase (EC 2.1.1.13). The mercaptopyruvate pathway is not regulated by cysteine in Hep3B cells. *↑, increase in enzymatic activity in oxidative stress; *↓, decrease in enzymatic activity in oxidative stress. *Modified fig. 1 from Nagahara and Sawada (2006).*

glutaredoxin-specific, depending on reducing agents, which show differences in their redox potential (Nagahara, 2011b, 2013).

In the switch containing single cysteine residue, the cysteine is oxidized to form sulfenate ($CysS\gamma O^-$), sulfinate ($CysS\gamma O_2^-$), and sulfonate ($CysS\gamma O_3^-$) (Lugo-Mas et al., 2006; Nagahara & Katayama, 2005;

Poole, Karplus, & Claiborne, 1989). Of these, only sulfenate can be reduced to cysteine. When a switch consists of two cysteine residues, they are oxidized to form a disulfide bond. Reducing reagents cleave the disulfide bond and hence, modulate the protein function based on the redox state. The protein function is regulated by direct and indirect mechanisms. Direct mechanism involves the so-called "locking and unlocking" of a cysteine residue essential for the protein function. Indirect mechanism involves allosteric-like "conformational change" and results in an oligomeric transformation.

Mammalian MSTs possess two or three exposed cysteines that serve as redox-sensing molecular switches. One switch is a catalytic site (Nagahara, 2008, 2011a, 2012, 2013; Nagahara & Katayama, 2005) while the other(s) is (are) present on the surface of the enzyme (Nagahara, 2008, 2011a, 2013; Nagahara et al., 2007). MST activity is regulated via redox-dependent changes at these molecular switches at post-translational level, and not at transcriptional level.

3.2. Catalytic site cysteine as a redox-sensing molecular switch

The catalytic site cysteine of MST is a single redox-sensitive cysteine residue. Matrix assisted laser desorption/ionization time-of-flight (MALDI-TOF) mass spectrometric analysis and protein chemical study using iodoacetate on rat MST revealed that stoichiometric concentration of hydrogen peroxide oxidizes the catalytic site cysteine (Cys^{247}) to form cysteine-sulfenate (Nagahara & Katayama, 2005). Interestingly, the sulfenate can be reduced not by reduced glutathione but by reduced thioredoxin, thioredoxin coupled to a reducing system containing thioredoxin reductase and NADPH, or dithiothreitol (Fig. 5; Nagahara, 2008, 2012, 2013; Nagahara & Katayama, 2005). This fact indicates that a reducing agent with midredox potential lower than that of reduced glutathione (−240 mV, Aslund, Berndt, & Holmgren, 1997) (and higher than that of reduced thioredoxin (−270 mV, Aslund et al., 1997)) can reduce the sulfenate (Fig. 6).

We also recognized thioredoxin peroxidase activity in the reduction process (Fig. 7; Nagahara & Katayama, 2005), which indicates that two electrons are transferred from two molecules of reduced thioredoxin to hydrogen peroxide via cysteine-sulfenate (Fig. 8). Excess molar concentration of hydrogen peroxide oxidizes the catalytic site cysteine to form cysteine-sulfinate and/or cysteine-sulfonate, which cannot be reduced by cellular reducing agents (Fig. 8; Nagahara, 2008, 2013; Nagahara & Katayama, 2005).

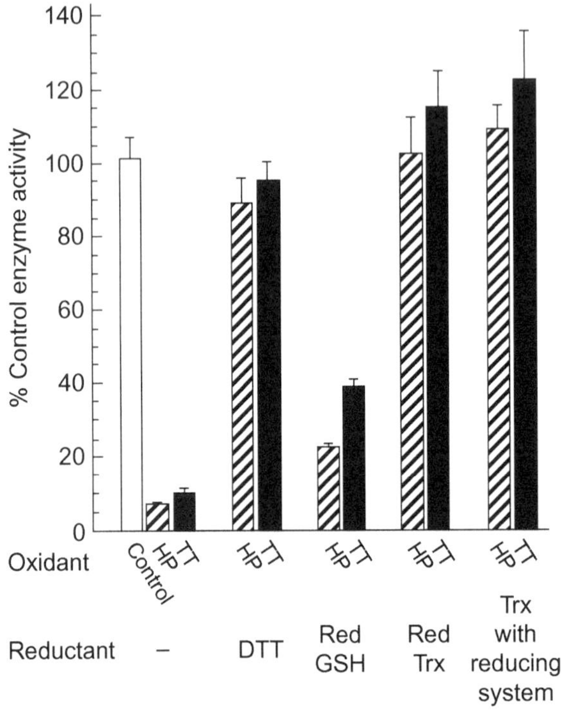

Figure 5 Oxidant-based inhibition of 3-mercaptopyruvate sulfurtransferase (MST) and reactivation by reductants. Each sample was assayed for rhodanese activity because MST assay mixture contained mercaptoethanol. Stoichiometric concentration of hydrogen peroxide (HP) ($E^{o\prime}=1.76$ mV) or tetrathionate (TT) ($E^{o\prime}=1.07$ mV, Das, Huie, & Neta, 1999) inhibited rat recombinant MST in a dose- and time-dependent manner. After removal of free oxidants with an NAP5 column, addition of dithiothreitol (DTT), *Escherichia coli* reduced thioredoxin (Trx), or *E. coli* thioredoxin with *E. coli* thioredoxin reductase and NADPH (Nagahara, 2005) completely restored the activity. Addition of yeast reduced glutathione (GSH) or yeast glutathione with glutathione reductase and NADPH (data not shown) restored the activity to less than 40%. The data are shown as a percentage of the oxidant-free control rhodanese activity (mean ± S.E., $n=3$). *Modified fig. 2 from Nagahara and Katayama (2005).*

Generally, a sulfenyl cysteine can be detected by 4-chloro-7-nitrobenzofurazan (NBD-Cl) as sulfenate-NBD adduct exhibiting maximal absorption at 347 nm in the spectrum (343 nm for free NBD-Cl and 420 nm for the cysteine-NBD adduct). In this case, the adduct with the sulfenyl cysteine cannot be formed probably because the midredox potential of the sulfenyl cysteine is lower than usual.

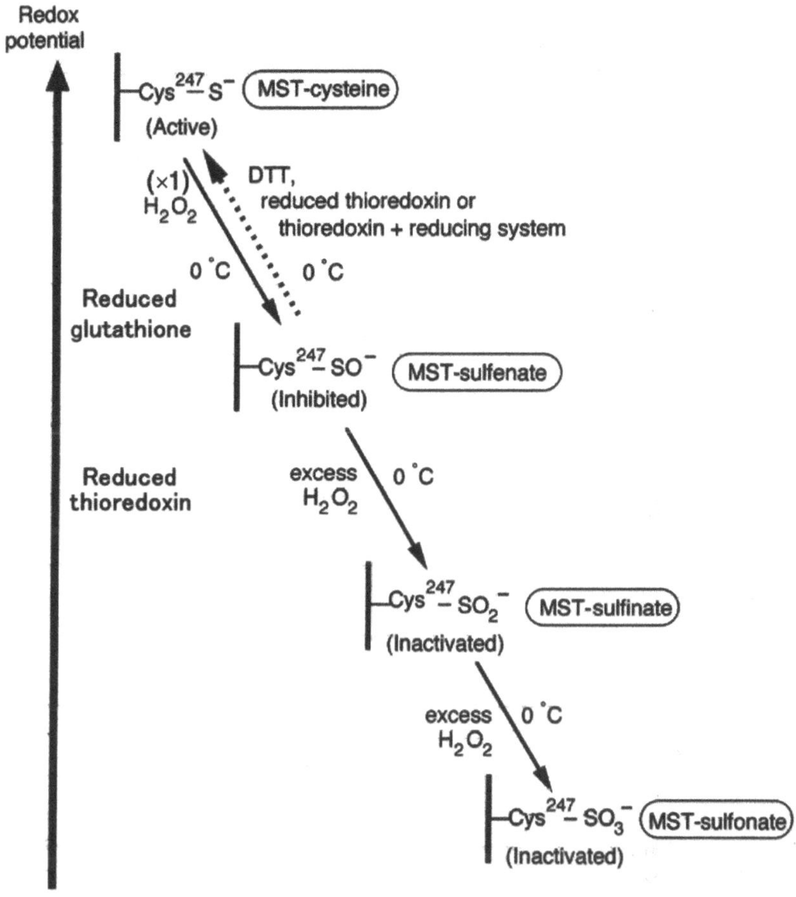

Figure 6 Midredox potential of modified Cys^{247} in the redox cycle. A stoichiometric concentration (×1) of hydrogen peroxide (H_2O_2) oxidizes the catalytic site Cys^{247} to form cysteine-sulfenate (Cys^{247}-SO^-), resulting in inhibition of MST. The midredox potential of the sulfenate is lower than that of reduced glutathione and higher than that of reduced thioredoxin. An excess molar dose of hydrogen peroxide reacts irreversibly to form cysteine-sulfinate ($Cys^{247}-SO_2^-$) and/or cysteine-sulfonate ($Cys^{247}-SO_3^-$). *Modified fig. 8 from Nagahara and Sawada (2006).*

3.3. Cysteine(s) residue(s) on the surface of MST as a redox-sensing molecular switch

Mammalian MSTs contain one or two cysteine residue(s) on the surface of the enzyme. Two cysteines from two different MST molecules together serve as a redox-sensing molecular switch (Nagahara & Katayama, 2005). Rat MST possesses two cysteines (Cys^{154} and Cys^{263}) of approximately

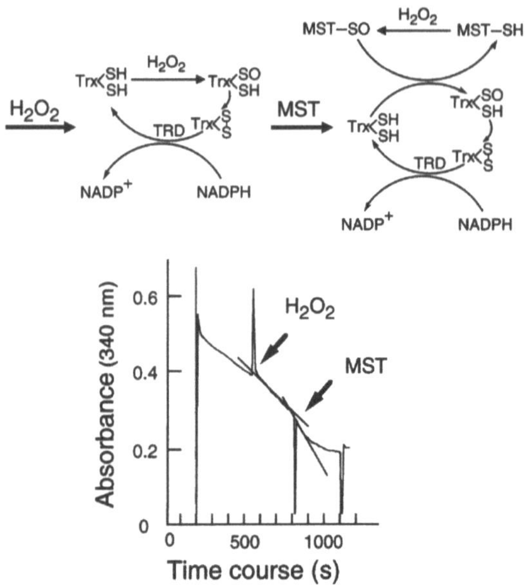

Figure 7 Thioredoxin peroxidase activity in the reduction of cysteine-sulfenate. Electrons are directly passed from reduced thioredoxin (Trx) to hydrogen peroxide (H_2O_2) without 3-mercaptopyruvate sulfurtransferase (MST). Thioredoxin is oxidized via sulfenyl intermediate and reduced by thioredoxin reductase (TRD) using NADPH (left-upper schema). Added MST is oxidized by hydrogen peroxide to form MST-sulfenate ($MST\text{-}SO^-$). Thus, hydrogen peroxide accepts electrons from MST. MST-sulfenate is reduced by reduced thioredoxin; reduced thioredoxin donates one electron to form a sulfenyl intermediate and is finally oxidized to thioredoxin (peroxide reaction; right-upper schema). C247S (Cys^{247} is replaced with Ser) does not possess peroxidase activity. *Modified fig. 4 from Nagahara and Katayama (2005).*

equivalent redox potential, and a dimeric MST is formed via disulfide bonds involving these cysteines from two MST molecules. Each cysteine is essential for regulation of a monomer–dimer equilibrium (Fig. 9; Nagahara, 2008, 2011b, 2013; Nagahara et al., 2007).

High pressure liquid chromatographic (HPLC) analysis using a gel filtration column was performed on rat recombinant MST and mutant enzymes (obtained by site-directed mutagenesis of cysteine(s), Cys^{154} and/or Cys^{263}, on the surface of MST) (Fig. 10; Nagahara et al., 2007). MST showed a monomer (inactive form)–dimer (active form) equilibrium in a redox cycle. On the other hand, the double mutant enzyme, C154/263S, exhibited a single peak that was assigned to the monomeric form. Oxidizing agents, such as, dissolved oxygen lead to formation of dimeric MST (an inactive form) with decrease in MST activity. Recombinant MST and authentic MST

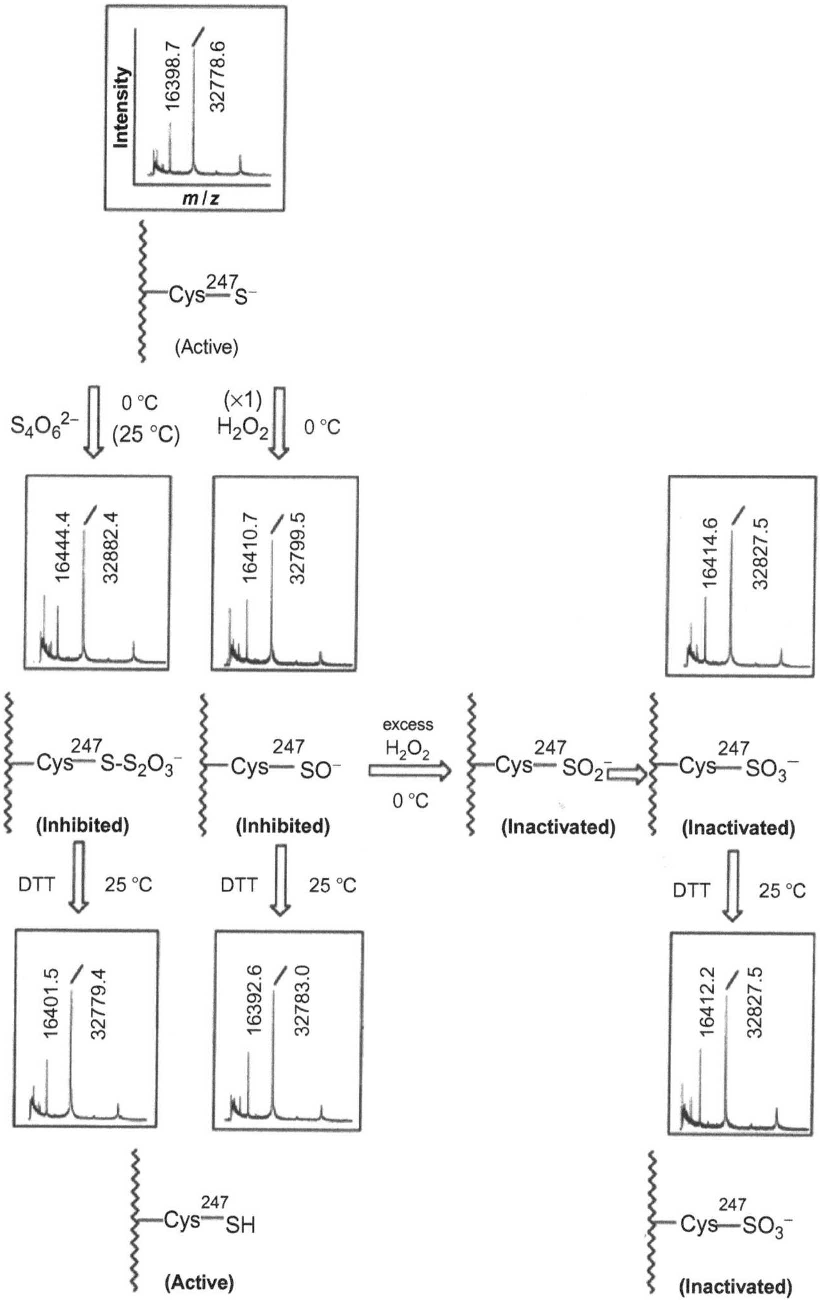

Figure 8 Mass spectrometric analysis of the redox cycle of 3-mercaptopyruvate sulfurtransferase (MST). MALDI-TOF mass spectrometric analysis was performed on a Reflex III (Bruker Daltonics) mass spectrometer equipped with a SCOUT 384 ion source laser, operating in the linear positive mode at a 20 kV acceleration voltage.

(Continued)

from rat liver exhibit monomer to dimer ratios of 92 to 8 and 2 to 1, respectively, under air-saturated conditions (Nagahara, 2013; Nagahara et al., 2007). The ratio of active form to inactive form in naturally occurring MST is higher than that in recombinant MST probably due to a large content of reducing agents in the liver. In contrast, reducing agents cleave the disulfide bond to form monomeric MST (active form) and restore the activity (Figs. 9 and 10; Nagahara et al., 2007).

Reducing agents cleave the disulfide bond to form a monomeric MST. When the intersubunit disulfide bond of MST was treated with reduced thioredoxin for 20 min, ~90% of the dimers were converted to their monomeric form. On the other hand, when MST was treated with dithiothreitol for 12 h, only 50% of the dimer was converted to its monomeric form. Thus, in this regard, reduced thioredoxin is more effective than dithiothreitol (Fig. 10; Nagahara, 2013; Nagahara et al., 2007). Rat MST does not contain both the "CFATC" and the "CXXXXC" motif sequences for binding site of thioredoxin. However, fractional sequences identical or related to thioredoxin reductase sequences from rat and *E. coli* are found in rat MST (Nagahara et al., 2007).

It is noteworthy that *E. coli* thioredoxin is more effective than rat thioredoxin in the reduction process. In one study, the rat MST activity increased to 4.5-fold that of the control when using the *E. coli* reducing system (*E. coli* thioredoxin coupled to a reducing system containing *E. coli* thioredoxin reductase and NADPH; [rat MST]:[*E. coli* thioredoxin]:[*E. coli* thioredoxin reductase]:[NADPH] = 1:5:0.02:12.5) (Fig. 11; Nagahara et al., 2007). However, its activity increased only

Figure 8—Cont'd External mass calibration was performed using the protein mixture of Protein calibration standard II. A stoichiometric concentration of hydrogen peroxide (H_2O_2) (12 m*M*) or tetrathionate ($S_2O_6^{2-}$) (12 m*M*) inhibits MST (12 m*M*) in 20 m*M* potassium phosphate buffer, pH 7.4 on ice for 20 min. Excess free-oxidants removed with gel filtration with a NAP5 column. The main peak from (*m/z* 32778.6; $[M+H]^+$ and 16398.7; $[M+2H]^{2+}$) of untreated control MST was shifted to (*m/z* 32799.5; $[M+H]^+$ and 16410.7; $[M+2H]^{2+}$) or (*m/z* 32882.4; $[M+H]^+$ and 164444.4; $[M+2H]^{2+}$), respectively. These results indicated the formation of sulfenate or sulfenyl thiosulfate at Cys^{247}, respectively. Further, 30 μl of each sample was treated with dithiothreitol (DTT) (0.6 m*M*) at 25 °C for 20 min and free DTT was removed with a NAP5 column. The activity was restored with a shift of the main peak to (*m/z* 32783.0, $[M+H]^+$ and 16392.6; $[M+2H]^{2+}$) or (*m/z* 327779.4, $[M+H]^+$ and 16401.5; $[M+2H]^{2+}$), respectively. An excess molar dose hydrogen peroxide inactivated MST with a shift of the main peak to (*m/z* 32827.5; $[M+H]^+$ and 16414.5 $[M+2H]^{2+}$), indicating the formation of sulfonate which cannot be reduced by dithiothreitol. *Modified fig. 3 from Nagahara & Katayama (2005).*

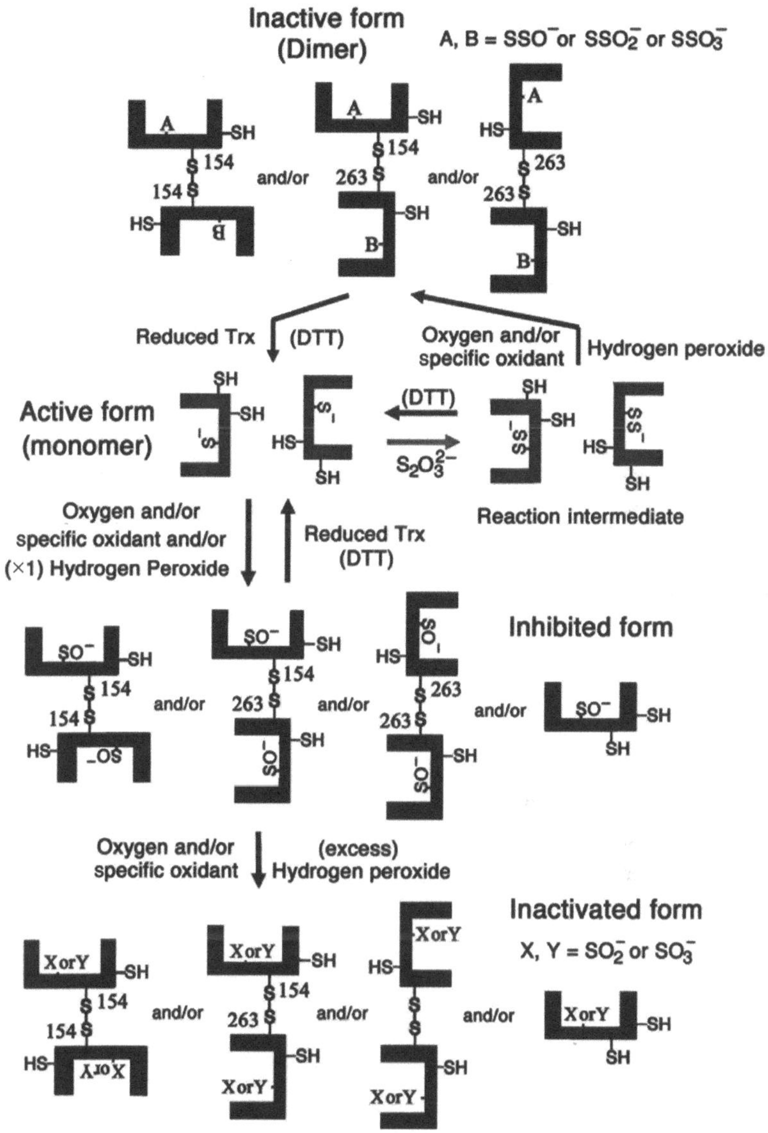

Figure 9 Redox changes in intermolecular and intramolecular redox-sensing switches of 3-mercaptopyruvate sulfurtransferase (MST), The intermolecular switch (Cys^{154} and Cys^{263}) is oxidized by hydrogen peroxide, oxygen, and/or specific oxidant to form an MST dimer (inactive form) via a disulfide bond between Cys^{154} and Cys^{154}, Cys^{154} and Cys^{263}, and/or Cys^{263} and Cys^{263}. The intramolecular switch, the catalytic site Cys^{247}, is oxidized to form cysteine-sulfenate (inhibited form) by a stoichiometric concentration (×1) of hydrogen peroxide, oxygen, and/or specific oxidant. An excess molar dose of the oxidant inactivates MST by forming cysteine-sulfinate and/or a cysteine-sulfonate. *Modified fig. 13 from Nagahara (2013).*

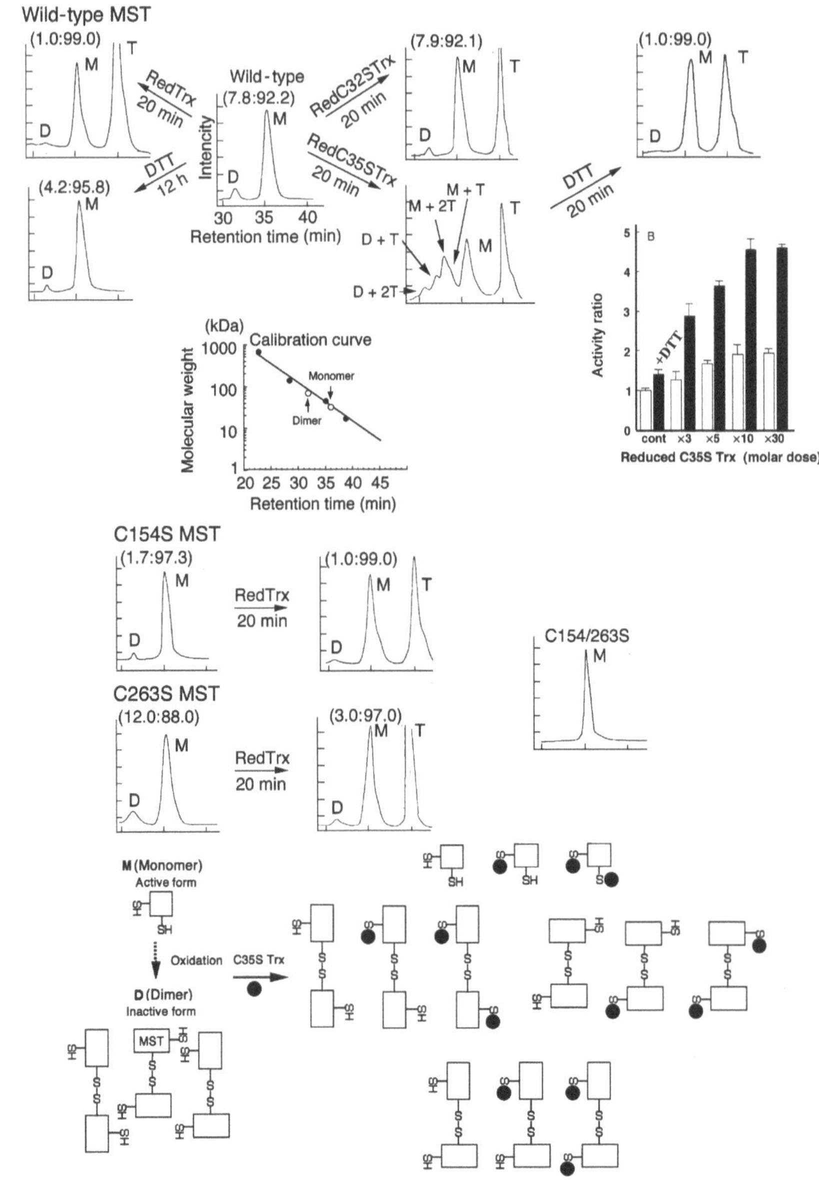

Figure 10 High performance liquid chromatography (HPLC) profiles for 3-mercaptopyruvate sulfurtransferase (MST) with thioredoxin treatment. Monomer–dimer equilibrium of MST was analyzed using an HPLC system equipped with two TSK gel filtrations (G3000SW$_{XL}$, 7.8 mm × 30 cm and G2000SW$_{XL}$, 7.8 mm × 30 cm connected in tandem). The mobile phase was composed of 0.2 *M* potassium phosphate buffer, pH 7.0, the flow rate was 0.5 ml/min, and detection was made at 280 nm. The equilibrium of wild-type MST (0.3 nmol) ([dimer]:[monomer] = 7.8:92.2) changed

threefold that of the control when using the rat reducing system (rat thioredoxin coupled to a reducing system containing rat thioredoxin reductase and NADPH; [rat MST]:[rat thioredoxin]:[rat thioredoxin reductase]:[NADPH] = 1:1:0.05:12.5) (Fig. 11; Nagahara et al., 2007).

During the molecular evolution of the redox-sensing molecular switch (Fig. 12; Nagahara, 2013), all terrestrial life forms that emerged after increase in the concentration of atmospheric oxygen acquired the redox-sensing switch in MST (Fig. 12; Nagahara, 2013).

4. MST KNOCKOUT MOUSE

We produced MST-knockout mice to investigate the pathogenesis of mercaptolactate-cysteine disulfiduria and to determine the biological function such as antioxidative function of MST in the embryonic and postnatal stages. Some organs in the mouse cannot be defensed against oxidative stress due to the impairment of regulation of MST activity.

Western blot analysis confirmed that there was no expression of MST (Nagahara et al., 2013) in the knockout mice. No morphological changes were observed; however, anxiety-like behavior was significantly increased in an elevated plus-maze test (Fig. 13; Nagahara et al., 2013). Levels of 5-hydroxytriptamine (5HT, serotonin) and the metabolite, 5-hydroxyindoleacetic acid, were significantly increased in the prefrontal cortex and hippocampus although dopamine level remained unchanged (Fig. 14; Nagahara et al., 2013).

to1.0:99.0 after incubation with *Escherichia coli* reduced (Red) thioredoxin (Trx) (1.5 nmol) in 20 μl of 20 m*M* potassium phosphate buffer, pH 7.0 on ice for 20 min. Dithiothreitol (DTT) (1.5 mmol DTT in 20 μl of 20 m*M* potassium phosphate buffer, pH 7.0 on ice) was less effective in changing this equilibrium to the same extent, even after12 h-incubation. *E. coli* reduced C35S (Cys^{35} replaced with Ser) (1.5 nmol) was ineffective. *E. coli* reduced C32S (Cys^{32} was replaced with Ser) (1.5 nmol) formed two types of adducts on reaction with MST. D, dimer MST (66.6 kDa, 30.9 min); M, monomer MST (32.8 kDa, 32.0 min); T, *E. coli* thioredoxin (11.8 kDa, 40.5 min); Mix, mixture of thioredoxin-MST complexes (D, D+T, D+2T, and M+2T). The equilibrium ratios of C154S ([dimer]:[monomer] = 1.7:97.3) and C263S ([dimer]:[monomer] = 12.0:88.0) changed to 1.0:99.0 and 3.0:97.0, respectively, after incubation with *E. coli* reduced Trx on ice for 20 min. It is noteworthy that C154/263S did not form any dimers. The calibration curve represents retention time for standard proteins (thyroglobulin [670 kDa], bovine gamma globulin [158 kDa], chicken ovalbumin [44 kDa], and equine myoglobin [17 kDa]). *Modified figs. 3 and 4 from Nagahara et al. (2007).*

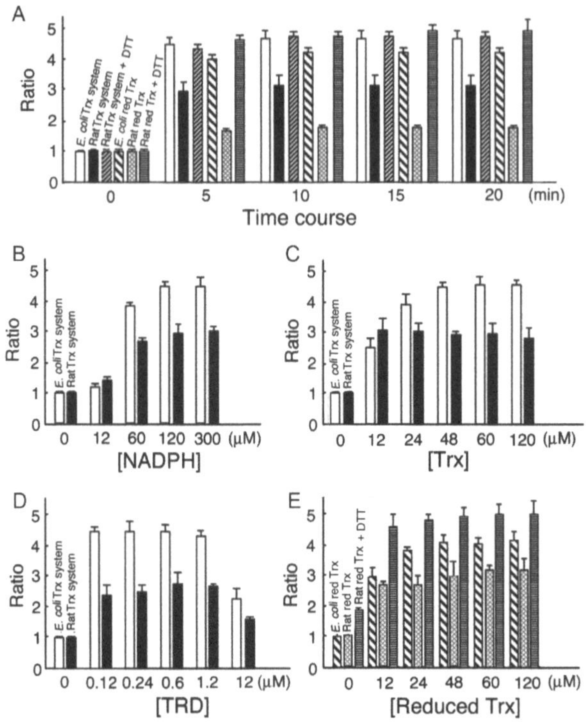

Figure 11 Time- and dose-dependent 3-mercaptopyruvate sulfurtransferase (MST) activation by thioredoxin (Trx) (A) Time-dependent activation of MST with reducing systems. The *Escherichia coli* system contained 12 μ*M* MST and 60 μ*M E. coli* reduced (red) thioredoxin. The rat system contained 60 μ*M* rat reduced thioredoxin with or without 1.2 m*M* dithiothreitol (DTT). The *E. coli* thioredoxin–thioredoxin reductase–NADPH system contained 12 μ*M* MST, 0.24 μ*M E. coli* thioredoxin reductase, and 150 μ*M* NADPH, and 60 μ*M E. coli* thioredoxin. The rat thioredoxin–thioredoxin reductase–NADPH system contains 12 μ*M* MST, 0.24 μ*M* rat thioredoxin reductase, and 150 μ*M* NADPH, and 12 μ*M* rat thioredoxin with or without 1.2 m*M* dithiothreitol. After incubation for a predetermined time, the rhodanese activity of each MST sample was assayed. The ratio of rhodanese activity in each sample to that in the control was calculated. (B) NADPH dose-dependent activation of MST using *E. coli* and rat thioredoxin. MST (12 μ*M*), various concentrations of NADPH (0, 12, 50, 120, and 300 μ*M*), *E. coli* or rat thioredoxin (60 or 12 μ*M*, respectively), and *E. coli* or rat thioredoxin reductase (0.24 or 0.6 μ*M*, respectively) were incubated on ice for 20 min, and the rhodanese activity of each MST sample was assayed. (C) Thioredoxin dose-dependent activation of MST using *E. coli* and rat thioredoxin. MST (12 μ*M*), various concentrations of *E. coli* or rat thioredoxin (0, 12, 24, 48, 60, and 120 μ*M*), NADPH (150 μ*M*), and *E. coli* or rat thioredoxin reductase (0.24 or 0.6 μ*M*, respectively) were incubated on ice for 20 min. (D) Thioredoxin reductase dose-dependent activation of MST using *E. coli* and rat thioredoxin. MST (12 μ*M*), various concentrations of *E. coli* or rat thioredoxin reductase (0, 0.12, 0.24, 0.6, 1.2, and 12 μ*M*), NADPH (150 μ*M*), and *E. coli* or rat thioredoxin (60 or 12 μ*M*, respectively) were incubated. (E) Dose-dependent activation of MST, using *E. coli* and rat reduced thioredoxin MST (12 μ*M*) and various concentrations of *E. coli* or rat reduced thioredoxin (0, 12, 24, 48, 60, and 120 μ*M*) were incubated. Further, 1.2 m*M* dithiothreitol was incubated with rat reduced thioredoxin. Data are shown as the mean ± S.E. (bar; $n = 3$). *Modified fig. 1 from Nagahara et al. (2007).*

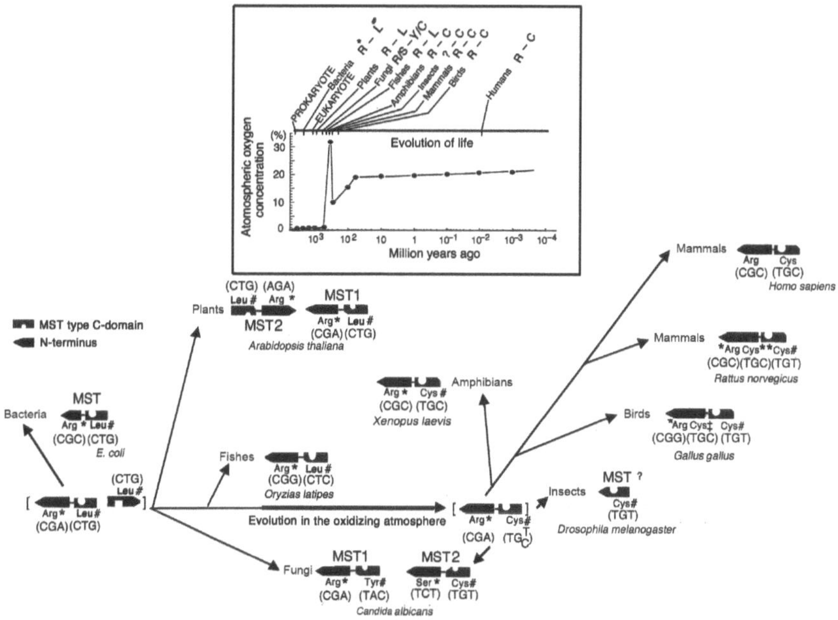

Figure 12 Molecular evolution of the redox-sensing switch in 3-mercaptopyruvate sulfurtransferase (MST) It is summarized that the codon [CTG], encoding a leucine residue in the C-terminal catalytically active domain of prokaryotic enzymes, was substituted with [TGT] or [TGC], encoding a cysteine residue (a redox-sensing switch) in eukaryotic enzymes. According to reports on the evolution of life depending on the Earth's atmospheric oxygen concentration (Brenner et al., 1993; Canfield, 2005; Kasting & Siefert, 2002), 1 billion years ago, plants diverged under a low atmospheric oxygen concentration (Brenner et al., 1993; Canfield, 2005; Kasting & Siefert, 2002); thus, the leucine residue [CTG] was conserved. Animals then began to diverge under an atmospheric oxygen concentration of less than 21% (Brenner et al., 1993; Canfield, 2005; Kasting & Siefert, 2002). Fish emerged around 570 million years ago. The codon encoding leucine [CTC] is conserved in the medaka enzyme, probably because the oxygen concentration is low compared to that in the atmosphere. Amphibians evolved around 370 million years ago. The leucine residue was substituted with cysteine residue [TGC] in the frog enzyme. Insects evolved around 360 million years ago. The leucine residue was replaced with cysteine residue [TGT] in the fruit fly enzyme. Mammals evolved around 250 million years ago. The leucine residue was replaced by cysteine in the rat enzyme. Birds evolved around 200 million years ago. The leucine residue was replaced by cysteine residue [TGT] in the chicken enzyme. The atmospheric oxygen concentration stabilized at approximately 21% around 170 million years ago. Humans, who evolved 10,000 years ago, also carry the cysteine residue [TGC]. Box: the intermolecular redox-sensing switch in MST evolved with changes in the oxygen concentration in the Earth's atmosphere. Data for codons and deduced amino acid residues are based on cDNA or incomplete genomic DNA data: *Aspergillus oryzae* (AP007175); *Candida albicans* (XM_709437); *Drosophila melanogaster* (BLAST data from FlyBase, National Center for Biotechnology Information, and Berkeley Drosophila Genome Project); *Gallus gallus* (D50564 or XM_001231690); *Oryzias latipes* (BLAST data from Medaka Expressed Sequence Tags data, the National Bio Resource Project Medaka Genome Project, and National Institute of Genetics, DNA Sequencing Center); *Xenopus laevis* (BC08421). C, Cysteine, which consists of a redox-sensing molecular switch; ., amino acid corresponding to Arg^{117} of rat MST; #, amino acid corresponding to Cys^{263} of rat MST. *Modified fig. 2 from Nagahara (2007) and modified fig. 3 from Nagahara (2011b).*

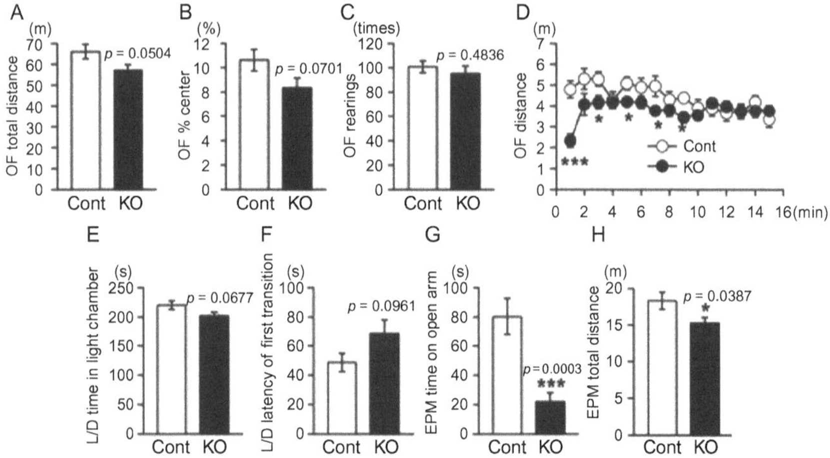

Figure 13 Anxiety-like behaviors in 3-mercaptopyruvate sulfurtransferase (MST)-knockout mice. All behavioral tests were performed using male MST-KO mice (between 10 and 12 weeks old; $n = 17$ per group). Tests for anxiety-like behaviors were performed on MST-knockout mice. (A–D) Open field test (OF) (the plastic open-field chamber was 50 (length) × 50 (width) × 50 cm (height). The field was illuminated at 40 lux. Behavior was monitored for 15 min.). (E and F) Light/dark transition test (L/D) (a cage (21 × 42 × 25 cm) was divided into two sections of equal size by a partition with a door. One chamber was brightly illuminated (600 lux), whereas the other chamber was dark (8 lux). Mice were placed on the dark side, and allowed to move freely between the two chambers with the door being open for 10 min.). (G and H) Elevated plus-maze test (EPM) (the apparatus was consisted of two open arms (25 × 5 cm) and two closed arms of the same size, with 15-cm-high transparent walls. The arms and central square were made of white plastic plates and were elevated to a height of 55 cm above the floor. Arms of the same type were arranged at opposite sides to each other. Each mouse was placed in the central square of the maze (5 × 5 cm), facing one of the closed arms. Mouse behavior was recorded during a 10-min test period.). Data represent the mean ± S.E. $^{*}p < 0.05$; $^{***}p < 0.001$; Cont, control mice; KO, homozygous (null) MST-knockout mice. *Figure 1 from Nagahara et al. (2013).*

The expression of serotonin receptor, 5HT type1A, was not altered significantly in these regions (Nagahara et al., 2013). On the other hand, 5HT type2A receptor was significantly increased in hippocampus (Fig. 15). There are two interesting reports: a knockout mouse in which 5HT type2A receptor was deleted, reduced anxiety-like behaviors (Weisstaub et al., 2006). When antagonist for 5HT type2A receptor was administered, anxiety-like behaviors were reduced (Kehne et al., 1996). These reports suggest that an abnormality in the 5HT type2A receptor expression might be related to the

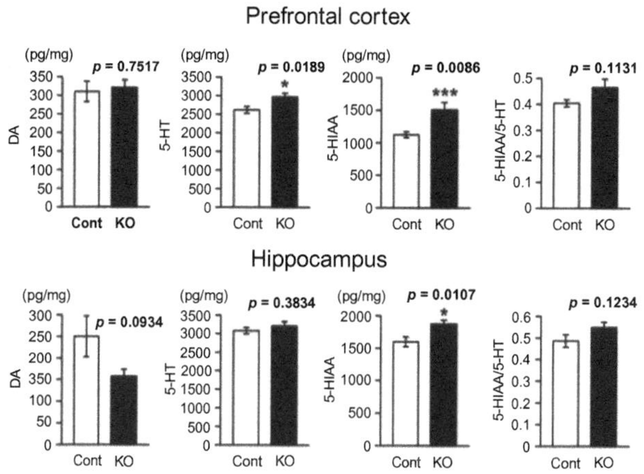

Figure 14 Region-wise levels of monoamines in the brain of 3-mercaptopyruvate sulfurtransferase (MST)-knockout mice. Brain tissue concentrations of monoamines were measured by a high performance liquid chromatography (HPLC) system. The brain tissues of 5 weeks old mice were used. (A and E) Dopamine (DA), (B and F) serotonin (5-hydroxytruptamine, 5-HT), and (C and G) 5-hydroxyindolacetic acid (5-HIAA) levels were assayed. All data represented as mean ± S.E. *$p < 0.05$; ***$p < 0.001$; Cont, control mice; KO, homozygous (null) MST-knockout mice. *Modified fig. 2 from Nagahara et al. (2013).*

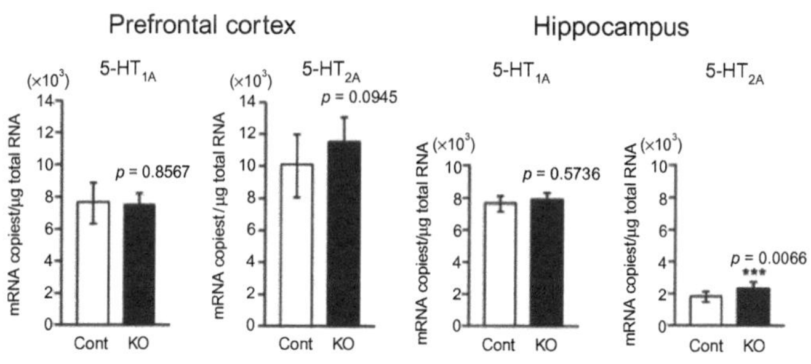

Figure 15 Quantitative analyses of brain $5HT_{1A}$ and $5HT_{2A}$ receptors in 3-mercaptopyruvate sulfurtransferase (MST)-knockout mice. Expression of $5HT_{1A}$ and $5HT_{2A}$ receptor mRNAs in the prefrontal cortex and hippocampus were quantified by reverse transcription polymerase chain reaction. All data represent the mean ± S.E. ***$p < 0.001$; Cont, control mice; KO, homozygous (null)-knockout mice. *Modified fig. 3 from Nagahara et al. (2013).*

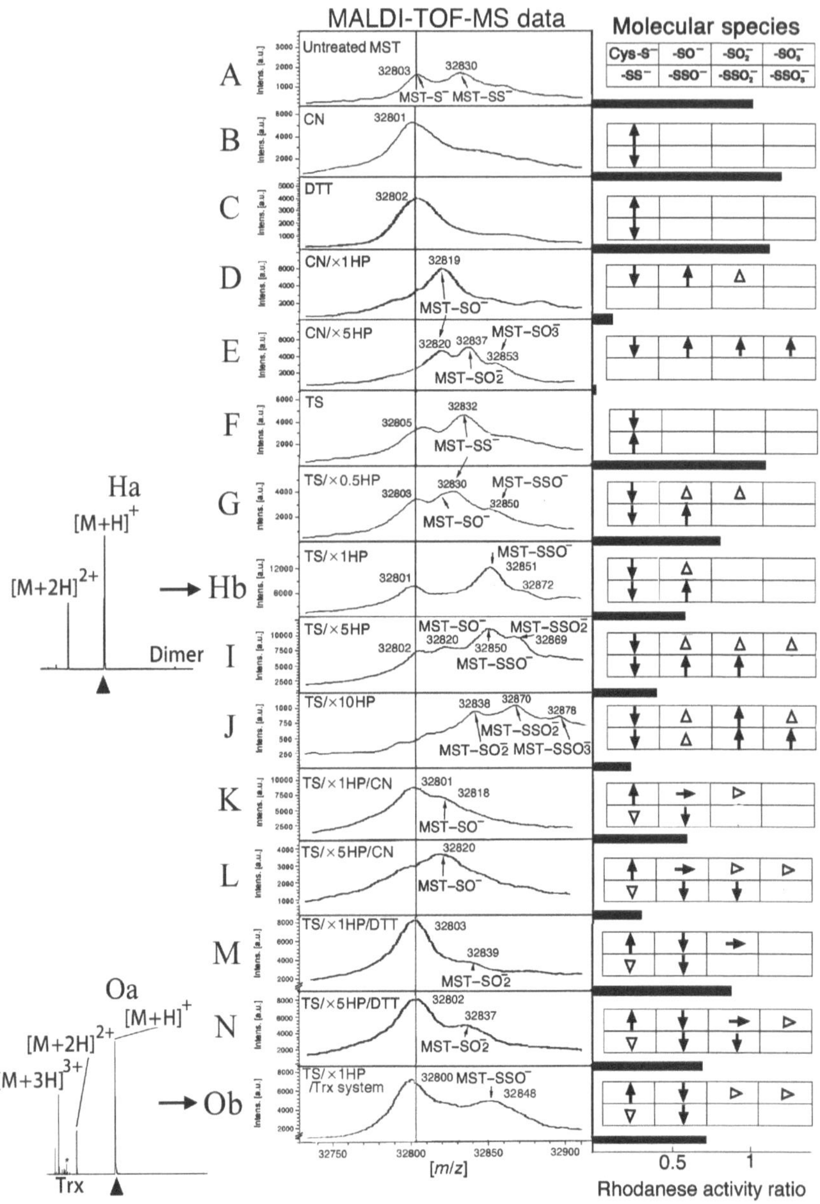

Figure 16 Matrix assisted laser desorption/ionization time-of-flight (MALDI-TOF) mass spectrometric analysis for persulfurated 3-mercaptopyruvate sulfurtransferase (MST) during the redox cycle MST (2.6 m*M*) was incubated with hydrogen peroxide (26 m*M* or 130 m*M*) in 20 m*M* potassium phosphate buffer, pH 7.4 on ice for 20 min. After excess of hydrogen peroxide was removed by gel filtration using a NAP5 column, MST was incubated with sodium thiosulfate at 100-fold molar concentration of MST on ice for

behavioral symptoms observed in MST-knockout mice. The relationship of hydrogen sulfide to this mechanism of pathogenesis has not been clarified.

5. OTHER INVESTIGATION

Recently, we indirectly demonstrated production of sulfur oxides (SO_x) during redox cycle of the persulfide (reaction intermediate) formed at the catalytic cysteine of MST, using MALDI-TOF mass spectrometry (Fig. 16; Nagahara et al., 2012). The persulfide at Cys^{247} is oxidized to thiosulfenyl MST ($MST\text{–}SSO^-$), thiosulfinyl MST ($MST - SSO_2^-$), and thiosulfonyl MST ($MST - SSO_3^-$) and MST is inhibited. Reducing agents can reduce them to form sulfur oxides. Interestingly, cyanide can also release

at least 30 min. Then excess of thiosulfate was removed by gel filtration with a NAP5 column, and the concentration was adjusted to 1 m*M*. The persulfurated MST was oxidized by 1, 5, and 10 m*M* hydrogen peroxide on ice for 20 min in 20 m*M* potassium phosphate buffer, pH 7.4. After excess hydrogen peroxide was removed by gel filtration with a NAP5 column, the sample was concentrated to 100 μ*M* with a VIVASPIN (10,000MWCO, PES). For a reduction experiment with dithiothreitol (DTT), a 50 μl aliquot of the oxidized sample was incubated with 10 μl of 500 m*M* DTT on ice for 30 min. For a reduction experiment with thioredoxin, 10 μl of 10 μ*M* oxidized MST was incubated with 10 μl of a thioredoxin combined with the reducing system was incubated on ice for 30 min. This reducing system contains 10 μ*M* rat thioredoxin, 0.5 μ*M* rat thioredoxin reductase, and 125 μ*M* NADPH. The ratio of these compounds (MST, thioredoxin, thioredoxin reductase, and NADPH) was 1:1:0.05:12.5. For cyanolysis of the disulfide bond with cyanide, a 50 μl aliquot of the oxidized sample was incubated with 10 μl of 500 m*M* potassium cyanide on ice for 30 min. MALDI-TOF mass spectrometric analysis was performed on an ultraflex III mass spectrometer equipped with a SCOUT 384 ion source laser, operating in the linear positive mode at a 25 kV acceleration voltage was performed. Spectra in Ha and Oa represent a reduced spectrum on the mass number scale and those in Hb and Ob represent an extended spectrum. ▲, monomeric MST; Dimer, dimeric MST; CN, cyanide; DTT, dithiothreitol; ×1HP, stoichiometric concentration of hydrogen peroxide; ×5HP, five times the dose of hydrogen peroxide; ×10HP, 10 times the dose of hydrogen peroxide; Trx; reduced thioredoxin; and Trx system, thioredoxin containing thioredoxin reductase and NADPH; TS, thiosulfate. In molecular species, $Cys\text{-}S^-$, non-modified MST; $\text{-}SO^-$, sulfenyl MST; $-SO_2^-$, sulfinyl MST; $-SO_3^-$, sulfonyl MST; $\text{-}SS^-$, persulfurated MST; $\text{-}SSO^-*$, thiosulfenyl MST; $-SSO_2^-*$, thiosulfinyl MST; and $-SSO_3^-*$, thiosulfonyl MST. ↑, definite increase in molecular content; ↓, definite decrease in molecular content; →, no change in molecular content; △, → possible increase in molecular content; ▽, possible decrease in molecular content; and possibly no change in molecular content. The ratio of rhodanese activity (the mean value; $n = 3$) for each sample to that (the mean value; $n = 3$) of untreated sample (sample A) was calculated. *Modified fig. 1 from Nagahara Nirasawa, Yoshii, and Niimura (2012).*

sulfur oxides from them via cyanolysis. We are now performing experiments to confirm the *in vivo* production of these sulfur oxides which may modulate MST activity.

REFERENCES

Alphey, M. S., Williams, R. A., Mottram, J. C., Coombs, G. H., & Hunter, W. N. (2003). The crystal structure of *Leishmania major* 3-mercaptopyruvate sulfurtransferase. A three-domain architecture with a serine protease-like triad at the active site. *Journal of Biological Chemistry, 278,* 48219–48227.

Aslund, F., Berndt, K. D., & Holmgren, A. (1997). Redox potentials of glutaredoxins and other thiol-disulfide oxidoreductases of the thioredoxin superfamily determined by direct protein-protein redox equilibria. *Journal of Biological Chemistry, 272,* 30780–30786.

Brenner, S., Elgar, G., Sandford, R., Macrae, A., Venkatesh, B., & Aparicio, S. (1993). Characterization of the puffer fish (Fugu) genome as a compact model vertebrate genome. *Nature, 366,* 265–268.

Canfield, D. E. (2005). The early history of atmospheric oxygen. *Annual Review of Earth and Planetary Science, 33,* 1–3.

Chen, Z., Chakraborty, S., & Banerjee, R. (1995). Demonstration that mammalian methionine synthases are predominantly cobalamin-loaded. *Journal of Biological Chemistry, 270,* 19246–19249.

Crawhall, J. C., Parker, R., Young, E. P., Ampola, M. G., Efron, M. L., & Bixby, E. M. (1968). Beta mercaptolactate-cysteine disulfide: Analog of cystine in the urine of a mentally retarded patient. *Science, 160,* 419–420.

Das, T. N., Huie, R. E., & Neta, P. (1999). Reduction potentials of SO_3^-, SO_5^-, and $S_4O_6^{3-}$ radicals in aqueous solution. *Journal of Physical Chemistry A, 103,* 3581–3588.

Dormer, U. H., Westwater, J., Stephen, D. W., & Jamieson, D. J. (2002). Oxidant regulation of the *Saccharomyces cerevisiae* GSH1 gene. *Biochimica et Biophysica Acta, 1576,* 23–29.

Jarabak, R., & Westley, J. (1978). Steady-state kinetics of 3-mercaptopyruvate sulfurtransferase from bovine kidney. *Archives of Biochemistry and Biophysics, 195,* 458–465.

Kasting, J. F., & Siefert, J. L. (2002). Life and the evolution of Earth's atmosphere. *Science, 296,* 1066–1068.

Kehne, J. H., Baron, B. M., Carr, A. A., Chaney, S. F., Elands, J., Feldman, D. J., et al. (1996). Preclinical characterization of the potential of the putative atypical antipsychotic MDL 100,907 as a potent 5-HT2A antagonist with a favorable CNS safety profile. *Journal of Pharmacology and Experimental Therapeutics, 277,* 968–981.

Kimura, H. (2013). Physiological role of hydrogen sulfide and polysulfide in the central nervous system. *Neurochemistry International, 63,* 492–497.

Lugo-Mas, P., Dey, A., Xu, L., Davin, S. D., Benedict, J., Kaminsky, W., et al. (2006). How does single oxygen atom addition affect the properties of an Fe-nitrile hydratase analogue? The compensatory role of the unmodified thiolate. *Journal of American Chemical Society, 128,* 11211–11221.

Meister, A. (1953). Conversion of α-keto analogue of cysteine to pyruvate and sulfur. *Federation Proceedings, 12,* 245.

Mosharov, E., Cranford, M. R., & Banerjee, R. (2000). The quantitatively important relationship between homocysteine metabolism and glutathione synthesis by the transsulfuration pathway and its regulation by redox changes. *Biochemistry, 39,* 13005–13011.

Nagahara, N. (2007). Molecular evolution of thioredoxin-dependent redox-sensing switch in mercaptopyruvate sulfurtransferase. In R. M. Mohan (Ed.), *Research advances in biological chemistry* (pp. 19–26). Kerala: Global Research Network.

Nagahara, N. (2008). A novel thioredoxin-dependent redox-sensing molecular switch of mercaptopyruvate sulfurtransferase: Maintenance of cellular redox equilibrium. *Mini-Reviews in Medical Chemistry, 8*, 585–589.

Nagahara, N. (2011a). Catalytic site cysteines of thiol enzymes: Sulfurtransferases. *Journal of Amino Acids, 2011*, 709404. http://dx.doi.org/10.4061/2011/709404.

Nagahara, N. (2011b). Intermolecular disulfide bond to modulate protein function as a redox-sensing switch. *Amino Acids, 41*, 59–72.

Nagahara, N. (2012). Redox-sensing cysteine dependent-molecular switches, intramolecular disulfide formation. In J. C. Taylor (Ed.), *Advances in chemistry research: Vol. 12* (pp. 243–256). New York: Nova Science Publishers, Inc.

Nagahara, N. (2013). Regulation of mercaptopyruvate sulfurtransferase activity via intrasubunit and intersubunit redox-sensing switches. *Antioxidants & Redox Signaling, 9*, 1792–1802.

Nagahara, N., Ito, T., Kitamura, H., & Nishino, T. (1998). Tissue and subcellular distribution of mercaptopyruvate sulfurtransferase in the rat: Confocal laser fluorescence and immunoelectron microscopic studies combined with biochemical analysis. *Histochemistry and Cell Biology, 110*, 243–250.

Nagahara, N., Ito, T., & Minami, M. (1999). Mercaptopyruvate sulfurtransferase as a defense against cyanide toxication: Molecular properties and mode of detoxification. *Histology and Histopathology, 14*, 1277–1286.

Nagahara, N., & Katayama, A. (2005). Post-translational regulation of mercaptopyruvate sulfurtransferase via a low redox potential cysteine-sulfenate in the maintenance of redox homeostasis. *Journal of Biological Chemistry, 280*, 34569–34576.

Nagahara, N., Nagano, M., Ito, T., Shimamura, K., Akimoto, K., & Suzuki, H. (2013). Antioxidant enzyme, 3-mercaptopyruvate sulfurtransferase-knockout mice exhibit increased anxiety-like behaviors: A model for human mercaptolactate-cysteine disulfiduria. *Scientific Reports, 3*, 1792–1802.

Nagahara, N., Nirasawa, T., Yoshii, T., & Niimura, Y. (2012). Is novel signal transducer sulfur oxide involved in the redox cycle of persulfide at the catalytic site cysteine in a stable reaction intermediate of mercaptopyruvate sulfurtransferase? *Antioxidants & Redox Signaling, 16*, 747–753.

Nagahara, N., & Nishino, T. (1996). Role of amino acid residues in the active site of rat liver mercaptopyruvate sulfurtransferase. cDNA cloning, overexpression, and site-directed mutagenesis. *Journal of Biological Chemistry, 271*, 27395–27401.

Nagahara, N., Okazaki, T., & Nishino, T. (1995). Cytosolic mercaptopyruvate sulfurtransferase is evolutionarily related to mitochondrial rhodanese. Striking similarity in active site amino acid sequence and the increase in the mercaptopyruvate sulfurtransferase activity of rhodanese by site-directed mutagenesis. *Journal of Biological Chemistry, 270*, 16230–16235.

Nagahara, N., & Sawada, N. (2006). The mercaptopyruvate pathway in cysteine catabolism: A physiologic role and related disease of the multifunctional 3-mercaptopyruvate sulfurtransferase. *Current Medical Chemistry, 13*, 1219–1930.

Nagahara, N., Sreeja, V. G., Li, Q., Shimizu, T., Tsuchiya, T., & Fujii-Kuriyama, Y. A. (2004). A point mutation in a silencer module reduces the promoter activity for the human mercaptopyruvate sulfurtransferase. *Biochimica et Biophysica Acta, 1680*, 176–184.

Nagahara, N., Yoshii, T., Abe, Y., & Matsumura, T. (2007). Thioredoxin-dependent enzymatic activation of mercaptopyruvate sulfurtransferase. An intersubunit disulfide bond serves as a redox switch for activation. *Journal of Biological Chemistry, 282*, 1561–1569.

Nakamura, T., Yamaguchi, Y., & Sano, H. (2000). Plant mercaptopyruvate sulfurtransferases: Molecular cloning, subcellular localization and enzymatic activities. *European Journal of Biochemistry, 267*, 5621–5630.

Poole, L. B., Karplus, P. A., & Claiborne, A. (1989). The non-flavin redox center of the streptococcal NADH peroxidase. II. Evidence for a stabilized cysteine-sulfenic acid. *Journal of Biological Chemistry*, *264*, 12330–12338.
Shibuya, N., Tanaka, M., Yoshida, M., Ogasawara, Y., Togawa, T., Ishii, K., et al. (2009). 3-Mercaptopyruvate sulfurtransferase produces hydrogen sulfide and bound sulfane sulfur in the brain. *Antioxidants & Redox Signaling*, *11*, 703–714.
Spallarossa, A., Forlani, F., Carpen, A., Armirotti, A., Pagani, S., Bolognesi, M., et al. (2004). The "rhodanese" fold and catalytic mechanism of 3-mercaptopyruvate sulfurtransferases: Crystal structure of SseA from *Escherichia coli*. *Journal of Molecular Biology*, *1335*, 583–593.
Taokam, S., Ohja, S., Shan, X., Kruger, W. D., & Banerjee, R. (1998). Evidence for heme-mediated redox regulation of human cystathionine beta-synthase activity. *Journal of Biological Chemistry*, *273*, 25179–25184.
Weisstaub, N. V., Zhou, M., Lira, A., Lambe, E., González-Maeso, J., Hornung, J. P., et al. (2006). Cortical 5-HT2A receptor signaling modulates anxiety-like behaviors in mice. *Science*, *313*, 536–540.
Wood, J. L., & Fiedler, H. (1953). Beta-Mercapto-pyruvate, a substrate for rhodanese. *Journal of Biological Chemistry*, *205*, 231–234.
Yadav, P. K., Yamada, K., Chiku, T., Koutmos, M., & Banerjee, R. (2013). Structure and kinetic analysis of H_2S production by human mercaptopyruvate sulfurtransferase. *Journal of Biological Chemistry*, *27*, 20002–20013.

CHAPTER FOURTEEN

Role of Human Sulfide:Quinone Oxidoreductase in H_2S Metabolism

Michael R. Jackson, Scott L. Melideo, Marilyn Schuman Jorns[1]
Department of Biochemistry and Molecular Biology, College of Medicine, Drexel University, Philadelphia, Pennsylvania, USA
[1]Corresponding author: e-mail address: marilyn.jorns@drexelmed.edu

Contents

Abstract

The first step in the mammalian metabolism of H_2S is catalyzed by sulfide:quinone oxidoreductase (SQOR). Human SQOR is an integral membrane protein, which presumably interacts with the inner mitochondrial membrane in a monotopic fashion. The enzyme is a member of a family of flavoprotein disulfide oxidoreductases (e.g., glutathione reductase) that utilize a Cys-S-S-Cys disulfide bridge as an additional redox center. SQOR catalyzes a two-electron oxidation of H_2S to sulfane sulfur using coenzyme Q as electron acceptor. The enzyme also requires a third substrate to act as the acceptor of the sulfane sulfur from a cysteine persulfide intermediate. Here, we describe a method for the bacterial expression of human SQOR as a catalytically active membrane-bound protein, procedures for solubilization and purification of the recombinant protein to >95% homogeneity, and spectrophotometric assays to monitor SQOR-mediated H_2S oxidation in reactions with different sulfane sulfur acceptors.

Methods in Enzymology, Volume 554
ISSN 0076-6879
http://dx.doi.org/10.1016/bs.mie.2014.11.037

1. INTRODUCTION

Hydrogen sulfide (H_2S) is the newest member of a small family of important biological signaling molecules, termed gasotransmitters, which also includes nitric oxide and carbon monoxide. H_2S biosynthesis in mammals is catalyzed by cystathionine β-synthase, cystathionine γ-lyase, and 3-mercaptopyruvate sulfurtransferase (Kabil & Banerjee, 2010; Yadav, Yamada, Chiku, Koutmos, & Banerjee, 2013). H_2S is the only gasotransmitter that is enzymatically metabolized and the only inorganic compound that can be used by mammalian mitochondria to generate ATP (Yong & Searcy, 2001). H_2S metabolism is particularly important because the gasotransmitter is a Janus-faced molecule that can exhibit toxic effects at supraphysiological concentrations. The first step in the mitochondrial metabolism of H_2S is catalyzed by sulfide:quinone oxidoreductase (SQOR), an inner mitochondrial membrane-bound flavoenzyme that catalyzes a two-electron oxidation of H_2S to sulfane sulfur (S^0) using coenzyme Q as electron acceptor (Hildebrandt, Grieshaber, & Westley, 2008; Jackson, Melideo, & Jorns, 2012; Lagoutte et al., 2010; Shahak & Hauska, 2008; Theissen & Martin, 2008). In addition to eukaryotes, SQORs are found in many bacteria where the oxidation of environmental H_2S provides an important source of energy and reducing equivalents.

SQORs have been classified into six types (Jones, Schaperdoth, & Macalady, 2014; Marcia, Ermler, Peng, & Michel, 2010). Type II SQORs include all known eukaryotic enzymes and a subset of prokaryotic enzymes. The other five types comprise only bacterial enzymes. Type II SQORs require a nucleophile (N:) to act as the acceptor of the sulfane sulfur from a persulfide intermediate ($E\text{-}CysSS^- + N: \Rightarrow E\text{-}CysS^- + NS$). Cyanide can act as an, albeit nonphysiological, acceptor of the sulfane sulfur, as judged by *in vitro* studies with mammalian, invertebrate, yeast, and bacterial type II enzymes (Hildebrandt et al., 2008; Jackson et al., 2012; Shahak & Hauska, 2008; Shibata & Kobayashi, 2006; Shibata, Suzuki, & Kobayashi, 2007; Theissen & Martin, 2008). Crystal structures are available for type 1 and type V bacterial SQORs. However, these enzymes catalyze an oxidative polymerization of H_2S to produce sulfane sulfur metabolites (polysulfide chains, cyclooctasulfur rings) in reactions that do not require an acceptor molecule (Brito et al., 2009; Cherney, Zhang, Solomonson, Weiner, & James, 2010; Marcia, Ermler, Peng, & Michel, 2009). Characterization of eukaryotic SQORs has proved far more challenging, as judged by difficulties encountered in attempts to isolate recombinant forms of the lugworm or yeast enzyme (Theissen & Martin, 2008; Weghe & Ow, 1999).

In this chapter, we described methods for expression of human SQOR in *Escherichia coli* and purification of the recombinant flavoenzyme, and the use of spectrophotometric assays to characterize the reaction of human SQOR with various acceptors of the sulfane sulfur. The significance of these findings with respect to the role of SQOR in mammalian H_2S metabolism is discussed.

2. EXPRESSION OF HUMAN SQOR IN *E. COLI*

We successfully expressed mature SQOR (47 kDa) as a catalytically active membrane-bound protein *in E. coli* by using a synthetic version of the human gene (*sqrdl*) and growth of the recombinant strain at low temperature in the presence of cold-adapted chaperonins, conditions that minimize inclusion body formation (Jackson et al., 2012). The synthetic gene contains a single Met in place of an N-terminal mitochondrial-targeting presequence, is optimized for expression in *E. coli*, and is flanked by unique *Nde1* and *Xho1* sites. To introduce a C-terminal $(His)_6$-tag, the synthetic gene was subcloned into plasmid pET23a (Novagen) to yield plasmid pET23a_matOPZSQOR. Plasmid pCPN10/60 is a pACYC-based plasmid that contains genes (*cpn10, cpn60*) for cold-adapted chaperonins from *Oleispira antarctica* (Cpn10, Cpn60) and a gentamycin-resistance gene. Plasmid pET23a_matOPZSQOR was used to transform *E. coli* BL21 (DE3) cells to ampicillin resistance. Plasmid pCPN10/60 was then used to transform BL21 (DE3)/pET23a_matOPZSQOR cells to gentamycin resistance. A starter culture is prepared by overnight growth of *E. coli* BL21 (DE3)/ pET23a_matOPZSQOR/pCPN10/60 cells at 37 °C in LB media containing gentamycin (20 μg/mL) and ampicillin (100 μg/mL). The starter culture is used to inoculate TB media containing the same two antibiotics. Cells are grown with shaking in 2 L flasks containing 500 mL of media at 15 °C. SQOR expression is induced with isopropyl β-D-thiogalactopyranoside (0.5 m*M*) when the cell density reaches $A_{595} \sim 1.1$. Cells are harvested 20 h after induction. The cell pellets (~100 g from 9 L of culture) are stored at −80 °C.

3. PURIFICATION OF RECOMBINANT HUMAN SQOR

All steps are conducted at 4 °C. Cell pellets (~50 g) are thawed and suspended in 75 mL of Tris–acetate buffer, pH 7.6, containing 0.5 *M* sucrose and 0.1 m*M* EDTA. The cell suspension is mixed with lysozyme (0.5 mg/mL) plus a cocktail of nucleases and protease inhibitors (DNAase, 20 μg/mL; RNAase, 20 μg/mL; 5 m*M* magnesium sulfate; soybean trypsin inhibitor, 12.6 μg/mL; aprotinin, 2 μg/mL; phenylmethanesulfonylfluoride, 25 μ*M*/mL; and tosyllysine chloromethylketone, 3 μg/mL). The suspension is incubated with stirring for 20 min and then sonicated (Branson Model 350, power setting = 6, duty cycle = 40%) for a total of 450 s in 30-s intervals, separated by 30-s cooling periods. Cell debris is removed by centrifugation at low speed (10 min at 10,000 × *g*). Membrane-bound SQOR is found in the low-speed supernatant which contains membrane fragments generated during sonication.

SQOR is optimally solubilized using 0.5% 1,2-diheptanoyl-*sn*-glycero-3-phosphocholine (DHPC) (Avanti Polar Lipids). Accordingly, the low-speed supernatant is diluted 1:1 with 50 m*M* potassium phosphate buffer, pH 7.4, containing 1% DHPC, 10% glycerol, and a cocktail of protease inhibitors and DNAase, as described above. The sample is incubated for 2 h on a rocking platform shaker to solubilize SQOR and then centrifuged at 120,000 × *g* for 1 h. The high-speed supernatant is collected and its buffer is modified to contain 200 m*M* Tris–HCl, pH 8.0, 200 m*M* sodium chloride and 40 m*M* imidazole. The high-speed supernatant is loaded onto a 5-mL HiTrap IMAC column (GE Healthcare), previously equilibrated with 40 m*M* Tris–HCl buffer, pH 8.0, containing 150 m*M* sodium chloride and 40 m*M* imidazole–HCl. The column is washed with 50 m*M* Tris–HCl buffer, pH 8.0, containing 150 m*M* sodium chloride, 10% glycerol, 80 m*M* imidazole–HCl, and 0.05% DHPC. SQOR is eluted with 50 m*M* Tris–HCl buffer, pH 8.0, containing 150 m*M* sodium chloride, 10% glycerol, 160 m*M* imidazole–HCl, and 0.1% DHPC and stored at −80 °C.

For the final step of the purification, eight batches of IMAC-purified SQOR obtained from ~400 g of cells are thawed, pooled, and dialyzed for 2 h versus a ~200-fold excess of Tris–HCl buffer, pH 8.0, containing 50 m*M* sodium chloride and 5% glycerol. The sample is centrifuged for 10 min at 30,000 × *g*. The supernatant is loaded onto a 50-mL HiLoad 26/10 Q Sepharose High Performance anion exchange column (GE Healthcare), previously equilibrated with 50 m*M* Tris–HCl buffer, pH 8.0, containing 2% glycerol and 0.03% DHPC. The column is washed with 50 m*M* Tris–HCl buffer, pH 8.0, containing 2% glycerol, 100 m*M* sodium chloride, and 0.06% DHPC (buffer A). SQOR is eluted with a 100 mL linear gradient formed with buffer A and buffer B (50 m*M* Tris–HCl buffer,

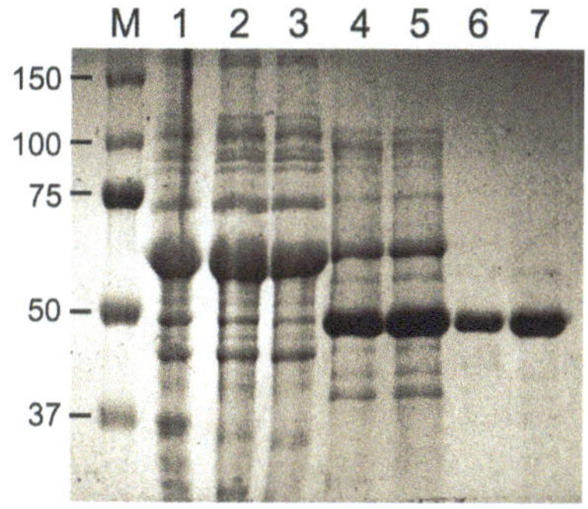

Figure 1 Purification of recombinant human SQOR. Molecular markers are shown in lane M. Lane 1, crude cell lysate; lane 2, high-speed supernatant; lane 3, IMAC wash; lane 4, IMAC eluate; lane 5, dialyzed IMAC eluate; lane 6, Q Sepharose eluate; lane 7, concentrated Q Sepharose eluate. The SDS-PAGE gel was stained for protein with ProSieve Blue Protein Staining Solution (Lonza). *Reprinted with permission from Jackson et al. (2012).*

Table 1 Purification of recombinant SQOR from *E. coli*

Purification step	Total activity (U)[a]	Total protein (mg)[b]	Specific activity (U/mg)	Yield (%)
High-speed supernatant	32,200	27,580	1.17	100
IMAC	11,100	89.8	123.6	34.5
Q Sepharose	8330	14.3	580.8	25.9

[a]A unit of activity is defined as the formation of 1 μmol of reduced CoQ_1 at 25 °C using a routine aerobic assay, as described in the text.
[b]Protein was determined using the Pierce BCA (Bicinchoninic Acid Assay) Protein Assay Kit.
Reprinted with permission from Jackson et al. (2012).

pH 8.0, containing 2% glycerol, 1.0 *M* sodium chloride, and 0.06% DHPC). SQOR-containing fractions are pooled, concentrated using a 10K Macrosep Advance Centrifugal Device (Pall Life Sciences), and stored in aliquots at −80 °C. The solubilized enzyme is purified to >95% homogeneity (Fig. 1) with an overall yield of 26% yield (Table 1).

4. CATALYTIC ASSAYS

Recombinant SQOR catalyzes a two-electron oxidation of H_2S to S^0 (sulfane sulfur). Enzyme activity is measured using CoQ_1 (Sigma-Aldrich), a water-soluble ubiquinone derivative, as the electron acceptor. Unless otherwise indicated, assays with sulfite, cyanide, or sulfide as the sulfane sulfur acceptor are conducted in 100 mM potassium phosphate, pH 7.5, 100 mM potassium pyrophosphate, pH 8.5, or 100 mM potassium phosphate, pH 7.0, respectively. All reaction buffers contain 0.5 mM EDTA. Except as noted below, assays are performed under anaerobic conditions at 25 °C using a cuvette (Spectrocell) with a screw-cap equipped with a Teflon-silicon membrane. Buffers used to prepare stock solutions of sodium sulfide, sodium sulfite, or potassium cyanide are bubbled with argon for at least 20 min. Sodium sulfite and sodium cyanide ($\leq$2 mM) stock solutions are prepared directly in the assay buffer. To prepare more concentrated sodium cyanide solutions, a 1 M solution of the substrate is neutralized to the desired pH with monobasic potassium phosphate and then diluted with assay buffer. Stock solutions of sodium sulfide are prepared in 50 mM potassium carbonate/bicarbonate buffer, pH 9.6, containing 250 mM EDTA. The sulfide concentration is determined based on its absorbance at 230 nm ($\varepsilon_{230} = 7200\ M^{-1}\ \text{cm}^{-1}$) (Hughes, Centelles, & Moore, 2009). Stock solutions of CoQ_1 are prepared in DMSO. Reagents are added to the cuvette using argon-purged gas-tight Hamilton syringes. Cuvettes containing buffer, CoQ_1, and, where indicated, sulfite or cyanide, are incubated at 25 °C for 2 min. An aliquot of SQOR is added, and the reaction is initiated immediately thereafter by addition of sodium sulfide. Reaction rates are determined by monitoring the reduction of CoQ_1 at 278 nm ($\Delta\varepsilon_{\text{ox-red}} = 12{,}000\ M^{-1}\ \text{cm}^{-1}$) (Fig. 2) and are corrected for the blank rate observed in the absence of SQOR.

A routine assay, which is used to monitor enzyme activity during purification, is conducted at 25 °C in 100 mM potassium phosphate buffer, pH 7.5, containing 0.5 mM EDTA, 80 μM CoQ_1, 600 μM sodium sulfite, and 200 μM sodium sulfide. Similar rates are observed at these substrate concentrations when assays are performed under aerobic or anaerobic conditions. Accordingly, routine assays are conducted using aerobic buffer and uncapped 1-mL cuvettes. Stock solutions of sodium sulfide and sodium sulfite are, however, prepared and stored on ice under anaerobic conditions.

5. SPECTRAL PROPERTIES OF RECOMBINANT HUMAN SQOR

The purified enzyme contains an approximately stoichiometric amount of noncovalently bound FAD (0.82 mol FAD/mol protein) and exhibits a typical flavoprotein visible absorption spectrum with peaks at 451 and 385 nm and a pronounced shoulder at 473 nm

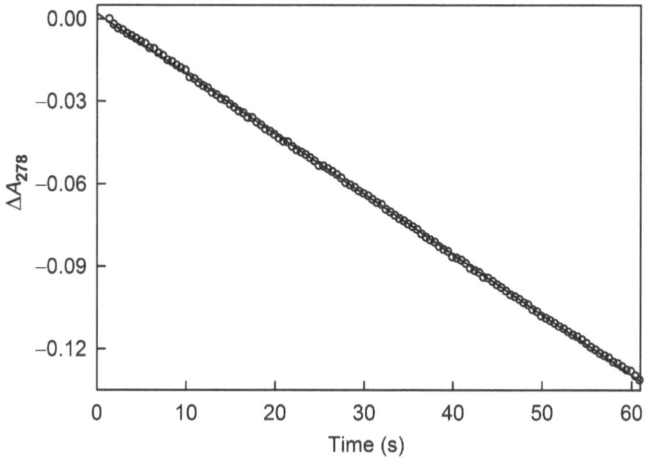

Figure 2 Measurement of SQOR activity using a continuous spectrophotometric assay. The trace shows the initial rate of reduction of CoQ_1 observed in an anaerobic assay conducted at 25 °C in 100 m*M* potassium phosphate buffer, pH 7.5, containing 0.5 m*M* EDTA, 80 μ*M* CoQ_1, 600 μ*M* sodium sulfite, 200 μ*M* sodium sulfide, and 0.75 n*M* SQOR. The reaction was initiated by addition of sodium sulfide. The black line was generated by a linear regression analysis of the data (red circles, black in the print version).

($A_{280}/A_{451} = 9.24$). The flavin is released into solution upon denaturation with guanidine hydrochloride (Fig. 3). The concentration of the purified native enzyme is estimated based on the flavin absorbance at 451 nm ($\varepsilon_{451} = 11{,}500\ M^{-1}\ \mathrm{cm}^{-1}$).

6. SURVEY OF POTENTIAL SULFANE SULFUR ACCEPTORS FOR HUMAN SQOR

H_2S oxidation by rat or lugworm SQOR is difficult to detect in the absence an acceptor for the sulfane sulfur, such as cyanide (Hildebrandt et al., 2008; Theissen & Martin, 2008). In contrast, H_2S oxidation by human SQOR is observed in assays containing only sulfide and CoQ_1. A 4.5- or 13.6-fold increase in the rate of H_2S oxidation is, however, observed upon addition of 1.0 m*M* cyanide or 600 μ*M* sulfite, respectively (Table 2). The reaction of human SQOR with cyanide or sulfite as acceptor is accompanied by the formation of a stoichiometric amount of thiocyanate or thiosulfate (Jackson et al., 2012).

It has been suggested that SQOR might use glutathione as the sulfane sulfur acceptor in a reaction that would produce the glutathione persulfide

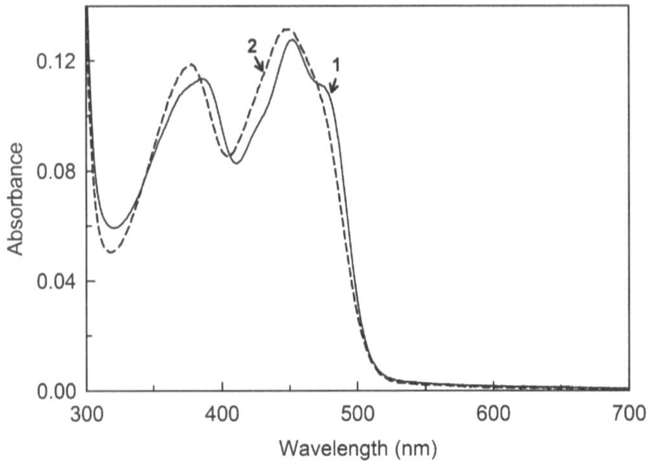

Figure 3 Absorption spectra of native and denatured recombinant human SQOR. Curve 1 is the absorption spectrum of SQOR in 100 m*M* potassium phosphate buffer, pH 7.4, containing 0.04% DHPC. Curve 2 was recorded after denaturation with 3 *M* guanidine hydrochloride. *Adapted with permission from Jackson et al. (2012).*

Table 2 Catalytic properties of recombinant human SQOR[a]

Reaction	$k_{cat\ app}$ (s^{-1})
$HS^- + CoQ_1$	18.5 ± 0.9
$HS^- + CoQ_1 + CN^-$ (1.0 m*M*)	82 ± 6
$HS^- + CoQ_1 + SO_3^{2-}$ (600 μ*M*)	251 ± 9
$HS^- + CoQ_1$ + GSH (1.0 m*M*)	19 ± 3
$HS^- + CoQ_1$ + SDO (907 n*M*)	19.2 ± 0.6

[a]Reaction rates were measured at 25 °C in anaerobic 100 m*M* potassium phosphate buffer, pH 8.0, containing 0.5 m*M* EDTA, 200 μ*M* sulfide, 80 μ*M* CoQ_1, and other indicated components at concentrations shown in parentheses.
Adapted with permission from Jackson et al. (2012).

(GSS^-) substrate for sulfur dioxygenase (SDO), an enzyme postulated to catalyze the second step in the pathway for H_2S metabolism (Hildebrandt et al., 2008; Kabil & Banerjee, 2010). However, a preliminary test of this hypothesis showed the rate of H_2S oxidation in the presence of 1 m*M* glutathione is, within experimental error, identical to that observed for the reaction with only sulfide and CoQ_1 (Table 2). A linear increase in the rate of H_2S oxidation is observed in studies conducted at higher glutathione concentrations (Fig. 4). The data provide no evidence for Michaelis complex formation at glutathione concentrations up to 40 m*M*. The rate observed at glutathione

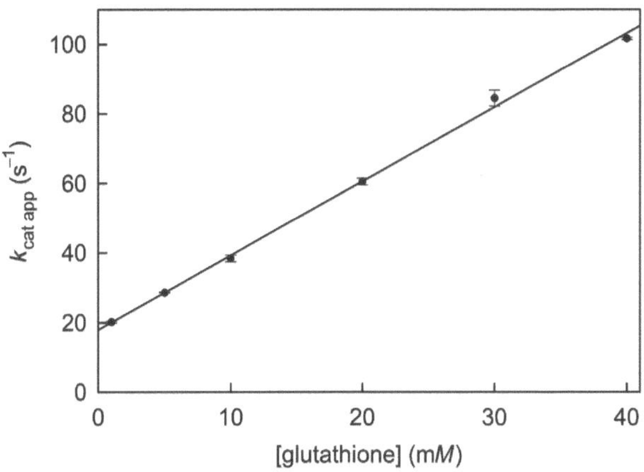

Figure 4 Effect of glutathione concentration on the observed rate of H_2S oxidation in assays conducted in 100 m*M* potassium phosphate buffer, pH 7.5, containing 0.5 m*M* EDTA, 80 μ*M* CoQ_1, 200 μ*M* sodium sulfide, and 2.84 n*M* SQOR. The black line was generated by a linear regression analysis ($r^2 = 0.9975$) of the data (red circles, black in the print version); the *Y*-intercept coincides with the rate of H_2S oxidation observed in the absence of glutathione.

concentrations found in cells (1–8 m*M*) (Griffith, 1999) is $\leq$ two-fold faster than observed in assays containing only sulfide and CoQ_1.

In an alternate scenario, it has been suggested that SDO might directly oxidize the sulfane sulfur in the SQOR persulfide intermediate ($E - CysSS^- + O_2 + H_2O \Rightarrow E - CysSS^- + SO_3{}^{2-} + 2H^+$) (Kabil & Banerjee, 2012). However, addition of human SDO to assays containing sulfide and CoQ_1 does not accelerate the rate of H_2S oxidation (Table 2).

7. SPECTRAL COURSE OF SQOR CATALYTIC ASSAYS WITH SULFITE, CYANIDE, OR SULFIDE AS SULFANE SULFUR ACCEPTOR

The SQOR reactions with sulfite or cyanide as acceptor yield products (thiosulfate or thiocyanate, respectively) that exhibit negligible absorption in the UV region ($\lambda > 240$ nm). The reaction with either acceptor results in a progressive loss of the intense absorption band of oxidized CoQ_1 at 278 nm ($\varepsilon_{278} = 14{,}500\ M^{-1}\ cm^{-1}$) and its tail of absorbance at longer wavelengths ($\lambda > 300$ nm). The end of the reaction is signaled by the appearance of a stable, relatively weak absorption band at 287 nm ($\varepsilon_{287} = 3340\ M^{-1}\ cm^{-1}$) due to reduced CoQ_1, as illustrated by the spectral course observed for the reaction with sulfite (Fig. 5A). A single minimum at

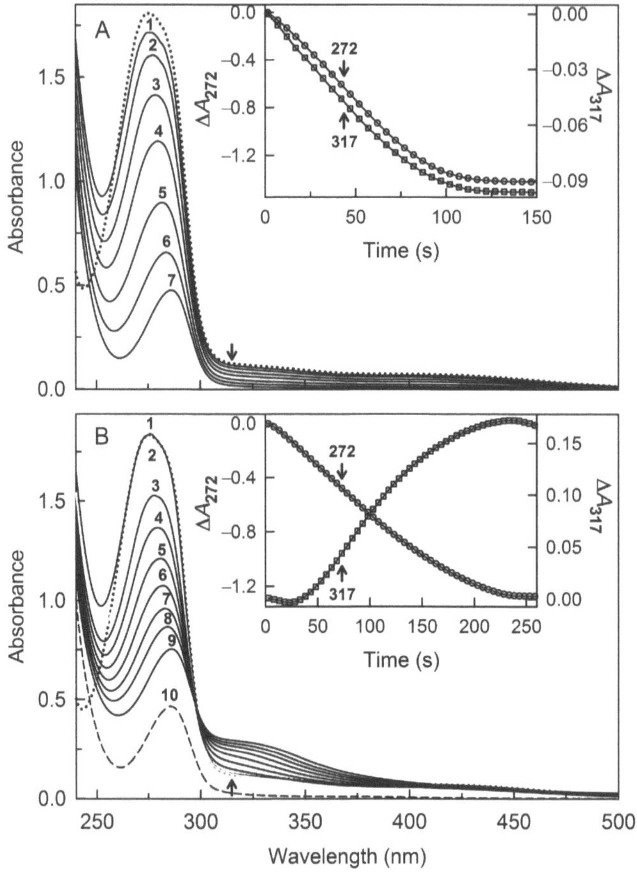

Figure 5 Spectral course of SQOR catalytic assays with sulfite or sulfide as sulfane sulfur acceptor. Reactions with sulfite (panel A) or sulfide (panel B) as acceptor were conducted in 100 m*M* potassium phosphate, pH 8.0, containing 0.5 m*M* EDTA at 25 °C. Red arrows indicate the direction of the spectral changes observed at $\lambda > 300$ nm. Panel A: The dotted black line was recorded before addition of 300 μ*M* sulfide to an assay mixture containing 160 μ*M* CoQ_1, 400 μ*M* sulfite, and 7.2 n*M* SQOR. Curves 2–7 were recorded 1.4, 12, 27, 42, 62, 82, and 132 s, respectively, after sulfide addition. Panel B: The dotted black line was recorded before addition of 300 μ*M* sulfide to an assay mixture containing 160 μ*M* CoQ_1 and 35.6 n*M* SQOR. Curves 1–9 were recorded 2.6, 33.2, 58.2, 88.2, 108, 133, 158, 183, and 233 s, respectively, after sulfide addition. Curve 10 was recorded 292 s after addition of 400 μ*M* sulfite. The inset in each panel shows the time course of absorbance changes at 272 and 317 nm plotted according to the left and right *y*-axes, respectively. *Adapted with permission from Jackson et al. (2012).* (See the color plate.)

272 nm is seen in difference spectra calculated by subtracting the final spectrum of reduced CoQ_1 from spectra observed during the reaction (data not shown). Similar kinetics are observed for the decrease in absorbance at 272 or 317 nm (Fig. 5A, inset).

A strikingly different spectral course is observed when assays are conducted in the absence of sulfite or cyanide, conditions under which sulfide acts as the apparent sulfane sulfur acceptor. In this case, the loss of the 278 nm absorption band of oxidized CoQ_1 is accompanied by the formation of a species absorbing in the 300 to 400 nm region (Fig. 5B, curve 9). The corresponding difference spectra exhibit a minimum at 272 nm and a maximum at 317 nm (data not shown). Except for a small initial lag, the increase in absorbance at 317 nm exhibits kinetics similar to that observed for the decrease in absorbance at 272 nm (Fig. 5B, inset).

The near-UV absorbing product observed in the absence of sulfite or cyanide is attributed to the formation of hydrogen disulfide monoanion (HS_2^-), via a reaction in which sulfide acts as the sulfane sulfur acceptor ($E-CysSS^- + HS^- \Rightarrow E-CysSS^- + HS_2^-$). Hydrogen disulfide is expected to readily react with nucleophiles or thiols, such as sulfite, cyanide, dithiothreitol, or glutathione. Indeed, addition of any of these reagents results the rapid loss of the absorbance attributed to HS_2^-, as illustrated by the spectrum obtained after addition of sulfite (Fig. 5B, curve 10), which is identical to that observed for reduced CoQ_1 by itself.

8. STEADY-STATE KINETIC PARAMETERS FOR H_2S OXIDATION BY SQOR WITH SULFITE, CYANIDE, OR SULFIDE AS SULFANE SULFUR ACCEPTOR

The reactions with sulfite or cyanide involve three different substrates. Steady-state kinetics parameters for such reactions can be estimated by varying the concentration of one substrate at a fixed, saturating concentration of the other two substrates (LaRonde-LeBlanc, Resto, & Gerratana, 2009). Reaction rates observed with sulfite or cyanide as acceptor exhibit an expected hyperbolic dependence on the concentration of the varied substrate ($k_{cat\ app} = k_{cat}[S]/(K_m + [S])$); values obtained for k_{cat} with either acceptor are independent of the nature of the varied substrate (Table 3). The turnover rate observed with sulfite at pH 7.5 is, within experimental error, identical to that observed with cyanide at pH 8.5, as judged by the average values obtained for k_{cat} (370 ± 14 and $345 \pm 11\ s^{-1}$, respectively). Similar apparent K_m values for sulfide and CoQ_1 are obtained with either

Table 3 Steady-state kinetic parameters for three reactions catalyzed by human SQOR[a]

Sulfide + sulfite + $CoQ_1 \Rightarrow$ thiosulfate + CoQ_1H_2				
Variable substrate	Fixed substrates[b]	K_m (μM)	k_{cat} (s^{-1})	k_{cat}/K_m ($M^{-1}\ s^{-1}$)
Sulfide	Sulfite, CoQ_1	13 ± 3	379 ± 20	$(2.9 \pm 0.6) \times 10^7$
Sulfite	Sulfide, CoQ_1	174 ± 20	368 ± 14	$(2.1 \pm 0.3) \times 10^6$
CoQ_1	Sulfide, sulfite	19 ± 2	364 ± 8	$(1.9 \pm 0.2) \times 10^7$
Sulfide + cyanide + $CoQ_1 \Rightarrow$ thiocyanate + CoQ_1H_2				
Variable substrate	Fixed substrates[c]	K_m (μM)	k_{cat} (s^{-1})	k_{cat}/K_m ($M^{-1}\ s^{-1}$)
Sulfide	Cyanide, CoQ_1	10.9 ± 0.7	343 ± 9	$(3.1 \pm 0.2) \times 10^7$
Cyanide	Sulfide, CoQ_1	650 ± 80	330 ± 12	$(5.1 \pm 0.7) \times 10^5$
CoQ_1	Sulfide, cyanide	14 ± 2	360 ± 13	$(2.7 \pm 0.3) \times 10^7$
2 Sulfide + $CoQ_1 \Rightarrow$ hydrogen disulfide + CoQ_1H_2				
Variable substrate	Fixed substrate[d]	K_m (μM)	k_{cat} (s^{-1})	k_{cat}/K_m ($M^{-1}\ s^{-1}$)
Sulfide	CoQ_1	315 ± 28	65 ± 2	$(2.1 \pm 0.2) \times 10^5$

[a]Apparent K_m values for the variable substrate were determined at saturating concentrations of the fixed substrates or at concentrations that yielded the maximum possible reaction rate in cases where excess substrate inhibition was observed. Reactions with sulfite, cyanide, or sulfide as the sulfane sulfur acceptor were conducted at the pH optimum determined for each acceptor (pH 7.5, 8.5, or 7.0, respectively) in phosphate or pyrophosphate buffers, as detailed in the text.
[b]Measurements were made at the following fixed substrate concentrations: 83 μM sulfide, 99 μM CoQ_1, and 2000 μM sulfite.
[c]Measurements were made at the following fixed substrate concentrations: 47 μM sulfide, 72.5 μM CoQ_1, and 6000 μM cyanide.
[d]Measurements were made at 66.2 μM CoQ_1.
Reprinted with permission from Jackson et al. (2012).

acceptor. The apparent K_m observed for sulfite ($174 \pm 20\ \mu M$) is fourfold smaller than observed for cyanide. The concentration of sulfite in human plasma has been measured after fasting ($4.87 \pm 2.49\ \mu M$) (Ji, Savon, & Jacobsen, 1995) but the relevant concentration within human cells is not known. Rat cardiac tissue contains 1.4 μmol sulfite/g protein (Liang et al., 2011). A value for the sulfite concentration in rat heart (440 μM) was calculated using this information and the cellular water space in rat cardiac tissue (3.16 mL/g protein) (Vinnakota, 2004).

$$V_{obs} = \frac{V_{max}}{1 + \frac{K_{m2}}{[S]}\left(1 + \frac{K_{m1}}{[S]}\right)} \tag{1}$$

The reaction with sulfide as the sulfane sulfur acceptor is essentially an oxidative dimerization of two identical substrates. According to a steady-state equation derived for this type of reaction (Eq. 1) (Jarret et al., 2000; Neuhaus, 1962), the velocity observed at a fixed saturating concentration of CoQ_1 should not exhibit a simple hyperbolic dependence on the concentration of sulfide, except under the limiting condition where [sulfide] $\gg K_{m1}$. In fact, a hyperbolic dependence is observed when the sulfide concentration is varied in the range from 40 to 2100 μM. (Measurements at higher sulfide concentrations are not possible owing to prohibitively high blank rates.) The apparent K_m obtained for sulfide ($315 \pm 28\ \mu M$) is about 25-fold larger than observed with sulfite or cyanide as the sulfane sulfur acceptor (Table 3). The high K_m value precludes studies at a fixed saturating sulfide concentration and a variable concentration of CoQ_1. The turnover rate observed for the reaction with sulfide as sulfane sulfur acceptor ($k_{cat} = 65 \pm 2\ s^{-1}$) is sixfold slower than the turnover rate observed for the reaction with sulfite or cyanide. The catalytic efficiency of the reaction with sulfide as acceptor is, however, more than 100-fold lower than observed with sulfite or cyanide, as judged by values calculated for the ratio, $k_{cat}/K_{m\ H_2S}$ (Table 3).

9. ROLE OF HUMAN SQOR IN H_2S METABOLISM

We propose that sulfite is the physiological acceptor of the sulfane sulfur generated during H_2S oxidation by SQOR. This reaction produces thiosulfate (Fig. 6, step 1), a known intermediate in the oxidation of H_2S to sulfate by animals or perfused liver (Bartholomew, Powell, Dodgson, &

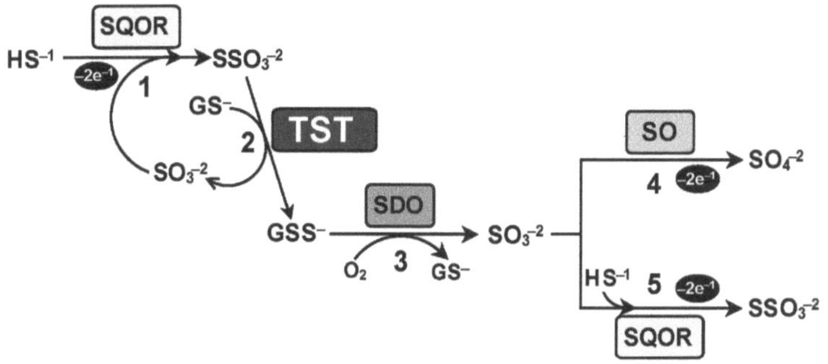

Figure 6 Proposed role of SQOR in mammalian H_2S metabolism. SQOR, sulfide:quinone oxidoreductase; TST, thiosulfate:glutathione sulfurtransferase; SDO, sulfur dioxygenase; SO, sulfite oxidase. *Reprinted with permission from Melideo, Jackson, and Jorns (2014).*

Curtis, 1980; Koj, Frendo, & Janik, 1967; Szczepkowski, Skarzynski, & Weber, 1961). Thiosulfate is also a major product of H_2S metabolism by colon, a tissue that must detoxify large amounts of H_2S produced by sulfate-reducing bacteria (Furne, Springfield, Koenig, DeMaster, & Levitt, 2001; Levitt, Furne, Springfield, Suarez, & DeMaster, 1999). Thiosulfate acts as a substrate for a postulated glutathione-dependent thiosulfate sulfurtransferase (TST) (Fig. 6, step 2). The TST reaction produces glutathione persulfide (GSS^-) and also regenerates the sulfite consumed in step 1. GSS^- is a known substrate for sulfur dioxygenase (SDO) (Hildebrandt et al., 2008; Kabil & Banerjee, 2012; Tiranti et al., 2009), an enzyme that catalyzes a four-electron oxidation of the sulfane sulfur in GSS^- to produce sulfite (Fig. 6, step 3). The last step of H_2S metabolism may proceed via one of two competing alternative reactions, a scenario that can account for observed tissues differences in the final product. In one path, the sulfite produced in step 3 undergoes a two-electron oxidation catalyzed by sulfite oxidase (SO) using cytochrome c as electron acceptor to produce sulfate (Fig. 6, step 4). This path results in an overall eight-electron oxidation of 1 mol of H_2S to sulfate, the major product of H_2S metabolism in liver. Alternatively, the sulfite produced in step 3 may be further metabolized by SQOR (Fig. 6, step 5). This path achieves an overall eight-electron oxidation of 2 mol of H_2S to 1 mol of thiosulfate, the major metabolic product observed in colon.

The proposed scheme is consistent with key features observed for mammalian metabolism of H_2S and the known catalytic properties of SQOR, SDO and SO. Recently, we used bioinformatic approaches to identify genes that encode human TST (*TSTD1*) and its yeast ortholog (*RDL1*) (Melideo et al., 2014). Human TSTD1 constitutes the hitherto missing link between the reactions catalyzed by SQOR and SDO. We obtained biochemical evidence for the functional interaction of the human TSTD1 with human SDO. Our discovery of bacterial Rosetta stone proteins that comprise fusions of SDO and TSTD1 provides phylogenetic evidence for the association of these enzymes. We found that many of the bacteria that express SDO–TSTD1 fusion proteins also contain an adjacent gene that encodes a human-like SQOR. The results strongly suggest that these prokaryotes metabolize H_2S via a pathway that is strikingly similar to the first three steps proposed for mammalian H_2S metabolism.

ACKNOWLEDGMENT

This work was supported in part by National Institutes of Health Grant R01 GM107389 (M. S. J.).

REFERENCES

Bartholomew, T. C., Powell, G. M., Dodgson, K. S., & Curtis, C. G. (1980). Oxidation of sodium sulphide by rat liver, lungs and kidney. *Biochemical Pharmacology*, *29*, 2431–2437.

Brito, J. A., Sousa, F. L., Stelter, M., Bandeiras, T. M., Vonrhein, C., Teixeira, M., et al. (2009). Structural and functional insights into sulfide:quinone oxidoreductase. *Biochemistry*, *48*, 5613–5622.

Cherney, M. M., Zhang, Y. F., Solomonson, M., Weiner, J. H., & James, M. N. G. (2010). Crystal structure of sulfide:quinone oxidoreductase from *Acidithiobacillus ferrooxidans*: Insights into sulfidotrophic respiration and detoxification. *Journal of Molecular Biology*, *398*, 292–305.

Furne, J., Springfield, J., Koenig, T., DeMaster, E., & Levitt, M. D. (2001). Oxidation of hydrogen sulfide and methanethiol to thiosulfate by rat tissues: A specialized function of the colonic mucosa. *Biochemical Pharmacology*, *62*, 255–259.

Griffith, O. W. (1999). Biologic and pharmacologic regulation of mammalian glutathione synthesis. *Free Radical Biology & Medicine*, *27*, 922–935.

Hildebrandt, T. M., Grieshaber, M. K., & Westley, J. (2008). Three enzymatic activities catalyze the oxidation of sulfide to thiosulfate in mammalian and invertebrate mitochondria. *FEBS Journal*, *275*, 3352–3361.

Hughes, M. N., Centelles, M. N., & Moore, K. P. (2009). Making and working with hydrogen sulfide: The chemistry and generation of hydrogen sulfide in vitro and its measurement in vivo: A review. *Free Radical Biology & Medicine*, *47*, 1346–1353.

Jackson, M. R., Melideo, S. L., & Jorns, M. S. (2012). Human sulfide:quinone oxidoreductase catalyzes the first step in hydrogen sulfide metabolism and produces a sulfane sulfur metabolite. *Biochemistry*, *51*, 6804–6815.

Jarret, C., Stauffer, F., Henz, M. E., Marty, M., Lüönd, R. M., Bobálová, J., et al. (2000). Inhibition of *Escherichia coli* porphobilinogen synthase using analogs of postulated intermediates. *Chemistry & Biology*, *7*, 185–196.

Ji, A. J., Savon, S. R., & Jacobsen, D. W. (1995). Determination of total serum sulfite by HPLC with fluorescence detection. *Clinical Chemistry*, *41*, 897–903.

Jones, D. S., Schaperdoth, I., & Macalady, J. L. (2014). Metagenomic evidence for sulfide oxidation in extremely acidic cave biofilms. *Geomicrobiology Journal*, *31*, 194–204.

Kabil, O., & Banerjee, R. (2010). Redox biochemistry of hydrogen sulfide. *The Journal of Biological Chemistry*, *285*, 21903–21907.

Kabil, O., & Banerjee, R. (2012). Characterization of patient mutations in human persulfide dioxygenase (ETHE1) involved in H_2S catabolism. *The Journal of Biological Chemistry*, *287*, 44561–44567.

Koj, A., Frendo, J., & Janik, Z. (1967). [^{35}S]-thiosulphate oxidation by rat liver mitochondria in the presence of glutathione. *Biochemical Journal*, *103*, 791.

Lagoutte, E., Mimoun, S., Andriamihaja, M., Chaumontet, C., Blachier, F., & Bouillaud, F. (2010). Oxidation of hydrogen sulfide remains a priority in mammalian cells and causes reverse electron transfer in colonocytes. *Biochimica et Biophysica Acta*, *1797*, 1500–1511.

LaRonde-LeBlanc, N., Resto, M., & Gerratana, B. (2009). Regulation of active site coupling in glutamine-dependent NAD^+ synthetase. *Nature Structural & Molecular Biology*, *16*, 421–429.

Levitt, M. D., Furne, J., Springfield, J., Suarez, F., & DeMaster, E. (1999). Detoxification of hydrogen sulfide and methanethiol in the cecal mucosa. *Journal of Clinical Investigation*, *104*, 1107–1114.

Liang, Y., Liu, D., Ochs, T., Tang, C., Chen, S., Zhang, S., et al. (2011). Endogenous sulfur dioxide protects against isoproterenol-induced myocardial injury and increases myocardial antioxidant capacity in rats. *Laboratory Investigation*, *91*, 12–23.

Marcia, M., Ermler, U., Peng, G., & Michel, H. (2009). The structure of *Aquifex aeolicus* sulfide:quinone oxidoreductase, a basis to understand sulfide detoxification and respiration. *Proceedings of the National Academy of Sciences of the United States of America*, *106*, 9625–9630.

Marcia, M., Ermler, U., Peng, G., & Michel, H. (2010). A new structure-based classification of sulfide:quinone oxidoreductases. *Proteins*, *78*, 1073–1083.

Melideo, S. L., Jackson, M. R., & Jorns, M. S. (2014). Biosynthesis of a central intermediate in hydrogen sulfide metabolism by a novel human sulfurtransferase and its yeast ortholog. *Biochemistry*, *53*, 4739–4753.

Neuhaus, F. C. (1962). The enzymatic synthesis of D-alanyl-D-alanine. I. Kinetic studies on D-alanyl-D-alanine synthetase. *The Journal of Biological Chemistry*, *237*, 3128–3135.

Shahak, Y., & Hauska, G. (2008). Sulfide oxidation from cyanobacteria to humans: Sulfide-quinone oxidoreductase (SQR). In R. Hell, C. Dahl, D. B. Knaff, & T. Leustek (Eds.), *Advances in photosynthesis and respiration* (pp. 320–335). Heidelberg: Springer.

Shibata, H., & Kobayashi, S. (2006). Characterization of a HMT2-like enzyme for sulfide oxidation from *Pseudomonas putida*. *Canadian Journal of Microbiology*, *52*, 724–730.

Shibata, H., Suzuki, K., & Kobayashi, S. (2007). Menaquinone reduction by an HMT2-like sulfide dehydrogenase from *Bacillus stearothermophilus*. *Canadian Journal of Microbiology*, *53*, 1091–1100.

Szczepkowski, T. W., Skarzynski, B., & Weber, M. (1961). The metabolic state of thiosulfate. *Nature*, *189*, 1007–1008.

Theissen, U., & Martin, W. (2008). Sulfide:quinone oxidoreductase from the lugworm *Arenicola marina* shows cyanide- and thioredoxin-dependent activity. *FEBS Journal*, *275*, 1131–1139.

Tiranti, V., Viscomi, C., Hildebrandt, T., Di Meo, I., Mineri, R., Tiveron, C., et al. (2009). Loss of ETHE1, a mitochondrial dioxygenase, causes fatal sulfide toxicity in ethylmalonic encephalopathy. *Nature Medicine*, *15*, 200–205.

Vinnakota, K. C. B. (2004). Myocardial density and composition: A basis for calculating intracellular metabolite concentrations. *American Journal of Physiology. Heart and Circulatory Physiology*, *286*, H1742–H1749.

Weghe, J. G. V., & Ow, D. W. (1999). A fission yeast gene for mitochondrial sulfide oxidation. *The Journal of Biological Chemistry*, *274*, 13250–13257.

Yadav, P. K., Yamada, K., Chiku, T., Koutmos, M., & Banerjee, R. (2013). Structure and kinetic analysis of H_2S production by human mercaptopyruvate sulfurtransferase. *The Journal of Biological Chemistry*, *288*, 20002–20013.

Yong, R., & Searcy, D. G. (2001). Sulfide oxidation coupled to ATP synthesis in chicken liver mitochondria. *Comparative Biochemistry and Physiology, Part B: Biochemistry & Molecular Biology*, *129*, 129–137.

CHAPTER FIFTEEN

H_2S Regulation of Nitric Oxide Metabolism

Gopi K. Kolluru[1], Shuai Yuan, Xinggui Shen, Christopher G. Kevil[1]
Department of Pathology, Louisiana State University Health Sciences Center–Shreveport, Shreveport, Louisiana, USA
[1]Corresponding authors: e-mail address: gkkolluru@gmail.com; ckevil@LSUHSC.edu

Contents

Abstract

Nitric oxide (NO) and hydrogen sulfide (H_2S) are two major gaseous signaling molecules that regulate diverse physiological functions. Recent publications indicate the regulatory role of H_2S on NO metabolism. In this chapter, we discuss the latest findings on H_2S–NO interactions through formation of novel chemical derivatives and experimental approaches to study these adducts. This chapter also addresses potential H_2S

Methods in Enzymology, Volume 554
ISSN 0076-6879
http://dx.doi.org/10.1016/bs.mie.2014.11.040

interference on various NO detection techniques, along with precautions for analyzing biological samples from various sources. This information will facilitate critical evaluation and clearer insight into H_2S regulation of NO signaling and its influence on various physiological functions.

1. INTRODUCTION

1.1. Nitric oxide and hydrogen sulfide: Key gaseous signaling molecules and their interactions

Nitric oxide (NO), the endothelium-derived relaxing factor, and the recently identified hydrogen sulfide (H_2S), which has been dubbed as endothelium-hyperpolarizing factor, are the gaseous signaling molecules with similar signaling functions. NO and H_2S regulate complex biological processes including cardiovascular, neuronal, cytoprotection, inflammation, and immune functions (Kolluru, Shen, Bir, & Kevil, 2013). These gasotransmitters have gained recent attention for their (1) unique regulation of physiological functions through common signaling cascades, (2) chemical interactions between the molecules leading to production of novel compounds that regulate physiological functions, and (3) interaction with thiols and other biological compounds. However, biochemical mechanisms that regulate these two gasotransmitter interactions have only recently been examined and require further investigation. This chapter describes the detection and quantification methods involved in identifying the effect of H_2S on NO and its metabolites mediated through nitric oxide synthases (NOSs) and nonenzymatic pathways (Kolluru, Shen, & Kevil, 2013; Liu & Huang, 2008; Fig. 1).

1.2. How does H_2S influence NOS and production of NO and its metabolites?

Pathophysiological effects of NO have been well established over many years (Hirst & Robson, 2011; Liu & Huang, 2008). Similar to NO, the biological effects of H_2S have been demonstrated in numerous publications (Kolluru, Shen, Bir, et al., 2013; Wang, 2012; Zamora, Vodovotz, & Billiar, 2000). However, only recent studies suggest that H_2S influences the bioavailability of NO and activity of its synthesis enzymes. The production of NO can be accomplished through either enzymatic or nonenzymatic pathway (Lundberg, Weitzberg, & Gladwin, 2008). It has been reported that H_2S induces phosphorylation of eNOS and also prevents its degradation (King et al., 2014; Kondo et al., 2013; Lei, Liu, Sheen, Chen, & Lii, 2010).

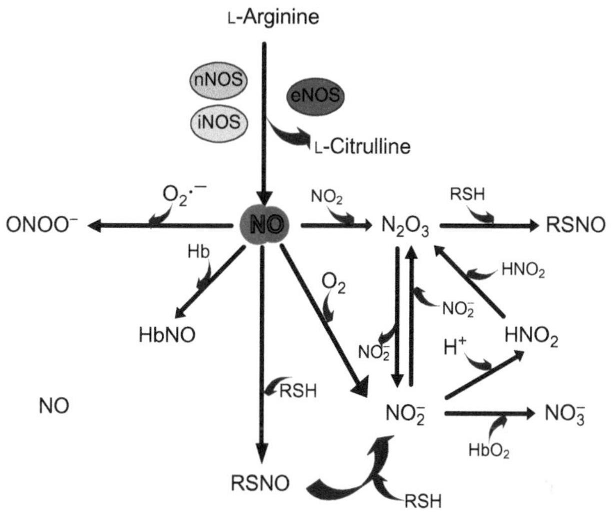

Figure 1 *Biosynthesis of NO and its metabolism*: NO is enzymatically synthesized by NOS isoforms—neuronal, inducible, and endothelial NO synthase (nNOS, iNOS, and eNOS) that catalyze the conversion of L-arginine to L-citrulline. NO undergoes oxidation to form nitrite (NO_2^-) and nitrate (NO_3^-). NO reacts with free thiol (RSH) to form nitrosothiol (RSNO). NO may react with either oxyhemoglobin (HbO_2) to form nitrate or with hemoglobin (Hb) to form nitrosylhemoglobin (HbNO). NO can also be formed nonenzymatically from reduction of nitrite. Nitrite upon reacting with NO or another nitrite forms dinitrogen trioxide (N_2O_3) that further reacts with free thiol to form RSNO or reduces to form nitrite. N_2O_3 can also be formed from nitrous acid (HNO_2) that in turn is formed from nitrite. Additionally, NO may also react with superoxide to produce peroxynitrite.

Conversely, exogenous H_2S donors, diallyl trisulfide (DATS) and NaHS, reduce iNOS expression and corresponding inflammation process (Benetti et al., 2013; Liu et al., 2006). Furthermore, non-NOS induction of NO production via sulfide-dependent nitrite reduction to NO under ischemic conditions has also been demonstrated (Bir et al., 2012). Together, these reports suggest that H_2S clearly influences NO production and its metabolites.

Studies with specific knockouts or heterozygous mutants of H_2S-producing enzymes, cystathionine-γ-lyase (CSE) and cystathionine-β-synthase (CBS), further substantiate H_2S influence on vascular functions including vasodilation that involves NO (Yang et al., 2008). However, no studies have been reported showing the influence of 3-mercaptopyruvate sulfurtransferase on NO metabolism. In a recent I/R study, the Lefer laboratory has demonstrated that the bioavailability of NO and eNOS expression decreases in CSE-knockout mice. However, acute H_2S therapy restored both eNOS function and NO bioavailability (King et al., 2014). Similarly, studies in animal models of hyperhomocysteinemia with heterozygous disruption of the $CBS^{+/-}$ gene resulted in impairment of vascular functions (Eberhardt et al., 2000). Further investigations in this model have demonstrated that this dysfunction is the result of decreased eNOS activity and bioavailability of NO and its metabolite *S*-nitrosothiol (Eberhardt et al., 2000; Upchurch et al., 1997; Zhang et al., 2000). Upregulation of other reactive oxygen species such as superoxide, peroxynitrite, and hydrogen peroxide in these models may also impair the biological activity of NO (Eberhardt et al., 2000; Upchurch et al., 1997).

Both H_2S and NO regulate several vascular functions through common signaling mechanisms (VEGF/Akt/PKB/p38 MAPK) (Altaany, Yang, & Wang, 2013; Bucci et al., 2010; Kondo et al., 2013; Lei et al., 2010; Yong et al., 2008; Yusof, Kamada, Kalogeris, Gaskin, & Korthuis, 2009). This is a relatively new area of research and is not without controversies. There are studies on H_2S–NO interactions that otherwise indicate H_2S-mediated downregulation of NOS expression and subsequently NOS/NO-mediated physiological functions (Geng et al., 2007; Kubo et al., 2007; Oh et al., 2006). However, it is increasingly clear that H_2S influences NOS/NO signaling in regulating various biological functions that requires further to understand specific mechanistic interactions.

1.3. Novel adducts formation from H_2S–NO interactions

Apart from influencing NO metabolism, H_2S has been recently identified to interact with NO metabolites (Kolluru, Shen, & Kevil, 2013). H_2S/HS^- (hydrosulfide ion), a strong nucleophile, can indirectly react with electrophilic NO through its oxidized or nitrosating forms to produce distinct biological compounds (Fig. 2). Recently, there have been several studies on novel H_2S–NO reactants. Whiteman et al. previously revealed that H_2S may reduce cytotoxicity by reacting with peroxynitrite ($ONOO^-$) forming a novel unknown compound (Whiteman et al., 2006). Later, Sebastian et al. reported that HS^- reacts with peroxynitrite to form sulfenic acid (HSOH) and NO_2^- (Carballal et al., 2011). Sulfenic acid may further form dihydrogen disulfide via reaction with HS^- (Carballal et al., 2011). HS^- may react with *S*-nitrosothiol (SNO) or *S*-nitrosoglutathione (GSNO) to form novel metabolites such as thionitrous acid (HSNO), sulfinyl nitrite (HS(O)NO or $HSNO_2$), or nitroxyl (HNO) (Filipovic et al., 2013; Filipovic, Miljkovic, Allgauer, et al., 2012; Filipovic, Miljkovic, Nauser, et al., 2012). However, a recent study reported

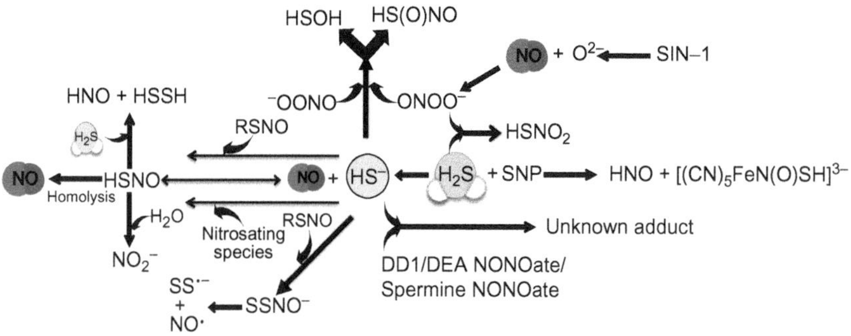

Figure 2 *Novel adducts from NO–H_2S interactions*: Sulfide radical HS^- can react with NO to form thionitrous acid (HSNO). Reaction of HS^- either with a thiol (RSNO) or with nitrosating species can also from HSNO. Alternatively, HS^- can react with RSNO to form nitrosopersulfide ($SSNO^-$), which further disintegrates into NO and disulfide radical ($SS^{\bullet -}$). HSNO can undergo hemolysis to form NO or react with H_2S to form a disulfide (HSSH) and HNO. It can also form nitrite upon hydrolysis (H_2O). NO donors can directly or indirectly involve to form new adducts. SIN-1 can generate both NO and O_2^- that leads to formation of peroxynitrite ($ONOO^-$). Peroxynitrite can react with either H_2S or its radical HS^- to form sulfinyl nitrite ($HSNO_2$ or HS(O)NO). HS^- can also form HSOH upon reacting with $ONOO^-$. Other NO donors such as DD1, DEA NONOate, or Spermine NONOate are known to react with HS^- to form an unknown adduct. Sodium nitroprusside (SNP) ($Na_2[Fe(CN)_5\ NO]$) reacts with H_2S to form nitroxyl (HNO) and an intermediate $[(CN)_5FeN(O)SH]^{3-}$.

that sulfide reacts with nitrosothiols rather to form nitrosopersulfide ($SSNO^-$) that may contribute to sulfide induced extended effects of NO mediated through nitrosothiols (Cortese-Krott et al., 2014). Nonetheless, a key conundrum remains in that the formation of these compounds has been reported using supra physiological concentrations of both H_2S and NO donors and their variants. Further studies are needed to investigate the formation of NO–H_2S adducts considering physiological conditions of pH and concentration.

2. TECHNIQUES DETERMINING ENZYMATIC ACTIVITY AND EXPRESSION OF NOS

NO is synthesized by three isoforms of NOSs—endothelial, neuronal, and inducible (Moncada & Higgs, 1993). Measurement of NOS activity and expression is essential to demonstrate the role of enzymatic synthesis of NO from various biological sources. This section describes currently available methods to quantify NOS activity and expression.

2.1. High-sensitive radiolabeled detection of NOS

Bredt and Snyder first reported the citrulline assay for quantification of NOS isoforms (Bredt & Snyder, 1990). The principle of this high-sensitive assay is based on the conversion of L-arginine to L-citrulline and NO formation as a byproduct. This assay is performed either on fresh samples such as cells and tissues or on biopsies (Bredt & Schmidt, 1996; Rachmilewitz et al., 1994; Steil, Garcia Rodriguez, Alonso, Crespo, & Bosca, 1995). The assay can also be performed on homogenates of frozen tissue or cells, provided the material is nonfixed, snap frozen in liquid nitrogen, and stored at −80 °C. Activity of NOS in comparison with that of control preparations indicates the level of induction of NOS activity.

Prior to starting the assay, samples are treated with addition of several cofactors and substrates, including calmodulin, tetrahydrobiopterin (BH_4), and NADPH. NOS activity is determined based on the amount of 1-mol radiolabeled L-arginine converted to radiolabeled L-citrulline. The samples are prepared by giving KCl wash to reduce nonspecific synthesis of L-citrulline. Cationic L-arginine is retained by AG 50 ion-exchange resin column in its Na^+ form, while L-citrulline is eluted. Either L-[^{3}H]arginine or L-[^{14}C]arginine is used as a precursor of L-citrulline. While ^{14}C-has a better radioisotope stability, L-[^{3}H]arginine has a better specific activity and economical. With few minor modifications, this assay can be performed to measure activity of all the three isoforms of NOS based on their dependence of Ca^{2+}-mediated activation (Salter, Knowles, & Moncada, 1991). Selective measurement of iNOS or nNOS activity is achieved by performing reactions with Ca^{2+}-chelating agent (1 m*M* EGTA) or in combination with NOS inhibitor (1 m*M* EGTA with 1 m*M* L-N^G-methylarginine).

2.1.1 Key points and limitations

1. Highly specific assay for determination of NOS enzymatic activity.
2. Works with biological materials containing high enzyme levels.
3. Additional metabolic pathways such as arginase and ornithine transcarbamylase can also indirectly synthesize L-citrulline from L-arginine in the absence of NOS.
4. Presence of endogenous L-arginine in the samples that may vary the results by affecting the specific activity of the enzymes.
5. Radioactive materials are hazardous and have to be handled with extreme caution.

2.2. Western blotting for detection of NOS expression

As mentioned earlier, sulfide influences NO bioavailability by upregulating the activity of its enzymes (King et al., 2014; Kondo et al., 2013; Lei et al., 2010). Levels of NOS proteins have been determined using western blot technique. For analysis, tissue/organs should be homogenized on ice using a tissue homogenizer in RIPA buffer (25 m*M* Tris–HCl, pH 7.4, and protease inhibitors) to yield a 10% (weight/volume) homogenate. The required amount of homogenate can be added to an equal volume of Laemmli buffer for SDS-PAGE analysis. Remaining sample can be snap frozen in liquid nitrogen for future analysis. Tissue/cellular extracts should be separated by centrifugation at 10,000 × *g* for 10 min at 4 °C. Supernatants are collected and measured for total protein concentrations will be determined by Bradford assay (Bio-Rad Laboratories, Hercules, CA, USA). Same amounts of proteins should be taken to separate on a 10%-SDS-polyacrylamide gel and transferred to PVDF membrane. Membranes should be probed with corresponding antibodies against eNOS, phospho-eNOS, nNOS, or iNOS (Abcam) (1:1000), respectively, with loading control β-tubulin or β-actin (cell signaling). The ECL western blot analysis system (Bio-Rad Laboratories or Amersham Life Systems) can be used for detection of the primary antibody signal and included a peroxidase-labeled secondary antibody. Our laboratory has achieved better specificity with primary antibodies purchased from Cell signaling or Abcam. The immunoreactivity of NOS is detected by changes in chemiluminescence. After transfer to an autoradiogram, signals of NOS isotypes can be quantified by densitometry using Image J software ver 1.48 (NIH).

2.3. Determination of mRNA expression of NOSs by qRT-PCR

To check the aspects of NOS gene expression mediated by sulfide, levels of NOS mRNA can be quantified using a RT-PCR reaction. The samples isolated from various tissue sources placed in RNA later 10% (weight/volume) should be homogenized using a tissue homogenizer. Total RNA will be obtained using the RNeasy Fibrous Tissue Kit (Qiagen). A total of 1 μg of DNase-I-treated RNA reverse should be transcribed into cDNA using the Bio-rad iscript cDNA supermix kit (Bio-rad) with random hexamers in a 20-μl reaction. PCR should be performed in duplicate for each sample using 1 μl of cDNA as a template, Bio-rad SYBR green assay kit (Bio-rad) in a 20-μl reaction. Specific primers should be used to probe for eNOS, nNOS, and iNOS with GAPDH controls should be designed (Beacon Designer 2.0 software). The sequences of the primers used for mouse as follows: eNOS accession number NM_008713 (sense-5′-CAGCAT CACCTACGACAC-3′; A-5′-GGCTCTGTAACTTCCTTGG-3′); nNOS accession number NM_008712 (sense-5′-GTCTTCCACCAGGAGATG-3′; A-5′-AAAGGCACAGAAGTGGGGGTA-3′); and iNOS, accession number NM_010927 (sense-5′-ACGGACGAGACGGATAG-3′; A-5′ GGGCTT CAAGATAGGGA-3′). The primers should be verified by a gel electrophoresis after obtaining a single PCR product with a conventional PCR. A reaction mixture lacking cDNA should be used as the negative control. Quantitative RT-PCR will be carried out in a Bio-Rad CFX96 RT-PCR using the following conditions: 2 min at 50 °C, 10 min at 95 °C followed by 40 cycles of 15 s at 95 °C and 1 min at 60 °C. The expression level of the three NOS isotypes should be calculated using the comparative threshold cycle method normalized to housekeeping gene GAPDH.

3. DETECTION OF NO AND ITS METABOLITES

NO can react with oxygen, hemoglobin and various other molecules to form various metabolites such as nitrite, nitrate, *S*-nitrosothiols (RSNOs), *S*-nitrosohemoglobin, etc. In view of difficulties inherent to direct NO detection, measurement of accumulated nitrite and nitrate, stable metabolites of NO, is often used to monitor cumulative NO synthesis, cells in culture, tissue homogenates, perfused organs, and organisms (Kevil, Kolluru, Pattillo, & Giordano, 2011). H_2S can regulate the bioavailability and generation of NO and its metabolites. H_2S also affect the release of NO through other species of NO such as nitrosothiol and metal nitrosyl complexes (Ondrias et al., 2008).

Detection of NO from biological samples is a challenge as this gaseous molecule is very unstable with a half-life in seconds. There are many discrepancies in the literature that reported the levels of NO often underestimated the actual NO produced from the biological samples. So, it is warranted to estimate accurate levels of NO by minimizing its loss through proper preservation/stabilization of the sample. Now there are several techniques available that determine the most accurate measurements of this gasotransmitter and its metabolites with minimum error (Bryan & Grisham, 2007; Cai, Dikalov, Griendling, & Harrison, 2007; Weissman & Gross, 2001). NO detection has evolved from the earlier measurements of Griess assay to the high-sensitive chemiluminescence and direct electrode detection methods.

This has reduced the erroneous detection levels of NO and its metabolites from millimolar to nanomolar over many years (Bryan & Grisham, 2007; Gomes, Fernandes, & Lima, 2006; Hetrick & Schoenfisch, 2009).

3.1. Griess assay: Classic biochemical assay for nitrite/nitrate/nitrosothiol detection

This is a spectrophotometric assay based on the formation of an aromatic amine, diazo derivative from nitrite ions in acidic conditions (Giustarini, Rossi, Milzani, & Dalle-Donne, 2008; Weissman & Gross, 2001). Oxidized NO or nitrite under acidified environment reacts with sulfanilamide to produce a diazonium ion, which further reacts with sulfanilamide and *N*-naphthyl ethylenediamine (NNED) to form a chromophoric azo product. For measuring NO synthesis, one potential drawback is that under normal conditions, only a fraction of NO is oxidized to NO_2^- (in the absence of hemoglobin, usually greater than half). Moreover, there are studies suggesting the involvement of sulfide on interfering with the assay (van der Zee, Lettinga, & Field, 2001; Wedmann et al., 2014). Possible interference of sulfide through decolorization of diazo derivatives, the prime end product (van der Zee et al., 2001), thereby affecting the detection assay.

3.1.1 Protocol

1. This assay can be performed on most of the samples including tissue extracts, cell culture medium, plasma, urine, solutions, etc.
2. Prepare working Griess reagent by mixing equal parts of A and B (NNED). Care has to be taken to avoid any dry material while mixing reagents A and B, which may react with Griess reagents to give increased background absorbance thereby affecting the sensitivity of the assay.
3. Prepare nitrite/nitrate standards from stock and add 25 μl to triplicate wells in a 96-well plate in different wells. On the same plate, add 25 μl volumes of unknowns and appropriate blanks to triplicate wells. A useful standard of nitrite/nitrate is prepared as: 0, 12.5, 25, 50, 75, and 100 μ*M*.
4. Dilute nitrate reductase (NR) to a final concentration of 0.1 U/ml and add 25 μl of NR buffer to all wells.
5. Add 100 μl of working Griess reagent to all wells. Quantify A540 or A550 in each well using a microtiter plate reader. A pink to purple color will develop within seconds, which is proportional to the combined concentrations of nitrite and nitrate in all samples.
6. Plot standard curves for nitrite and nitrate from all unknown samples nitrite/nitrate concentrations using a combined standard curve.

Some modified Griess reactions further reduce nitrate to nitrite using chemical reductants including cadmium (Cd^{2+}), zinc (Zn^{2+}), mercuric chloride ($HgCl_2$), and Vanadium chloride or enzymes such as bacterial NR. However, some of these methods are not reliable as they further reduce nitrite (zinc) or limited shelf life of the compounds and, additionally, slow the reaction kinetics. Cadmium in presence of copper is a most favorable choice (Cortas & Wakid, 1990; Green et al., 1982), where the conversion of nitrate to nitrite can be achieved rapidly as represented below:

$$NO^{3-} + H_2O + Cd \rightarrow NO^{2-} + Cd^{2+} + 2NH_2OH$$

It has been suggested that prior treatment of biological samples with Cd^{2+} and Zn^{2+} can reduce the interference of sulfide with the Griess reaction (Wedmann et al., 2014).

3.1.2 Key points and limitations

1. Less sensitive than any other assay such as chemiluminescence and does not perform a direct detection of NO.
2. Only a fraction of NO is oxidized to NO_2^-, the remainder goes to NO_3^-.
3. It is a simple chemical/enzymatic assay for quantifying the sum of nitrate and nitrite in biological samples with colorimetric read out.
4. It can be performed in 96 well plates.
5. Does not require specialized equipment, following diazotization with the Griess reagent.
6. Reduced azo dye signal by a possible interference of sulfide with the Griess reagents.
7. Reactivity of different compounds varies with NO among various biological samples. So the measurement of NO must be interpreted with caution considering the variations in the settings. Wang et al. reported no changes in plasma NO levels upon H_2S therapy during tissue ischemia (Wang et al., 2010). However, this study only examined plasma nitrite/nitrate levels with the Griess reaction methodology, which does not account for NO, nitrosothiols, and nitrosoheme.

3.2. Chemiluminescent detection of NO metabolites

Advancement in techniques for sensitive and specific detection of NO species has further revealed the physiological and biochemical role of NO and its metabolites including RSNOs and HbNO. Detection of NO metabolites using chemiluminescence technique is a widely used approach.

The principle of this detection method is based on specific reaction of NO with ozone (O_3) to form excited-state NO_2*. Spontaneous return to ground-state NO_2 results in photon emission in the range of red to infrared energy.

As the reaction between NO and O_3 proceeds at a near-diffusion-limited rate and in the absence of a significant activation barrier, this reaction goes to rapid completion at room temperature. Conventional photomultiplier tubes (PMTs) are sensitive to the lower end of this range of emitted light and afford an enormous amplification (>106-fold) of the current arising from a single-photon-induced displacement of an electron from the photosensitive PMT surface. Thus, chemiluminescent detection provides a very sensitive method for direct quantification of NO in picomolar quantities.

3.2.1 Protocol

Stabilization Solution preparation: Biological samples should be collected and placed in the stabilization buffer as homogenized in a solution designed to stabilize NO and all its derivations (Dejam et al., 2005; Kumar et al., 2008; Lang et al., 2007). Composition of stabilization buffer varies based on the sample to be processed.

(a) Stabilization solution 1 (SS1) comprising of *N*-ethylmaleimide (NEM) (5-m*M* final concentration) and DTPA (100-μ*M* final concentration) along with an anticoagulant sodium citrate (1.5% w/v) or Heparin (1 unit/ml 10% v/v).

(b) Stabilization solution 2 (SS2) comprising of NEM (1-m*M* final concentration) and DTPA (100-μ*M* final concentration) along with an anticoagulant sodium citrate (1.5% w/v) or Heparin (1 unit/ml 10% v/v), NEM = 1 m*M*, and DTPA = 100 μ*M*.

(c) Stabilization solution 3 (SS3) comprising of Potassium Ferricyanide ($K_3(FeCN)_6$) (133-m*M* final concentration), NEM (1.67-m*M* final concentration), and detergent Nonidet NP-40 (1% v/v), prepared in PBS.

Collection and storage of the samples: Whole blood (500 μl) has to be immediately (~15–30 s) mixed with SS3 (100 μl); Cell/tissue samples should be collected and homogenized in SS3. Blood is centrifuged at 5000 rpm for 5 min at 4 °C to separate the plasma and RBC. Plasma is stored in SS1 (1:1 vol), while RBC is stored in SS2 (1:5 vol). Samples should be mixed vigorously and incubated on ice for 5 min to allow reaction with $K_3(FeCN)_6$ to reach completion and then snap frozen in liquid N_2 and stored at −80 °C until analysis on a Siever's 280i Nitric Oxide Analyzer (General Electric, CA).

3.2.2 Preparation of samples for analysis

Samples are allowed to completely thaw on ice in the dark. Samples are vortex mixed briefly and centrifuged at 15,000 × *g* for 2 min. Supernatants are to be taken for further measurements as described. Aliquots of samples will be tested for sulfanilamide resistance or mercuric chloride ($HgCl_2$), by addition of an acidic sulfanilamide solution or $HgCl_2$ to a final concentration of 0.5% v/v, and allowed to sit in the dark for 15 min prior to injection into the analyzer.

3.2.3 Process for analyses of nitrate, nitrite, S-nitrosothiols, and XNO

1. Assemble the purge vessel with 1 ml of sodium iodide (NaI) catalyst and 7 ml of glacial acetic acid, for nitrite/nitrate detection. For nitrite/nitrate, use 8 ml of acidic vanadium chloride and 100 μl Antifoam B in the purge vessel.
2. Connect the purge vessel to a helium tank and begin gassing through the porous glass frit at a rate that causes the liquid level to approach the top of the purge vessel (~ 10 ml/min). A continuous supply of water is provided into the purge vessel. The outflow of the purge vessel should be connected through a trap containing 1 *M* NaOH. Connect the trap to the inflow valve of the chemiluminescence detector. If trap is not used, the purge vessel directly connects to the chemiluminescence detector.
3. The vacuum, ozone generator, and PMT are turned ON. The vacuum introduces the sample into the reaction chamber and also purges the gasses that may quench the light generated by NO_2^*. The vacuum should be preferably between 6 and <200 torr.
4. The PMT output signal can be seen on a monitor integrated to the chemiluminescence detector via the data-analysis system. Once the PMT output signal stabilizes, the system is ready for measurements.
5. Before measuring the samples, it is important to generate a calibration curve with known volume for known standard concentrations. Prepare a calibration curve with appropriate sodium nitrite standards and take the readings in duplicate or triplicate to establish linearity and reproducibility of peak areas.
6. Samples are introduced into the purge vessel with acidified NaI buffer by injection using a gas-tight Hamilton syringe through a septum on top of the apparatus. Helium rapidly sweeps the aqueous sample in the gaseous form into the chemiluminescence detector. The relatively small space of the purge vessel provides sharper peaks and high sensitivity. All the levels of NO metabolites have to be normalized to protein concentrations.

3.2.4 Experimental protocol to identify interference of H_2S on NO detection

To address possible effects of H_2S on chemiluminescent detection of NO, we have experimentally demonstrated using a cell free sample with donors of NO and H_2S using reductive acidic triodiode (KI/I3) as described below:

1. Prepare various working concentrations (0.1, 1, 10, 100, 200, 500 μ*M*, and 1 m*M*) of NO donors—sodium nitrite ($NaNO_2$; Sigma Chemicals) and diethylenetriamine NONOate (DETA NONOate; Sigma chemicals); sulfide donors—sodium sulfide (Na_2S) and GYY4137 (Cayman chemicals). Prepare another set of samples with a pretreatment of $CuSO_4$ to reduce any reactive thiols present in the samples.
2. Assemble the purge vessel of the chemiluminescence detector and prepare sodium nitrite calibration curve as mentioned in the protocol above.
3. Inject the 50 μl samples, as indicated above in step 1 into the purge vessel containing a solution of potassium iodide and glacial acetic acid. Samples are run by replacing the purge vessel with fresh reaction solution between each individual experiments.
4. Signals generated from the samples are integrated and compared to sodium nitrite standard curves to calculate concentrations. Signals from the samples reflect total NOx concentrations (nitrite + *S*-nitrosothiol + XNO).

Figure 3 indicates distinct peaks of NO levels obtained from various concentrations (0.1, 1, 10, and 100 μ*M*) of NO donors sodium nitrite ($NaNO_2$) and DETA NONOate (Fig. 3A and B). In contrast, no signal peaks were observed using sulfide donors Na_2S at 0.1, 1.0, and 10 μ*M* (Fig. 3C) or GYY4137 at 0.1, 1.0, 10, 100, 500 μ*M*, and 1 m*M* (Fig. 3D). However, very small signal peaks were observed using 200, 500 μ*M*, and 1 m*M* concentrations of Na_2S but no signals from GYY4137 at the same concentrations (inset Fig. 3C and D) using chemiluminescent detection. Together, these data demonstrate that chemiluminescent detection is 50 times more sensitive in detection of 10 μ*M* $NaNO_2$ than an equal molar concentration of Na_2S. Nonetheless, high concentrations of H_2S/HS^- (>500 μ*M*) might possibly interfere with measurements of low NO concentrations using chemiluminescence. Teng et al. have also reported similar observations, but under alkaline pH using higher concentrations of H_2S (1–10 m*M*; Teng et al., 2008). Thus, care should be taken to employ proper controls and appreciate the potential concentration parameters of possible sulfide levels when measuring NO using chemiluminescence.

3.2.5 Key points and limitations

1. Chemiluminescence assay is a highly sensitive detection method for measurement of NO metabolites.

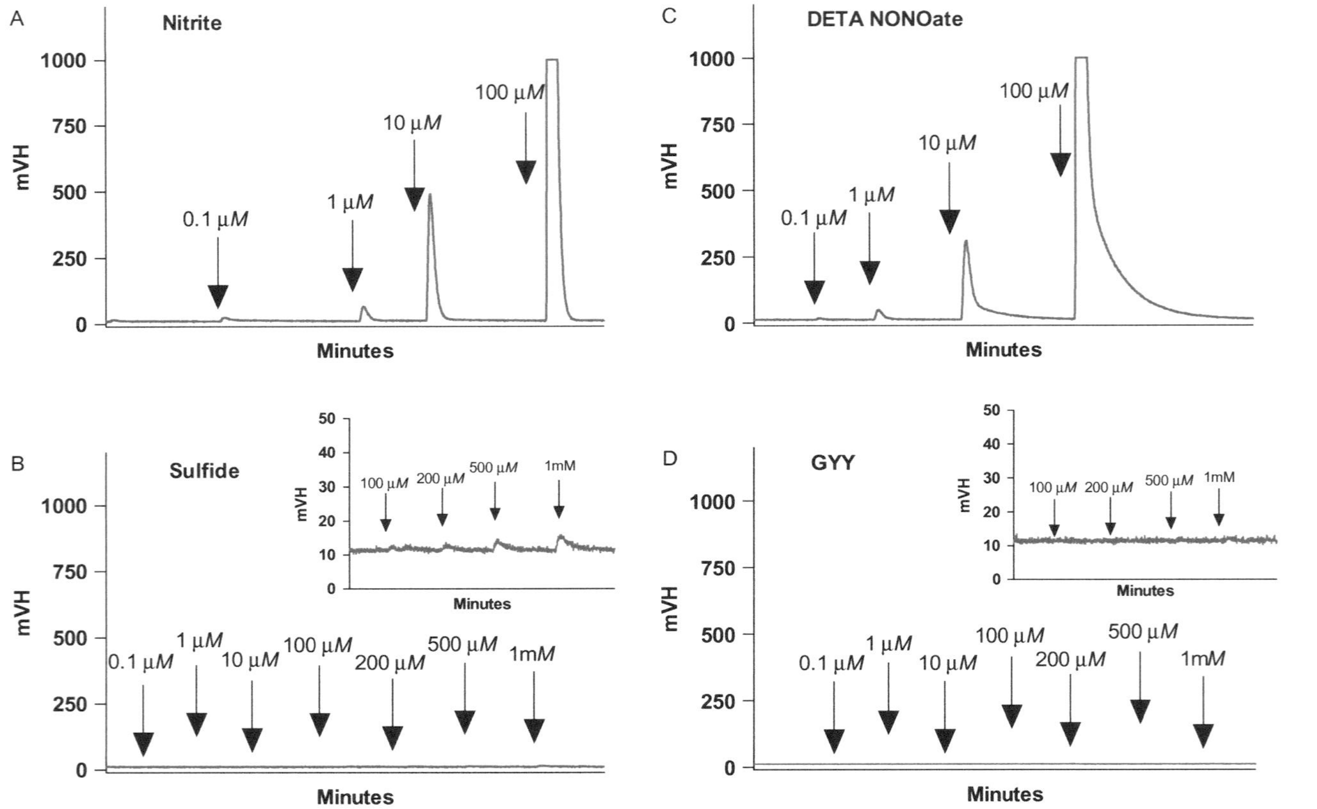

Figure 3 *H_2S interference in chemiluminescence detection of NO*: Using the tri-iodide method in the Sievers 280i chemiluminescence detector, 50 μl of increasing concentrations (0.1 μ*M*–1 m*M*) of either (A) sodium nitrite, (B) DETA NONOate, (C) sodium sulfide, or (D) the sulfide donor GYY were injected. As shown in (A) and (B), NO donors resulted in increasing NO*x* generation in a dose-dependent manner. Panels C and D illustrate little to no NO*x* production with increasing concentrations up to 1 m*M* when sulfide donors were injected. The insets in panels C and D show a magnified section of the higher concentrations. The small peaks shown in the magnified inset with sulfide (C) could indicate the limitation of filter system used in the Sievers 280i to eliminate sulfide chemiluminesence. These small peaks are not visible in the GYY samples (D inset).

2. Chemiluminescence assay can measure multiple forms of NO metabolites like free nitrite, nitrate, nitosothiols, and nitosoheme upon simple pretreatment of the samples with certain chemicals/inhibitors.
3. One of its disadvantages is that it cannot measure real-time NO levels directly emanating from intact tissues.
4. Minimal interference of signals with higher concentrations $H_2S > 1\ mM$ (Fig. 3).

3.3. Real-time detection of NO by electrode probe

Malinski et al. has first devised a carbon fiber "microsensor" in amperometric or voltammetric mode for detection of NO gas in biological systems, with a detection limit of 10 n*M* (Malinski & Taha, 1992). Presently, there are World precision instruments and other commercial companies like Innovative Instruments Inc. offer this electrode probe for NO detection with sensitivity as low as 0.03 pA/n*M*. Amperometric detection of NO is based on oxidation of NO on the surface of an electrode and subsequent generation of current. Gaseous NO diffuses across the NO selective membrane and into sensor containing Ag/AgCl electrode pair. Upon applying of potential equivalent to 900 mV on the working electrode, small redox current is generated due to the oxidation of NO at the electrode, which is measured by an amplifier and recorded. The principle of this detection is represented in the equation below:

$$\begin{aligned} NO + e^- &\rightarrow NO^+ + OH^- \rightarrow HNO_2 \rightarrow H^+ + NO_2 \\ &\rightarrow NO_3^- + 2H_2O + 3e^- \end{aligned}$$

NO in the presence of OH^-, it is converted into nitrite (NO_2^-) further converted to nitrate.

However, a recent study indicates the response exhibited by this electrode upon addition of H_2S (Wedmann et al., 2014). This study revealed that the NO specific electrode displays signals as response to addition of H_2S (1 m*M*). It has to be further investigated with various concentrations of H_2S at physiological range. However, the usage of this direct NO detection has to be carried out with caution considering the electrode response to other radicals/gaseous molecules such as H_2S.

3.3.1 Calibration

Like most other electrodes, NO electrode should be calibrated before quantitative measurements can be made. Calibration based on a chemical reaction of nitrite with iodide in acid that generates known concentrations of NO as the following equation:

$$2KNO_2 + 2KI + 2H_2SO_4 \rightarrow 2NO + I_2 + 2H_2O + 2K_2SO_4$$

A calibration curve demonstrating changes in current or peak height as a function of NO concentration can be produced. The generation of NO is based on the equation mentioned above, where a known amount of KNO_2 is used.

The electrode probe has to be gently placed on to the surface of the cells with Hank's balanced salt solution (supplemented with 2-m*M* $CaCl_2$ and 20-m*M* Hepes, pH 7.4) or can be used on the surface of tissues to measure the at variable concentrations the peaks in NO.

3.3.2 Key points and limitations

1. Usage of this electrode probes for sensitive NO detection real time; it is very simple and easy to perform.
2. Studies looking for H_2S effects on real-time NO variations can be performed using this probe.
3. However, this method can measure NO levels on the periphery and not suitable for detection of NO metabolites in deeper tissues.

3.4. ESR detection of NO

ESR (or electron paramagnetic resonance (EPR)—spin trapping) spectroscopy is a useful tool to detect NO from various sources—*in vitro, ex vivo,* and *in vivo*. Detection of NO using ESR has been detailed in the literature (Cai et al., 2007; Hogg, 2010). NO is a relatively stable radical and not directly detectable by EPR spectroscopy and should be conjugated with compounds such as "spin traps." Upon reacting with spin traps, NO becomes detectable by EPR. Different spin traps can used for NO detection by EPR, such as Iron (Fe)-based hydrophobic diethyldithiocarbamate ($Fe^{2+}(DETC)_2$), which is more compatible with biological fluids such as blood and hydrophilic *N*-methyl-D-glucamine dithiocarbamate ($Fe^{2+}(MGD)_2$) spin traps for extracellular detection of NO such as cells and tissues (Cai et al., 2007; Vanin & Poltorakov, 2009). Nitronyl nitroxides (NNOs) or imidazolinoxyl-*N*-oxides have been identified as both NO detectors and scavengers (Akaike et al., 1993; Joseph, Kalyanaraman, & Hyde, 1993). One of the disadvantages using with dithiocarbamates is copper interference with the Fe/NO signal.

One of the commonly used probe in this category is 2-(4-carboxyphenyl)-4,5-dihydro-4,4,5,5, tetramethyl-1*H*-imidazoyl-1-oxyl-3-oxide (CPTIO). Recent study demonstrated that CPTIO directly reacts with H_2S to form hydroxylamine (Wedmann et al., 2014). Usage of NNOs for NO detection for biological samples with H_2S has to be performed with caution. However, the concentrations used (CPTIO 1 m*M* and Na_2S 10 m*M*) to understand this interaction requires further examination using physiological ranges of sulfide and relevant levels of CPTIO used for experimental purposes.

3.5. Fluorescent detection of NO

Several fluorescence indicators for direct NO-imaging *in vitro* or *in vivo* have been commercially developed based on aromatic diamino derivatives of fluorescent chromophores. These fluorescent probes have distinct excitation/emission (ex/em) wavelengths. Some of the popular fluorescent probes and their ex/em include 2,3-diaminonaphthalene (365/415 nm), 4,5-diaminofluorescein (DAF-2), and 4-amino-5-methylamino-2′,7′-difluorofluorescein (DAF-FM) both with same 495/515 nm, whereas diaminorhodamine-4M (560/575) and 1,2-diaminoanthraquinone (Lacza et al., 2005; von Bohlen und Halbach, 2003). These substances interact with NO to form a fluorescent triazole complex. However, these fluorescent probes have issues with the specificity (detection of other thiols, nitrosating species) and autofluorescence based on the biological substances such as tissues. Therefore, use of these fluorescent probes have to be carried out with caution as there is a possibility of H_2S influence that may produce erroneous NO levels.

4. NOVEL ADDUCTS FROM H_2S–NO INTERACTIONS

The sulfur of H_2S/HS^- has the lowest valence (−2) and therefore serves as a reductant. NO, on the other hand, is a poor one-electron oxidant, which acts as monomer in biological conditions (Fukuto et al., 2012). Additionally, the paramagnetism of NO makes it even less possible for direct interaction with the diamagnetic H_2S/HS^- (Bruce, 2013). Although recent studies suggested direct reaction between H_2S and NO is possible (Eberhardt et al., 2014), mounting evidence supports NO adducts as important intermediates of H_2S and NO interactions. We discuss here various –NO adducts that are discussed in recent research publications.

4.1. Peroxynitrite ($ONOOH/ONOO^-$)

NO and superoxide react with a near-diffusion rate constant (Huie & Padmaja, 1993) to give rise to peroxynitrite ($ONOOH/ONOO^-$), a potent oxidant with important roles in neurodegenerative diseases, inflammation, and ischemia reperfusion. Moore et al. first reported H_2S scavenges $ONOO^-$ and prevents $ONOO^-$-induced cytotoxicity, similarly to GSH (Whiteman et al., 2004). Carballal et al. proposed a nucleophilic attack by H_2S on $ONOO^-$ to form HSOH and NO_2^- (Carballal et al., 2011). Filipovic et al., on the other hand, observed H_2S scavenges $ONOO^-$ with a second-order rate constant of $8 \pm 2 \times 10^3$. The major product is sulfinyl nitrite (mainly HS(O)NO) as the major product, which further decomposes to NO (Milos et al., 2012). The antioxidant effects of H_2S have been partially attributed to the direct scavenging of oxidants. However, this is questionable considering the relatively low physiological concentration of H_2S (Carballal et al., 2011).

4.2. H_2S interactions with NO donors

Bian's group and Moore's group studied the interaction of H_2S and NO using NaHS and a variety of NO donors. In the study of Yong et al., while NaHS (50 μ*M*) does not affect myocyte contractility, it reversed the negative inotropic effect introduced by NO donors (L-arginine, SNP, and DEA/NO) (Yong, Hu, Wang, Huang, & Bian, 2010), suggesting formation of a product other than NO. They suggested that this product to be HNO/NO^- by using Angeli's salt ($Na_2N_2O_3$) to mimic observed effects and using HNO/NO^- scavengers (NAC, L-cysteine, and GSH) to block response mediated by NaHS and SNP. Similarly, Filipovic et al. have found that SNP blocks H_2S-induced contractility of rat uteri and increased cellular level of HNO/NO^- in human umbilical vein endothelial cells (HUVECs) treated with 100 μ*M* Na_2S and 100 μ*M* SNP. However, they indicated that this is a multistep direct reaction between H_2S and SNP, involving formation of $[(CN)_5FeN(O)SH]^{3-}$, reduction of $HSNO/SNO^-$, and reaction between polysulfides and $[(CN)_5Fe(HNO)]^{3-}$ (Filipovic et al., 2013).

Multiple endogenous sources of HNO/NO^- have been proposed. An early work suggested that *S*-nitrosothiols may react with thiols to yield HNO/NO^- (Wong et al., 1998). HNO was evaluated according to the product of the reaction between HNO cyclohexyl mercaptan, by HPLC, LC–MS, and GC–MS. However, the study used GSNO and GSH at m*M* concentrations, which may lack physiological relevance. It is also shown that NOS may generate HNO/NO^- instead of NO when electron transfer to H_4B is hindered (Wei, Wang, Hemann, Hille, & Stuehr, 2003). Oxidative heme proteins have shown to oxidize hydroxylamine derivatives to release HNO/NO measured by methemoglobin$^-$ (Reisz, Bechtold, & King, 2010). Recently, improved analytical methods using phosphine ligation have also been reported (Reisz, Klorig, Wright, & King, 2009; Reisz, Zink, & King, 2011). A novel fluorescent probe has also been developed to measure the cellular HNO/NO^- level. Improved analytical methods using phosphine ligation have also been reported (Filipovic et al., 2013; Rosenthal & Lippard, 2010; Wrobel, Johnstone, Deliz Liang, Lippard, & Rivera-Fuentes, 2014; Zhou et al., 2011).

In another study, Moore et al. found that $HgCl_2$ and $CuCl_2$ restore the increase of cGMP by 100 μ*M* SNP, which is inhibited by 100 μ*M* NaHS (Whiteman et al., 2006). Since $HgCl_2$ enhances the decomposition of *S*-nitrosothiol to nitrite, they proposed *S*-nitrosothiol as a product of H_2S and NO reaction. Despite the fact that SNP can react with H_2S directly, another NO donor, SIN-1, releases O_2^- along with NO (Feelisch, Ostrowski, & Noack, 1989), leaving the reaction to be between H_2S and oxidative nitrogen species. Moreover, as mentioned previously (Qian & Jack, 2013), Moore's study used various NO donors, including SNP, SIN-1, spermine NONOate, DD1 (4-bromo-3,3,4-trimethyl-1-oxo-1,2-diazetidin-1-ium-2-olate), DEA NONOate, and EDTA NONOate; yet only nitrite introduced by spermine NONOate and DEA NONOate have been inhibited by NaHS. Also, in the case of SNP, SIN-1 and DD1, nitrite level "restored" by $HgCl_2$ has been much higher than NO donor alone.

4.3. *S*-nitrosothiols

Unlike the unfavorable reaction between H_2S and NO, H_2S reacts with *S*-nitrosothiol to yield a "yellowish product," which has been observed by different groups (Cortese-Krott et al., 2014; Filipovic, Miljkovic, Nauser, et al., 2012; Munro & Williams, 2000; Seel & Wagner, 1988). The mechanism of this reaction remains unclear. Munro and others have assigned this yellow product to $HSSNO/SSNO^-$ (Munro & Williams, 2000). On the contrary, Filipovic and his colleagues have demonstrated using ultrahigh-resolution ESI-TOP mass spectrometry, in which $HSSNO/SSNO^-$ has been absent from the product of the reaction between Na_2S and GSNO (Filipovic, Miljkovic, Allgauer, et al., 2012; Filipovic, Miljkovic, Nauser, et al., 2012), and identified HSNO as the product.

Moreover, using FTIR spectrometry, they also observed the disappearance of characteristic N=O vibration with retaining of the yellow color. Therefore, it has been concluded that HSNO mediates H_2S and RSNO interaction instead of HSSNO/$SSNO^-$.

On the contrary, Feelisch and his group suggested that the mobile hydrogen in HSNO makes the molecule to be present as HNSO, HOSN, and HONS as well (Cortese-Krott et al., 2014). Instead of releasing NO, these isomers tend to form gas or polymers (Müller, Nonella, & Huber, 1984). Although, the complexity of UV–vis spectra implies the existence of multiple reactions, the formation of HSNO and its long-term presence is debated. It has been proposed that HSNO should react with HS^- to yield HSSNO/$SSNO^-$ eventually (Cortese-Krott et al., 2014). However, more direct evidence is needed to prove these potential mechanisms.

4.4. Experimental procedures

4.4.1 UV–vis and stopped-flow spectroscopy

Stopped-flow is a spectroscopic technique used for kinetics study in solutions. Test reagents are mixed rapidly and contained in the sample cell for absorption and fluorescence measurements at a specific wavelength. Recordings over time can help analyze reaction rate and mechanism. The following protocol, adopted from Filipovic's study (Milos et al., 2012), uses Na_2S and ONOOH/$ONOO^-$ as examples to illustrate how to use stopped-flow spectroscopy to study H_2S interaction with NO adducts.

4.4.1.1 Protocol

Na_2S powder is kept in a desiccator at 4 °C and dissolved in degased ultrapure water for experiment freshly. Working stock solution of Na_2S concentration is verified with monobromobimane and RP-HPLC as we have published (Shen, Peter, Bir, Wang, & Kevil, 2012). ONOOH/$ONOO^-$ is prepared as reviewed elsewhere (Robinson & Beckman, 2004). The measurements are achieved by μSFM-20 Bio-Logic stopped-flow module, Huber CC90 cryostat, and J&M TIDAS high-speed diode array spectrometer with deuterium and tungsten lamps. The 10-ml Hamilton syringes sealed with Isolast perfluoroelastomer O-rings are controlled by different drives. Data are analyzed by Bio-Kine software and Specfit/32™. Na_2S and $ONOO^-$ solutions are diluted in 300-m*M* phosphate buffer (pH 7.4). While $ONOO^-$ concentration is fixed at 140 μ*M*, Na_2S is kept excessive (0.5–3 m*M*).

4.4.1.2 Key points

1. UV–vis light spectroscopy is one of the oldest forms of spectroscopy. It is simple to use and able to study reaction kinetics with stopped-flow setup.
2. Stopped-flow spectroscopy is especially useful to capture transient intermediate in reactions and reveal a whole spectrum of products. However, it does not confirm the identity of an intermediate or product, but only electronic and vibrational properties.
3. Absorption of the structure has to be known to make an intelligent speculation. Interpretations of the data may vary due to the complexity of the reaction.

4.4.2 Mass spectrometry

Mass spectrometry is one of the most powerful techniques for probing biological samples. Assays based on mass spectrometry have been used to analyze a variety of NO and H_2S derivatives. The following is an example using LC–MS for direct GSNO measurement.

4.4.2.1 Protocol

Desalt the biological samples by eluting through a C18 column (90% formic acid [0.1% v/v], 10% methanol, 1.5 ml/min). Lyophilize the eluted samples and reconstitute. Add ^{15}N-glutathione as a control to avoid artificial on-column formation of GSNO. Photolyze the samples or pretreat with $HgCl_2$ samples to use as a negative control. Analyze the eluted samples by an electrospray ionization MS (Finnigan LCQ Deca XP, Xcalibur software). GSNO cations are detected by selective ion monitoring at 336.9 (m/z).

4.4.2.2 Key comments

1. SNO species are liable to photolysis and oxidation. Biological samples have to be processed rapidly in cool and inert solution.
2. If samples are processed in acidic condition, a corresponding reduced thiol (in this example ^{15}N-glutathione) has to be used as a control for artificially increase SNO signal. Other SNO species can be measured with this protocol at appropriate m/z values.
3. The detection limit of this method for SNO is reported to be ~200 nM (Gow, Doctor, Mannick, & Gaston, 2007). Protein SNO can be biotin labeled before MS measurement. However, the biotin switch assay can be unreliable if samples are not processed properly, as discussed above.
4. Other NO and H_2S derivatives can be measured using isotope labeled NO and H_2S donors, such as $GS^{15}NO$ and $Na2^{34}S$ (Filipovic, Miljkovic,

Nauser, et al., 2012; Shen, Chakraborty, Dugas, & Kevil, 2014). It is also reported that HNO can be measured by phosphine-based labeling *in vitro* and *in vivo*. However, only AS derived and enzymatically synthesized HNO is tested. Its application in biological samples remains to be investigated.

4.4.3 HNO/NO^- imaging with CuBOT1

CuBOT1 is the first fluorescent probe detecting HNO/NO^- over other ROS/RNS in live cells. It has absorption and excitation wavelength at 518 nm and emission wavelength at 526 nm (Rosenthal & Lippard, 2010).

4.4.3.1 Protocol

Subculture human umbilical vein endothelial cells (HUVECs, passage 3–7) in 35 μ dishes to confluence in low serum medium overnight. Incubate the cells with 10-μ*M* CuBOT1 in cell culture medium for 20 min (Eberhardt et al., 2014; Filipovic et al., 2013; Filipovic, Miljkovic, Nauser, et al., 2012). Wash the cells with fresh medium three times before the treatments. Real-time image can be captured with an inverted epifluorescent microscope with green fluorescent filters.

4.4.3.2 Key points

1. CuBOT1 is shown to be highly specific to HNO, with no interference by NO_3^-, NO_2^-, ONO_2^-, OCl^-, O_2^-, or NO. Angeli's salt activates CuBOT1, so does the combination of DEA NONOate and Na_2S in cell culture (Eberhardt et al., 2014). Interestingly, neither DEA NONOate nor H_2S itself increases fluorescence.
2. Exogenous cysteine at 200 μ*M* increased fluorescence of the probe rapidly, while the normal intracellular cysteine level did not activate the probe (Rosenthal & Lippard, 2010). Consistently, recent study showed that depletion of either arginine or cysteine decreases fluorescence significantly (Eberhardt et al., 2014). These evidence suggest that NO and H_2S may interact to give rise to HNO.
3. Moreover, as a fluorescence probe, CuBOT1 is able to do real-time imaging and reveal subcellular location. However, it does not measure absolute concentration of HNO.

5. CONCLUSION

It is beyond contention from available literature that there is an interaction between H_2S and NO. Proposed interactions between these two molecules involve formation of various adducts. The information provided in this chapter details the precautions to be taken while performing NO measurements, in lieu with the possible H_2S interference.

ACKNOWLEDGMENTS

C. G. K. is the recipient of NIH Grant HL113303, and G. K. K. and S. Y. are funded by fellowships from the Malcom Feist Cardiovascular Research Endowment, LSU Health Sciences Center–Shreveport. Authors would like to acknowledge Ms. Sibile Pardue for her suggestions in designing experiments with chemiluminescence detection.

REFERENCES

Akaike, T., Yoshida, M., Miyamoto, Y., Sato, K., Kohno, M., Sasamoto, K., et al. (1993). Antagonistic action of imidazolineoxyl N-oxides against endothelium-derived relaxing factor/NO through a radical reaction. *Biochemistry, 32*, 827–832.

Altaany, Z., Yang, G., & Wang, R. (2013). Crosstalk between hydrogen sulfide and nitric oxide in endothelial cells. *Journal of Cellular and Molecular Medicine, 17*, 879–888.

Benetti, L. R., Campos, D., Gurgueira, S. A., Vercesi, A. E., Guedes, C. E., Santos, K. L., et al. (2013). Hydrogen sulfide inhibits oxidative stress in lungs from allergic mice in vivo. *European Journal of Pharmacology, 698*, 463–469.

Bir, S. C., Kolluru, G. K., McCarthy, P., Shen, X., Pardue, S., Pattillo, C. B., et al. (2012). Hydrogen sulfide stimulates ischemic vascular remodeling through nitric oxide synthase and nitrite reduction activity regulating hypoxia-inducible factor-1alpha and vascular endothelial growth factor-dependent angiogenesis. *Journal of the American Heart Association, 1*, e004093.

Bredt, D., & Schmidt, H. (1996). The citrulline assay. In J. Stamler & M. Feelisch (Eds.), *Methods in nitric oxide research* (pp. 249–255). New York, NY: John Wiley & Sons.

Bredt, D. S., & Snyder, S. H. (1990). Isolation of nitric oxide synthetase, a calmodulin-requiring enzyme. *Proceedings of the National Academy of Sciences of the United States of America, 87*, 682–685.

Bruce, S. (2013). Potential biological chemistry of hydrogen sulfide (H_2S) with the nitrogen oxides. *Free Radical Biology & Medicine, 55*, 1–7.

Bryan, N. S., & Grisham, M. B. (2007). Methods to detect nitric oxide and its metabolites in biological samples. *Free Radical Biology & Medicine, 43*, 645–657.

Bucci, M., Papapetropoulos, A., Vellecco, V., Zhou, Z., Pyriochou, A., Roussos, C., et al. (2010). Hydrogen sulfide is an endogenous inhibitor of phosphodiesterase activity. *Arteriosclerosis, Thrombosis, and Vascular Biology, 30*, 1998–2004.

Cai, H., Dikalov, S., Griendling, K. K., & Harrison, D. G. (2007). Detection of reactive oxygen species and nitric oxide in vascular cells and tissues: Comparison of sensitivity and specificity. *Methods in Molecular Medicine, 139*, 293–311.

Carballal, S., Trujillo, M., Cuevasanta, E., Bartesaghi, S., Möller, M. N., Folkes, L. K., et al. (2011). Reactivity of hydrogen sulfide with peroxynitrite and other oxidants of biological interest. *Free Radical Biology & Medicine, 50*, 196–205.

Cortas, N. K., & Wakid, N. W. (1990). Determination of inorganic nitrate in serum and urine by a kinetic cadmium-reduction method. *Clinical Chemistry, 36*, 1440–1443.

Cortese-Krott, M. M., Fernandez, B. O., Santos, J. L., Mergia, E., Grman, M., Nagy, P., et al. (2014). Nitrosopersulfide (SSNO(-)) accounts for sustained NO bioactivity of S-nitrosothiols following reaction with sulfide. *Redox Biology, 2*, 234–244.

Dejam, A., Hunter, C. J., Pelletier, M. M., Hsu, L. L., Machado, R. F., Shiva, S., et al. (2005). Erythrocytes are the major intravascular storage sites of nitrite in human blood. *Blood, 106*, 734–739.

Eberhardt, M., Dux, M., Namer, B., Miljkovic, J., Cordasic, N., Will, C., et al. (2014). H_2S and NO cooperatively regulate vascular tone by activating a neuroendocrine HNO-TRPA1-CGRP signalling pathway. *Nature Communications, 20*(5), 747–758.

Eberhardt, R. T., Forgione, M. A., Cap, A., Leopold, J. A., Rudd, M. A., Trolliet, M., et al. (2000). Endothelial dysfunction in a murine model of mild hyperhomocyst(e)inemia. *The Journal of Clinical Investigation, 106*, 483–491.

Feelisch, M., Ostrowski, J., & Noack, E. (1989). On the mechanism of NO release from sydnonimines. *Journal of Cardiovascular Pharmacology, 14*(Suppl. 11), S13–S22.

Filipovic, M. R., Eberhardt, M., Prokopovic, V., Mijuskovic, A., Orescanin-Dusic, Z., Reeh, P., et al. (2013). Beyond H_2S and NO interplay: Hydrogen sulfide and nitroprusside react directly to give nitroxyl (HNO). A new pharmacological source of HNO. *Journal of Medicinal Chemistry, 56*, 1499–1508.

Filipovic, M. R., Miljkovic, J., Allgauer, A., Chaurio, R., Shubina, T., Herrmann, M., et al. (2012). Biochemical insight into physiological effects of H (2)S: Reaction with peroxynitrite and formation of a new nitric oxide donor, sulfinyl nitrite. *The Biochemical Journal, 441*, 609–621.

Filipovic, M. R., Miljkovic, J. L., Nauser, T., Royzen, M., Klos, K., Shubina, T., et al. (2012). Chemical characterization of the smallest S-nitrosothiol, HSNO; cellular cross-talk of H_2S and S-nitrosothiols. *Journal of the American Chemical Society, 134*, 12016–12027.

Fukuto, J. M., Carrington, S. J., Tantillo, D. J., Harrison, J. G., Ignarro, L. J., Freeman, B. A., et al. (2012). Small molecule signaling agents: The integrated chemistry and biochemistry of nitrogen oxides, oxides of carbon, dioxygen, hydrogen sulfide, and their derived species. *Chemical Research in Toxicology, 25*, 769–793.

Geng, B., Cui, Y., Zhao, J., Yu, F., Zhu, Y., Xu, G., et al. (2007). Hydrogen sulfide downregulates the aortic L-arginine/nitric oxide pathway in rats. *American Journal of Physiology. Regulatory, Integrative and Comparative Physiology, 293*, R1608–R1618.

Giustarini, D., Rossi, R., Milzani, A., & Dalle-Donne, I. (2008). Nitrite and nitrate measurement by Griess reagent in human plasma: Evaluation of interferences and standardization. *Methods in Enzymology, 440*, 361–380.

Gomes, A., Fernandes, E., & Lima, J. L. (2006). Use of fluorescence probes for detection of reactive nitrogen species: A review. *Journal of Fluorescence, 16*, 119–139.

Gow, A., Doctor, A., Mannick, J., & Gaston, B. (2007). S-Nitrosothiol measurements in biological systems. *Journal of Chromatography. B, Analytical Technologies in the Biomedical and Life Sciences, 851*, 140–151.

Green, L. C., Wagner, D. A., Glogowski, J., Skipper, P. L., Wishnok, J. S., & Tannenbaum, S. R. (1982). Analysis of nitrate, nitrite, and [15 N]nitrate in biological fluids. *Analytical Biochemistry, 126*, 131–138.

Hetrick, E. M., & Schoenfisch, M. H. (2009). Analytical chemistry of nitric oxide. *Annual Review of Analytical Chemistry (Palo Alto, California), 2*, 409–433.

Hirst, D. G., & Robson, T. (2011). Nitric oxide physiology and pathology. *Methods in Molecular Biology, 704*, 1–13.

Hogg, N. (2010). Detection of nitric oxide by electron paramagnetic resonance spectroscopy. *Free Radical Biology & Medicine, 49*, 122–129.

Huie, R. E., & Padmaja, S. (1993). The reaction of no with superoxide. *Free Radical Research Communications, 18*, 195–199.

Joseph, J., Kalyanaraman, B., & Hyde, J. S. (1993). Trapping of nitric oxide by nitronyl nitroxides: An electron spin resonance investigation. *Biochemical and Biophysical Research Communications, 192*, 926–934.

Kevil, C. G., Kolluru, G. K., Pattillo, C. B., & Giordano, T. (2011). Inorganic nitrite therapy: Historical perspective and future directions. *Free Radical Biology & Medicine*, *51*, 576–593.

King, A. L., Polhemus, D. J., Bhushan, S., Otsuka, H., Kondo, K., Nicholson, C. K., et al. (2014). Hydrogen sulfide cytoprotective signaling is endothelial nitric oxide synthase-nitric oxide dependent. *Proceedings of the National Academy of Sciences of the United States of America*, *111*, 3182–3187.

Kolluru, G. K., Shen, X., Bir, S. C., & Kevil, C. G. (2013). Hydrogen sulfide chemical biology: Pathophysiological roles and detection. *Nitric Oxide: Biology and Chemistry/Official Journal of the Nitric Oxide Society*, *35*, 5–20.

Kolluru, G. K., Shen, X., & Kevil, C. G. (2013). A tale of two gases: NO and H_2S, foes or friends for life? *Redox Biology*, *1*, 313–318.

Kondo, K., Bhushan, S., King, A. L., Prabhu, S. D., Hamid, T., Koenig, S., et al. (2013). H(2)S protects against pressure overload-induced heart failure via upregulation of endothelial nitric oxide synthase. *Circulation*, *127*, 1116–1127.

Kubo, S., Kurokawa, Y., Doe, I., Masuko, T., Sekiguchi, F., & Kawabata, A. (2007). Hydrogen sulfide inhibits activity of three isoforms of recombinant nitric oxide synthase. *Toxicology*, *241*, 92–97.

Kumar, D., Branch, B. G., Pattillo, C. B., Hood, J., Thoma, S., Simpson, S., et al. (2008). Chronic sodium nitrite therapy augments ischemia-induced angiogenesis and arteriogenesis. *Proceedings of the National Academy of Sciences of the United States of America*, *105*, 7540–7545.

Lacza, Z., Horvath, E. M., Pankotai, E., Csordas, A., Kollai, M., Szabo, C., et al. (2005). The novel red-fluorescent probe DAR-4M measures reactive nitrogen species rather than NO. *Journal of Pharmacological and Toxicological Methods*, *52*, 335–340.

Lang, J. D., Jr., Teng, X., Chumley, P., Crawford, J. H., Isbell, T. S., Chacko, B. K., et al. (2007). Inhaled NO accelerates restoration of liver function in adults following orthotopic liver transplantation. *The Journal of Clinical Investigation*, *117*, 2583–2591.

Lei, Y. P., Liu, C. T., Sheen, L. Y., Chen, H. W., & Lii, C. K. (2010). Diallyl disulfide and diallyl trisulfide protect endothelial nitric oxide synthase against damage by oxidized low-density lipoprotein. *Molecular Nutrition & Food Research*, *54*(Suppl. 1), S42–S52.

Liu, K. L., Chen, H. W., Wang, R. Y., Lei, Y. P., Sheen, L. Y., & Lii, C. K. (2006). DATS reduces LPS-induced iNOS expression, NO production, oxidative stress, and NF-kappaB activation in RAW 264.7 macrophages. *Journal of Agricultural and Food Chemistry*, *54*, 3472–3478.

Liu, V. W., & Huang, P. L. (2008). Cardiovascular roles of nitric oxide: A review of insights from nitric oxide synthase gene disrupted mice. *Cardiovascular Research*, *77*, 19–29.

Lundberg, J. O., Weitzberg, E., & Gladwin, M. T. (2008). The nitrate-nitrite-nitric oxide pathway in physiology and therapeutics. *Nature Reviews. Drug Discovery*, *7*, 156–167.

Malinski, T., & Taha, Z. (1992). Nitric oxide release from a single cell measured in situ by a porphyrinic-based microsensor. *Nature*, *358*, 676–678.

Milos, R. F., Jan, M., Andrea, A., Ricardo, C., Tatyana, S., Martin, H., et al. (2012). Biochemical insight into physiological effects of H_2S: Reaction with peroxynitrite and formation of a new nitric oxide donor, sulfinyl nitrite. *The Biochemical Journal*, *441*, 609–621.

Moncada, S., & Higgs, A. (1993). The L-arginine-nitric oxide pathway. *The New England Journal of Medicine*, *329*, 2002–2012.

Müller, R. P., Nonella, M., & Huber, J. R. (1984). Spectroscopic investigation of HSNO in a low temperature matrix. UV, vis, and IR-induced isomerisations. *Chemical Physics*, *87*, 351–361.

Munro, A. P., & Williams, D. L. H. (2000). Reactivity of sulfur nucleophiles towards S-nitrosothiols. *Journal of the Chemical Society, Perkin Transactions*, *2*, 1794–1797.

Oh, G. S., Pae, H. O., Lee, B. S., Kim, B. N., Kim, J. M., Kim, H. R., et al. (2006). Hydrogen sulfide inhibits nitric oxide production and nuclear factor-kappaB via heme oxygenase-1 expression in RAW264.7 macrophages stimulated with lipopolysaccharide. *Free Radical Biology & Medicine, 41*, 106–119.

Ondrias, K., Stasko, A., Cacanyiova, S., Sulova, Z., Krizanova, O., Kristek, F., et al. (2008). H(2)S and HS(-) donor NaHS releases nitric oxide from nitrosothiols, metal nitrosyl complex, brain homogenate and murine L1210 leukaemia cells. *Pflügers Archiv, 457*, 271–279.

Qian, L., & Jack, R. L. (2013). Chemical foundations of hydrogen sulfide biology. *Nitric Oxide: Biology and Chemistry/Official Journal of the Nitric Oxide Society, 35*, 21–34.

Rachmilewitz, D., Karmeli, F., Eliakim, R., Stalnikowicz, R., Ackerman, Z., Amir, G., et al. (1994). Enhanced gastric nitric oxide synthase activity in duodenal ulcer patients. *Gut, 35*, 1394–1397.

Reisz, J. A., Bechtold, E., & King, S. B. (2010). Oxidative heme protein-mediated nitroxyl (HNO) generation. *Dalton Transactions, 39*, 5203–5212.

Reisz, J. A., Klorig, E. B., Wright, M. W., & King, S. B. (2009). Reductive phosphine-mediated ligation of nitroxyl (HNO). *Organic Letters, 11*, 2719–2721.

Reisz, J. A., Zink, C. N., & King, S. B. (2011). Rapid and selective nitroxyl (HNO) trapping by phosphines: Kinetics and new aqueous ligations for HNO detection and quantitation. *Journal of the American Chemical Society, 133*, 11675–11685.

Robinson, K. M., & Beckman, J. S. (2004). Synthesis of peroxynitrite from nitrite and hydrogen peroxide. *Methods in Enzymology, 396*, 207–214.

Rosenthal, J., & Lippard, S. J. (2010). Direct detection of nitroxyl in aqueous solution using a tripodal copper(II) BODIPY complex. *Journal of the American Chemical Society, 132*, 5536–5537.

Salter, M., Knowles, R. G., & Moncada, S. (1991). Widespread tissue distribution, species distribution and changes in activity of Ca(2+)-dependent and Ca(2+)-independent nitric oxide synthases. *FEBS Letters, 291*, 145–149.

Seel, F., & Wagner, M. (1988). Reaction of sulfides with nitrogen monoxide in aqueous solution. *Zeitschrift für Anorganische und Allgemeine Chemie, 558*, 189–192.

Shen, X., Chakraborty, S., Dugas, T. R., & Kevil, C. G. (2014). Hydrogen sulfide measurement using sulfide dibimane: Critical evaluation with electrospray ion trap mass spectrometry. *Nitric Oxide: Biology and Chemistry/Official Journal of the Nitric Oxide Society, 41*, 97–104.

Shen, X., Peter, E. A., Bir, S., Wang, R., & Kevil, C. G. (2012). Analytical measurement of discrete hydrogen sulfide pools in biological specimens. *Free Radical Biology & Medicine, 52*, 2276–2283.

Steil, A. A., Garcia Rodriguez, M. C., Alonso, A., Crespo, M. S., & Bosca, L. (1995). Platelet-activating factor: The effector of protein-rich plasma extravasation and nitric oxide synthase induction in rat immune complex peritonitis. *British Journal of Pharmacology, 114*, 895–901.

Teng, X., Scott Isbell, T., Crawford, J. H., Bosworth, C. A., Giles, G. I., Koenitzer, J. R., et al. (2008). Novel method for measuring S-nitrosothiols using hydrogen sulfide. *Methods in Enzymology, 441*, 161–172.

Upchurch, G. R., Jr., Welch, G. N., Fabian, A. J., Freedman, J. E., Johnson, J. L., Keaney, J. F., Jr., et al. (1997). Homocyst(e)ine decreases bioavailable nitric oxide by a mechanism involving glutathione peroxidase. *The Journal of Biological Chemistry, 272*, 17012–17017.

van der Zee, F. P., Lettinga, G., & Field, J. A. (2001). Azo dye decolourisation by anaerobic granular sludge. *Chemosphere, 44*, 1169–1176.

Vanin, A., & Poltorakov, A. (2009). NO spin trapping in biological systems. *Frontiers in Bioscience (Landmark Edition), 14*, 4427–4435.

von Bohlen und Halbach, O. (2003). Nitric oxide imaging in living neuronal tissues using fluorescent probes. *Nitric Oxide: Biology and Chemistry/Official Journal of the Nitric Oxide Society*, *9*, 217–228.

Wang, R. (2012). Physiological implications of hydrogen sulfide: A whiff exploration that blossomed. *Physiological Reviews*, *92*, 791–896.

Wang, M. J., Cai, W. J., Li, N., Ding, Y. J., Chen, Y., & Zhu, Y. C. (2010). The hydrogen sulfide donor NaHS promotes angiogenesis in a rat model of hind limb ischemia. *Antioxidants & Redox Signaling*, *12*, 1065–1077.

Wedmann, R., Bertlein, S., Macinkovic, I., Boltz, S., Miljkovic, J. L., Munoz, L. E., et al. (2014). Working with "H_2S": Facts and apparent artifacts. *Nitric Oxide: Biology and Chemistry/Official Journal of the Nitric Oxide Society*, *41*, 85–96.

Wei, C.-C. C., Wang, Z.-Q. Q., Hemann, C., Hille, R., & Stuehr, D. J. (2003). A tetrahydrobiopterin radical forms and then becomes reduced during Nomega-hydroxyarginine oxidation by nitric-oxide synthase. *The Journal of Biological Chemistry*, *278*, 46668–46673.

Weissman, B. A., & Gross, S. S. (2001). Measurement of NO and NO synthase. *Current Protocols in Neuroscience*, Chapter 7: Unit 7.13. http://dx.doi.org/10.1002/0471142301.ns0713s05.

Whiteman, M., Armstrong, J. S., Chu, S. H., Jia-Ling, S., Wong, B. S., & Cheung, N. S. (2004). The novel neuromodulator hydrogen sulfide: An endogenous peroxynitrite 'scavenger'? *Journal of Neurochemistry*, *90*, 765–768.

Whiteman, M., Li, L., Kostetski, I., Chu, S. H., Siau, J. L., Bhatia, M., et al. (2006). Evidence for the formation of a novel nitrosothiol from the gaseous mediators nitric oxide and hydrogen sulphide. *Biochemical and Biophysical Research Communications*, *343*, 303–310.

Wong, P. S., Hyun, J., Fukuto, J. M., Shirota, F. N., DeMaster, E. G., Shoeman, D. W., et al. (1998). Reaction between S-nitrosothiols and thiols: Generation of nitroxyl (HNO) and subsequent chemistry. *Biochemistry*, *37*, 5362–5371.

Wrobel, A. T., Johnstone, T. C., Deliz Liang, A., Lippard, S. J., & Rivera-Fuentes, P. (2014). A fast and selective near-infrared fluorescent sensor for multicolor imaging of biological nitroxyl (HNO). *Journal of the American Chemical Society*, *136*, 4697–4705.

Yang, G., Wu, L., Jiang, B., Yang, W., Qi, J., Cao, K., et al. (2008). H_2S as a physiologic vasorelaxant: Hypertension in mice with deletion of cystathionine gamma-lyase. *Science*, *322*, 587–590.

Yong, Q. C., Hu, L. F., Wang, S., Huang, D., & Bian, J. S. (2010). Hydrogen sulfide interacts with nitric oxide in the heart: Possible involvement of nitroxyl. *Cardiovascular Research*, *88*, 482–491.

Yong, Q. C., Lee, S. W., Foo, C. S., Neo, K. L., Chen, X., & Bian, J. S. (2008). Endogenous hydrogen sulphide mediates the cardioprotection induced by ischemic postconditioning. *American Journal of Physiology Heart and Circulatory Physiology*, *295*, H1330–H1340.

Yusof, M., Kamada, K., Kalogeris, T., Gaskin, F. S., & Korthuis, R. J. (2009). Hydrogen sulfide triggers late-phase preconditioning in postischemic small intestine by an NO- and p38 MAPK-dependent mechanism. *American Journal of Physiology Heart and Circulatory Physiology*, *296*, H868–H876.

Zamora, R., Vodovotz, Y., & Billiar, T. R. (2000). Inducible nitric oxide synthase and inflammatory diseases. *Molecular Medicine*, *6*, 347–373.

Zhang, X., Li, H., Jin, H., Ebin, Z., Brodsky, S., & Goligorsky, M. S. (2000). Effects of homocysteine on endothelial nitric oxide production. *American Journal of Physiology Renal Physiology*, *279*, F671–F678.

Zhou, Y., Liu, K., Li, J.-Y., Fang, Y., Zhao, T.-C., & Yao, C. (2011). Visualization of nitroxyl in living cells by a chelated copper(II) coumarin complex. *Organic Letters*, *13*, 1290–1293.

AUTHOR INDEX

Note: Page numbers followed by "*f*" indicate figures, "*t*" indicate tables and "*np*" indicate footnotes.

C

D

E

F

G

L

P

SUBJECT INDEX

Note: Page numbers followed by "*f*" indicate figures and "*t*" indicate tables.

R

S

T

W

Z

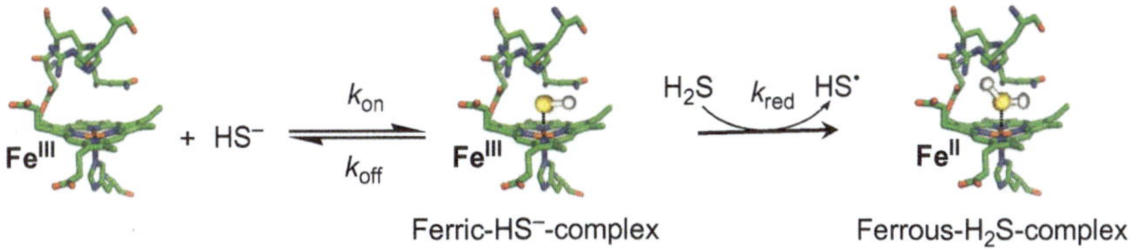

Péter Nagy, Figure 3 Reaction of sulfide with heme Fe^{3+} centers. Coordination of sulfide to a high spin Fe^{3+} center of a heme protein (k_{on}) results in a low-spin Fe^{3+}-sulfide complex. Sulfide can be liberated from this complex (k_{off}) but redox reaction with another sulfide (k_{red}) can generate the ferrous(Fe^{2+})-sulfide complex and $HS^•$.

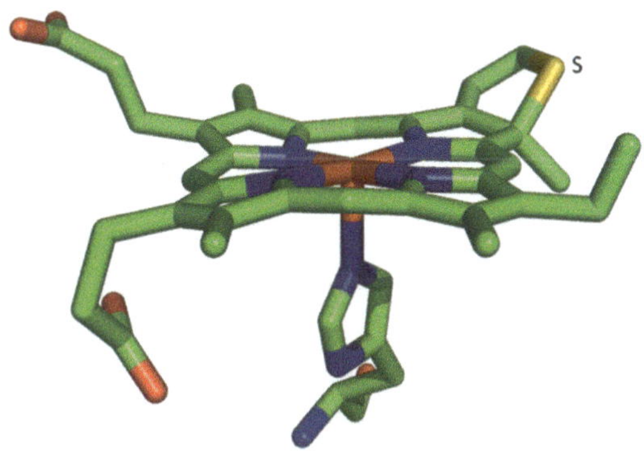

Péter Nagy, Figure 4 Sulfheme moiety from the crystal structure of sulfmyoglobin using PDB file 1YMC.

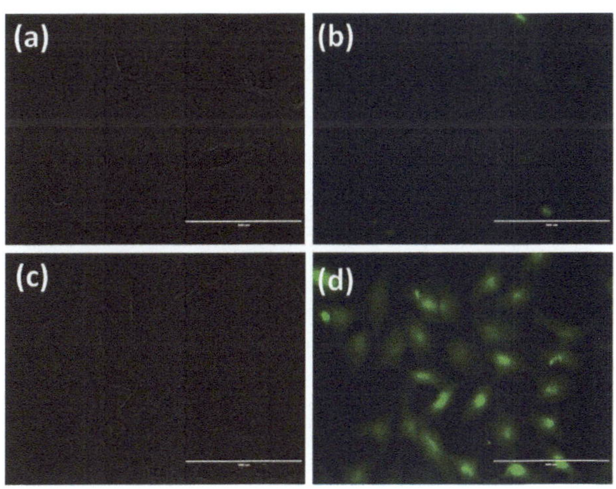

Bo Peng and Ming Xian, Figure 4 Fluorescence images of H_2S in HeLa cells using WSP5. Cells on a 24-well plate were incubated with WSP5 (50 μ*M*) for 30 min and then washed and subjected to different treatments: (A, B) Control (no NaHS was added) and (C, D) treated with 100 μ*M* NaHS. Scale bar: 100 nm. *Reprinted with permission from Peng et al. (2014). Copyright 2014 John Wiley & Sons, Inc.*

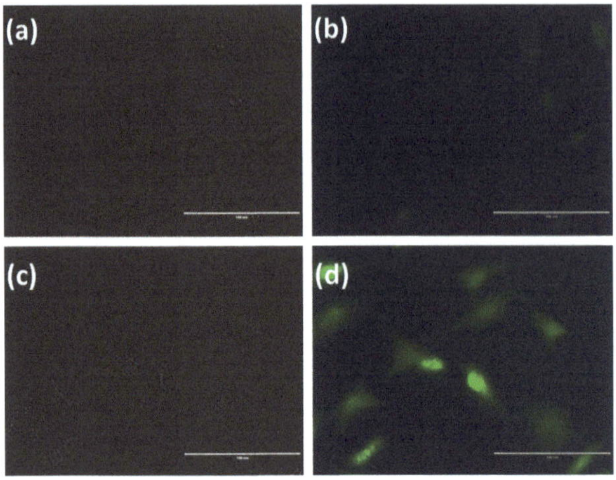

Bo Peng and Ming Xian, Figure 5 Fluorescence images of H_2S production from a H_2S donor YZ-4-074 in HeLa cells. Cells on a 24-well plate were incubated with or without YZ-4-074 for 30 min and then washed and incubated with 30 μ*M* WSP4 for 30 min. (A, B) Controls (no donor was added) and (C, D) treated with 100 μ*M* YZ-4-074. Scale bar: 100 nm. *Reprinted with permission from Peng et al. (2014). Copyright 2014 John Wiley & Sons, Inc.*

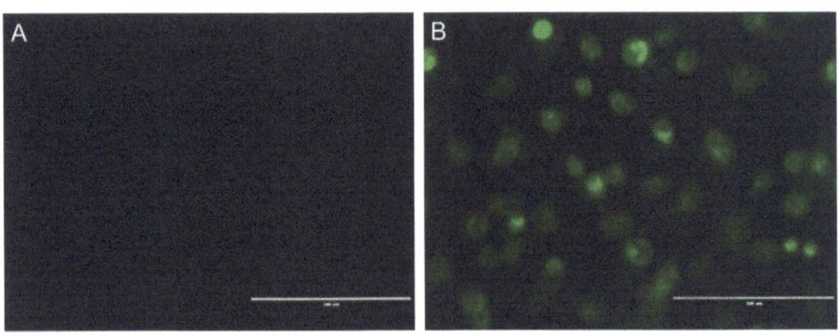

Bo Peng and Ming Xian, Figure 6 H_2S release from NOD-1d in HeLa cells: (A) cells treated with donor (200 μ*M*) and WSP-1 (50 μ*M*), no UV irradiation; (B) cells treated with donor (200 μ*M*) and WSP1 (50 μ*M*), under UV irradiation. *Reprinted with permission from Devarie-Baez et al. (2013). Copyright 2013 American Chemical Society.*

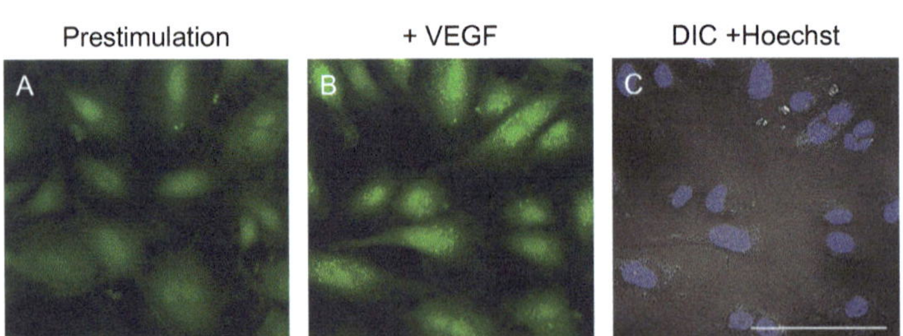

Vivian S. Lin *et al.*, Figure 4 (A) HUVECs incubated with 5 μ*M* SF7-AM for 30 min at 37 °C, washed, and then imaged. (B) The same field of HUVECs in (A) was treated on stage with 2 μL of 10 μg/mL VEGF for 30 min at 37 °C, and then imaged. (C) Brightfield images of the same field of cells in (B) overlaid with Hoechst 33342 stain (1 μ*M*) at 37 °C. Images in (A) and (B) are the maximum intensity projections of 8 × 2 μm *z*-stacks. Scale bar represents 100 μm. *Adapted with permission from Lin et al. (2013). Copyright 2013 Proceedings of the National Academy of Sciences.*

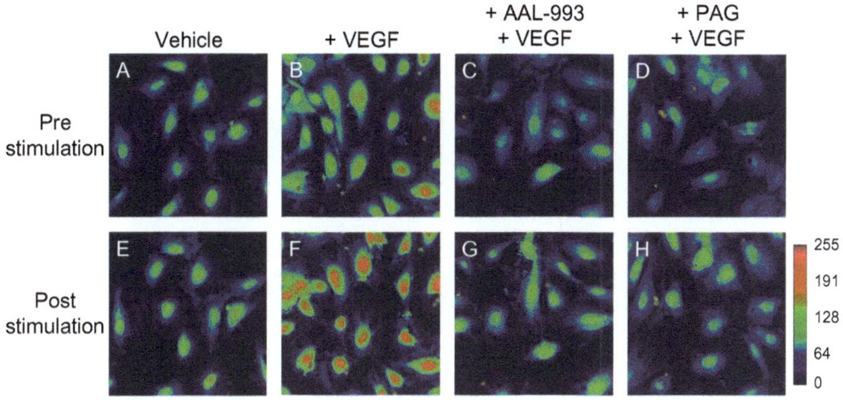

Vivian S. Lin *et al.*, Figure 5 Representative confocal fluorescence images of H_2S signaling in live HUVECs. HUVECs were incubated with 2.5 μ*M* SF7-AM, washed, and imaged before (A) and after (E) treatment with 0.1% BSA in H_2O as a vehicle control. HUVECs were incubated with 2.5 μ*M* SF7-AM, washed, and imaged before (B) and after (F) treatment with VEGF. SF7-AM-labeled HUVECs pretreated with 30 μ*M* AAL-993 (AAL) for 40 min before (C) and after (G) treatment with VEGF. SF7-AM-labeled HUVECs pretreated with 100 μ*M* PAG for 10 min before (D) and after (H) treatment with VEGF. *Adapted with permission from Lin et al. (2013). Copyright 2013 Proceedings of the National Academy of Sciences.*

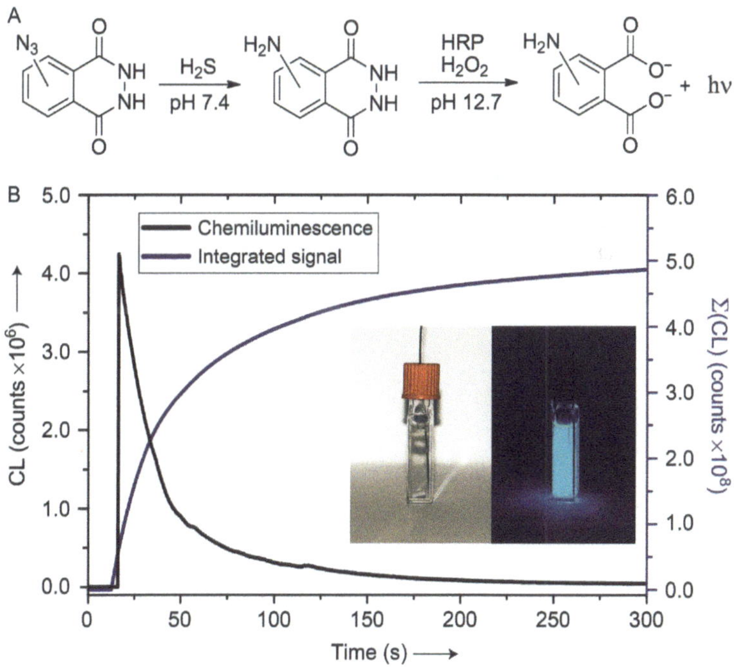

T. Spencer Bailey and Michael D. Pluth, Figure 3 Chemiluminescence response profile for CLSS-1 and CLSS-2. (A) H_2S-mediated reduction of CLSS-1 and CLSS-2 followed by luminol visualization using HRP (horseradish peroxidase) and H_2O_2. (B) Chemiluminescence profile of 50 μ*M* CLSS-1 after incubation for 1 h with 33 equiv. of NaSH. Visual detection (inset) at 10 × concentration with 5 s camera exposure. Samples were incubated in pH 7.4 PIPES buffer at 37 °C prior to analysis. *Reprinted with permission from Bailey and Pluth (2013). Copyright 2013 American Chemical Society.*

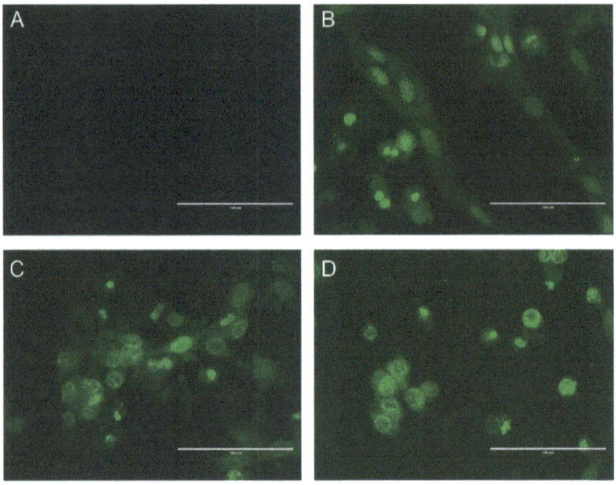

Chung-Min Park and Ming Xian, Figure 4 H_2S production of **PDD-1** in H9c2 cells. Cells were incubated with vehicle (A), 100 μ*M* (B), 200 μ*M* (C), and 400 μ*M* (D) of donor for 24 h. After removal of excess donor, 100 μ*M* of WSP-1 was added. Images were taken afterwards. *Reprinted with permission from Park et al. (2013). Copyright 2013 Molecular BioSystems.*

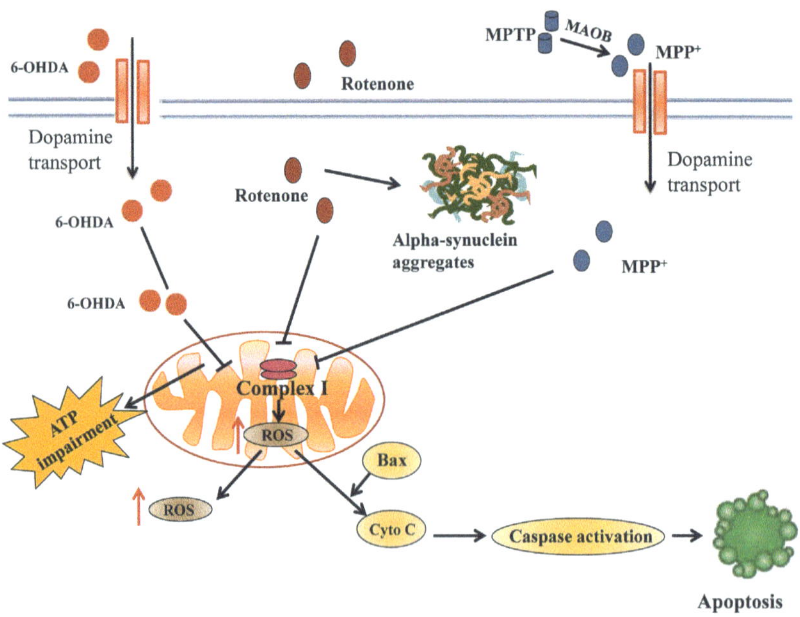

Xue Xue and Jin-Song Bian, Figure 1 Schematic illustration showing the mechanisms for different neurotoxinsinduced PD models. All neurotoxins-induced neuronal damages are mainly caused by the massive oxidative stress. Rotenone is highly lipophilic and readily enters all cells, where it blocks complex I activity, causing massive formation of reactive ROS. Different from Rotenone, both 6-OHDA and MPP+ have a similar affinity as dopamine to the dopamine transporter (DPT) and are therefore uptaken into dopaminergic neurons to generate ROS such as superoxide radicals. This explains why and how these two neurotoxins selectively kill dopaminergic neurons.

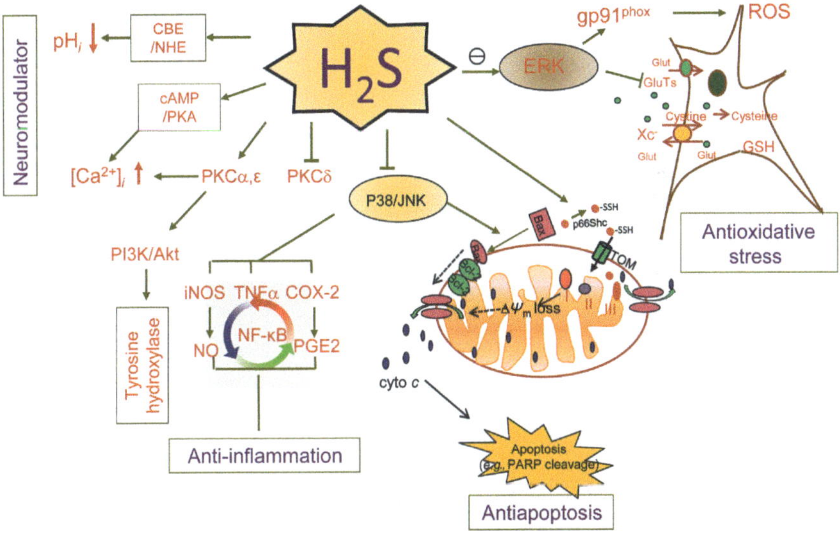

Xue Xue and Jin-Song Bian, Figure 3 Schematic illustration showing the neuromodulatory and -protective effects of H_2S. Physiologically, H_2S regulates intracellular calcium, pHi, and cAMP level in brain cells. In pathological situations, H_2S produces anti-inflammatory, -oxidant, and -apoptotic effects via multiple mechanisms. This can be mediated by suppression or activation of different protein kinases including PKC, PI3K/Akt, p38, JNK, and ERK–MAPKs. Direct effect is mediated by S-sulfhydration of p66SHC at Cys-59, which prevents oxidative stress-induced phosphorylation and translocation of this protein to mitochondria. This effect further contributes to the inhibitory effect of H_2S on ROS production in mitochondria.

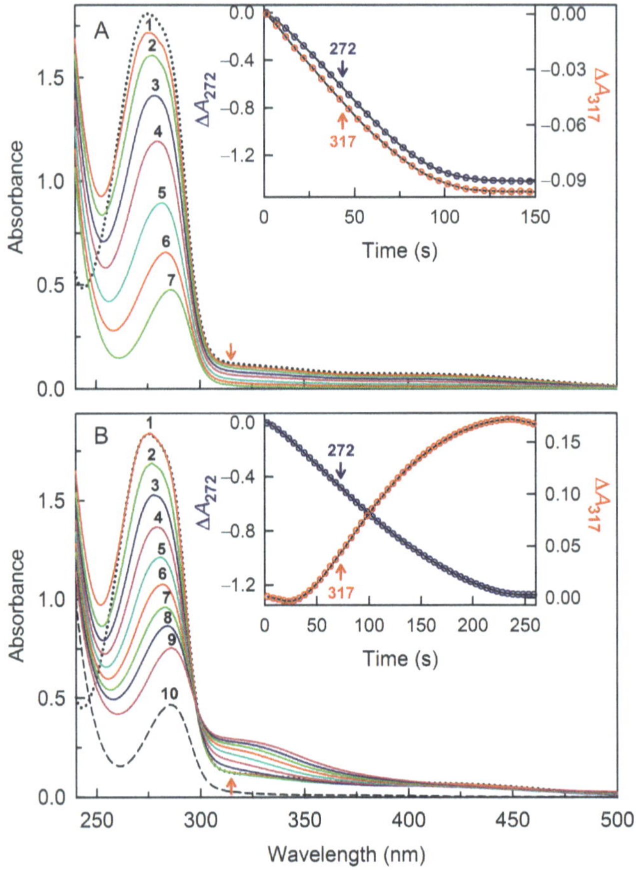

Michael R. Jackson *et al.*, Figure 5 Spectral course of SQOR catalytic assays with sulfite or sulfide as sulfane sulfur acceptor. Reactions with sulfite (panel A) or sulfide (panel B) as acceptor were conducted in 100 m*M* potassium phosphate, pH 8.0, containing 0.5 m*M* EDTA at 25 °C. Red arrows indicate the direction of the spectral changes observed at $\lambda > 300$ nm. Panel A: The dotted black line was recorded before addition of 300 μ*M* sulfide to an assay mixture containing 160 μ*M* CoQ_1, 400 μ*M* sulfite, and 7.2 n*M* SQOR. Curves 2–7 were recorded 1.4, 12, 27, 42, 62, 82, and 132 s, respectively, after sulfide addition. Panel B: The dotted black line was recorded before addition of 300 μ*M* sulfide to an assay mixture containing 160 μ*M* CoQ_1 and 35.6 n*M* SQOR. Curves 1–9 were recorded 2.6, 33.2, 58.2, 88.2, 108, 133, 158, 183, and 233 s, respectively, after sulfide addition. Curve 10 was recorded 292 s after addition of 400 μ*M* sulfite. The inset in each panel shows the time course of absorbance changes at 272 and 317 nm plotted according to the left and right *y*-axes, respectively. *Adapted with permission from Jackson et al. (2012).*

FSC
www.fsc.org
MIX
Papier | Fördert gute Waldnutzung
FSC® C083411

Kolibri 360 GmbH
Ferdinand-Jühlke-Straße 7
99095 Erfurt, Deutschland
produktsicherheit@kolibri360.de